Posthuman Management

Creating Effective Organizations in an Age of Social Robotics, Ubiquitous AI, Human Augmentation, and Virtual Worlds

Second Edition

Matthew E. Gladden

Posthuman Management: Creating Effective Organizations in an Age of Social Robotics, Ubiquitous AI, Human Augmentation, and Virtual Worlds (Second Edition)

Published in the United States of America
by Defragmenter Media in association with West Pole & Larkspur,
imprints of Synthypnion Press LLC

Synthypnion Press LLC
Indianapolis, IN 46227
http://www.synthypnionpress.com

Defragmenter Media
http://defragmenter.media

ISBN 978-1-944373-05-4 (hardcover print edition)
ISBN 978-1-944373-13-9 (paperback print edition)
ISBN 978-1-944373-06-1 (ebook)
10 9 8 7 6 5 4 3
August 2016

Chapters One, Two, and Three of this volume were originally published as Parts One (pp. 31-91), Two (pp. 93-131), and Three (pp. 133-201) of Gladden, Matthew E., *Sapient Circuits and Digitalized Flesh: The Organization as Locus of Technological Posthumanization*. Indianapolis: Defragmenter Media, 2016. This volume's Introduction has been adapted from "The Posthumanized Organization as a Synergism of Human, Synthetic, and Hybrid Agents," published in that same text (pp. 17-30).

Chapter Five in this volume was originally published as Gladden, Matthew E., "Neural Implants as Gateways to Digital-Physical Ecosystems and Posthuman Socioeconomic Interaction," in *Digital Ecosystems: Society in the Digital Age*, edited by Łukasz Jonak, Natalia Juchniewicz, and Renata Włoch, pp. 85-98. Warsaw: Digital Economy Lab, University of Warsaw, 2016.

Chapter Eight has been expanded and adapted from Gladden, Matthew E., "The Social Robot as CEO: Developing Synthetic Charismatic Leadership for Human Organizations," in *Sociable Robots and the Future of Social Relations: Proceedings of Robo-Philosophy 2014*, edited by Johanna Seibt, Raul Hakli, and Marco Nørskov, pp. 329-339. Frontiers in Artificial Intelligence and Applications, vol. 273. Amsterdam: IOS Press, 2014.

Chapter Nine is reprinted from Gladden, Matthew E., "The Diffuse Intelligent Other: An Ontology of Nonlocalizable Robots as Moral and Legal Actors," in *Social Robots: Boundaries, Potential, Challenges*, edited by Marco Nørskov, pp. 177-98. Farnham: Ashgate, 2016.

Chapter Ten was first published as Gladden, Matthew E., "Leveraging the Cross-Cultural Capacities of Artificial Agents as Leaders of Human Virtual Teams," in *Proceedings of the 10th European Conference on Management Leadership and Governance*, edited by Visnja Grozdanić, pp. 428-435. Reading: Academic Conferences and Publishing International Limited, 2014.

Chapter Eleven was originally published as Gladden, Matthew E., "A Fractal Measure for Comparing the Work Effort of Human and Artificial Agents Performing Management Functions," in *Position Papers of the 2014 Federated Conference on Computer Science and Information Systems*, edited by Maria Ganzha, Leszek Maciaszek, and Marcin Paprzycki, pp. 219-226. Annals of Computer Science and Information Systems vol. 3. Warsaw: Polskie Towarzystwo Informatyczne, 2014.

Chapter Twelve was originally published as Gladden, Matthew E., "Managerial Robotics: a Model of Sociality and Autonomy for Robots Managing Human Beings and Machines," *International Journal of Contemporary Management* vol. 13, no. 3 (2014): 67-76.

Chapter Thirteen is a revised and expanded version of Gladden, Matthew E., "The Artificial Life-Form as Entrepreneur: Synthetic Organism-Enterprises and the Reconceptualization of Business," published in *Proceedings of the Fourteenth International Conference on the Synthesis and Simulation of Living Systems*, edited by Hiroki Sayama, John Rieffel, Sebastian Risi, René Doursat and Hod Lipson, pp. 417-18. Cambridge, MA: The MIT Press, 2014.

To all the managers, leaders, and executives
with whom I have had the opportunity to work and study –
who have given me an understanding and experience
of organizational life at its most meaningful and worthwhile

Brief Table of Contents

Detailed Table of Contents

Preface

The purpose of this book is to offer a posthumanist perspective on organizational management that is largely lacking in the current literature but which, I believe, will grow increasingly important in the coming years. The impact of accelerating technological change on contemporary organizations is undeniable, and in recent decades posthumanist thought has developed a rich array of approaches for identifying, understanding, and predicting the future path of such forces of 'technologization' in other spheres of human activity. However, there is a relative dearth of scholarship that seeks to make posthumanist methodologies and insights relevant and accessible for theorists and practitioners of contemporary organizational management. The leading works on posthumanism have generally appeared within fields like literary criticism or bioethics that have little direct application to organizational management. Conversely, the numerous management texts exploring the impact of emerging technologies on organizations typically display a functional or strategic perspective that is very practical; rarely do they robustly address (or even acknowledge) the deep-seated transformation of the nature of human agency and our relationship to our environment that is being driven by ongoing technological change and whose ontological, phenomenological, psychological, aesthetic, ethical, legal, and political dimensions are being analyzed in a constructive fashion by the many varieties of posthumanist thought.

In its own way, each of the texts gathered in this volume seeks to build a bridge between posthumanism and the world of organizational theory and management. The first three chapters originally appeared in my book *Sapient Circuits and Digitalized Flesh: The Organization as Locus of Technological Posthumanization*, and they provide a systematic theoretical framework that undergirds and ties together this entire volume. The remainder of the book comprises chapters that were originally published as articles in academic journals, as chapters in edited volumes, or in other collections or were prepared especially for this book. Each of these chapters employs a qualitative or quantitative methodology to investigate some particular aspect of management in a posthuman context. My hope is that this volume will prove useful for management scholars and students who are interested in the posthumanizing aspect of emerging technologies, as well as for managers whose decisions affect the lives of countless people and who are seeking to better understand the rapidly evolving sociotechnological dynamics that are reshaping their organizations – and in particular, the advent of new forms of artificial agency, technologically augmented human agency, and hybrid human-synthetic agency.

The preparation of this volume would have been impossible were it not for the many organizations and individuals who allowed me to share and receive feedback on my research over the last couple of years by publishing texts in journals and books and delivering lectures and conference presentations. I am grateful to everyone who made possible those opportunities for dialogue. In particular, I am thankful to the faculty, staff, and students of the universities and other research institutions where ideas contained in this volume were first presented, including those of Aarhus University, VERN′ University of Applied Sciences, the Jagiellonian University, the Facta Ficta Research Center, the Warsaw University of Technology, the University of Silesia in Katowice, the Centrum Informacji Naukowej i Biblioteka Akademicka (CINiBA) in Katowice, the Faculty of Humanities of the AGH University of Science and Technology, the Institute of Computer Science of the Polish Academy of Sciences, and the Digital Economy Lab of the University of Warsaw. I am also thankful to the editors who have made possible the publication of my previous research into these topics, including those at the MIT Press, IOS Press, Ashgate Publishing, *Creatio Fantastica*, the *Annals of Computer Science and Information Systems*, *Informatyka Ekonomiczna / Business Informatics*, *Annales: Ethics in Economic Life*, the *International Journal of Contemporary Management*, and *Fronda Lux*. In particular, I offer my deepest gratitude to Krzysztof Maj, Ksenia Olkusz, Mateusz Zimnoch, Magdalena Szczepocka, Marco Nørskov, Johanna Seibt, Helena Dudycz, Natalia Juchniewicz, Renata Włoch, Mateusz Matyszkowicz, and Jerzy Kopański.

I am also thankful to many individuals who have been involved with the management and governance of organizations of diverse types from whom I have learned so much, including Mariann M. Payne, Rita M. Rodriguez, Matthew R. Hummer, Harry A. Rissetto, Anthony F. Essaye, John G. Quigley, George R. Houston, Jr., Timothy Brown, S.J., Gerald J. Chojnacki, S.J., John C. Haughey, S.J., Thomas E. Scheye, Nancy L. Swartz, Rob Eby, Jillian Gobbo, Joseph A. McCartin, James T. Lamiell, Brad Kelly, Aisulu Raspayeva, Maria Ferrara, Keelan Downton, Laura Michener, Marley Moynahan, Kathryn Tucker, David Wilkins, John McKelvie, Kevin Flannery, Christian Lambert, Tessa Pulaski, Madeline Howard, Pamela Paniagua, Tom Rijntjes, Paweł Urbański, and Nathan Fouts. I am also deeply thankful to everyone at Georgetown University's School of Continuing Studies, especially Edwin

Schmierer and Douglas M. McCabe; to Serge Pukas, Paulina Krystosiak, Robert Pająk, Jacek Koronacki, and everyone affiliated with the Institute of Computer Science of the Polish Academy of Sciences; to Nicole Cunningham for her friendship and inspirational example as an author; and to Sarah Stern for shaping me as a writer and editor.

I am especially thankful to those scholars who are fashioning innovative links between posthumanist thought and various spheres of human existence whom I have had the opportunity to hear present their research and with whom I have enjoyed valuable conversations about these topics – especially Bartosz Kłoda-Staniecko, Agata Kowalewska, Krzysztof Maj, and Magdalena Szczepocka.

As always, I am deeply grateful to my parents, my brother, and other relatives and friends for the support that they have provided me in my research, and I offer my heartfelt thanks to my wife for her sound counsel and continual encouragement. And, especially, I am grateful to Terry R. Armstrong and Gasper Lo Biondo, S.J., without whom I would never have acquired the managerial experience, the knowledge, or – most importantly – the inspiration that made it possible for me to embark upon the work of preparing this text.

The individuals mentioned above – along with many other conference attendees and anonymous reviewers who provided feedback on my papers – have contributed greatly to developing whatever insights and value are to be found within this book. Whatever flaws or limitations remain within the text are due not to any failures on their part but are my responsibility alone.

Matthew E. Gladden
Pruszków, July 5, 2016

Introduction

Technologization and the Evolution of Intelligent Agency: Toward Posthuman Models of Organizational Management[1]

Abstract. We live in an era of accelerating technological posthumanization in which the form and capacities of human and artificial agents are converging in ways that might be understood as either exciting or unsettling. Ongoing developments in fields like biocybernetics, neuroprosthetics, wearable computing, virtual reality, and genetic engineering are yielding technologically augmented human beings who possess physical components and behaviors resembling those traditionally found in electronic computers. Meanwhile, developments in artificial intelligence, social robotics, artificial life, nanotechnology, and ubiquitous computing are creating synthetic entities whose structures and processes ever more closely resemble those of living organisms. Such human and nonhuman agents exist and interact within increasingly sophisticated digital-physical ecosystems in which entities shift continually between actual and virtual worlds. Insofar as such agents constitute the building-blocks of contemporary organizations, the processes of technological posthumanization that are transforming them are also reshaping the theory and practice of organizational management.

Posthuman Management provides a wide-ranging and systematic investigation of these issues by collecting relevant texts recently published in academic journals along with original content prepared for this volume. This introductory chapter to *Posthuman Management* presents an overview of the major issues explored within

[1] This chapter is an expanded version of "The Posthumanized Organization as a Synergism of Human, Synthetic, and Hybrid Agents," originally published as the introduction to Gladden, Matthew E., *Sapient Circuits and Digitalized Flesh: The Organization as Locus of Technological Posthumanization*, pp. 17-30. Indianapolis: Defragmenter Media, 2016.

> the volume's three parts and the methodologies employed. In Part I of the book, a general theoretical and practical framework for the field of posthuman management is developed. Each chapter approaches this task at a different level, moving from the more abstract sphere of a basic exploration of the nature of posthumanism to the more concrete sphere of formulating tools for posthumanized business analysis and considering the implications of posthumanization for a specific management discipline (in this case, organization development). The three chapters in Part II take a closer look at look at the ways in which organizational management will be affected by the posthumanizing augmentation of human beings through technologies such as neuroprosthetics, virtual reality, and genetic engineering. Finally, the six chapters of Part III explore in more depth the ways in which increasingly advanced technologies for robotics, artificial intelligence, and artificial life will impact organizational management.

The realities of organizational life are quickly catching up with the visions long explored by science fiction writers and futurists. Many of us can now expect to experience during our working lives a world in which 'ordinary' human beings labor alongside artificial general intelligences, social robots, sapient networks, nanorobotic swarms, and human beings with genetically engineered capacities and neurocybernetic implants. What strange new features might such a world contain? A robot boss embraced by its human subordinates because it is more empathetic, fair, honest, intelligent, and creative than its human predecessor. A customized product and marketing campaign designed for a single human consumer by an AI that can deduce the consumer's deepest fears and desires. Artificial life-forms that function as self-contained 'businesses' by gathering resources from the environment, transforming them into products, and selling them to consumers, all without the involvement of any human beings. Intelligent, evolvable bioelectronic viruses that can infect an organization's information infrastructure by moving effortlessly back and forth between human employees and their computers. Corporate espionage conducted by hacking into the video stream of a rival CEO's robotic eye. An office building or manufacturing facility or orbiting satellite or tropical resort where an organization's employees gather every day to work but which exists only as a persistent virtual world experienced using immersive multisensory VR. Employees who engage their colleagues as avatars within virtual environments, without knowing or caring whether a particular coworker is a 'normal' human being, uploaded human mind, social robot, or artificial general intelligence. Different classes and subclasses of 'metahuman' and 'parahuman' employees and customers who have been genetically engineered to possess radically nonhuman minds and bodies. Human workers who no longer control the intellectual property rights to their own thoughts, dreams, or memories, because they were produced with the assistance of neuroprosthetic implants or cognitive enhancement software provided by their employer. Human beings who are unable to quit their jobs because they rely on their employers for a lifetime supply of antivirus updates, immunosuppressive drugs, or physical maintenance for their full cyborg body. Human workers whose invasive neural interfaces allow them to dwell permanently within virtual worlds and whose physical bodies must be cared for by their employer's biomedical support division. Neurocybernetically linked human workers who lose their personal identity and merge to form a hive mind whose members share collective sensations, emotions, and volitions. A vast, tangled, digital-physical ecosystem in which an organization's human and synthetic employees, buildings, vehicles, manufacturing equipment, databases, products, and customers are all cybernetically linked through their participation in the 'Internet of Being.'

Such possibilities terrify some of us just as they exhilarate others. Because of the ongoing rapid technological developments taking place in many fields, these hypothetical scenarios present all who are involved with the study or management of organizations with complex ethical, legal, and operational questions whose thoughtful consideration cannot easily be further delayed.

The Ongoing Posthumanization of Organizations

It is widely acknowledged that the nature of human organizations is undergoing a profound transformation. Historic approaches to long-term strategic planning are increasingly being rendered obsolete by intensifying forces of globalized competition, rising worker mobility, and the breathtaking pace of technological change that is driving organizations of all types to devote growing resources to activities like online commerce, social media, cloud computing, data mining, and the development of artificially intelligent tools.[2] Rich bodies of scholarship and best practices have been formulated to guide organizations in grappling with such change. However, while such analyses are of great practical value for informing decision-making in areas like marketing, sales, logistics, and finance, they have barely begun to plumb the deeper forces which – at an ontological and phenomenological level – are reshaping human beings' capacity and desire to join with one another in the organized pursuit of shared goals.

---

[2] For a discussion of the ways in which multidimensional and synergistic 'business models' of the sort pioneered by technology firms are now supplementing or supplanting previous types of linear 'business plans,' see, e.g., Magretta, "Why Business Models Matter" (2002); Casadesus-Masanell & Ricart, "How to Design a Winning Business Model" (2011); and DaSilva & Trkman, "Business Model: What It Is and What It Is Not" (2014). Regarding the increasing difficulty – or even futility – of attempting to secure a competitive advantage of a lasting structural nature for an organization, see McGrath, *The End of Competitive Advantage: How to Keep Your Strategy Moving as Fast as Your Business* (2013).

Among the more noteworthy forces driving such change are those which can be collectively described as processes of *posthumanization*.[3] The dynamics of posthumanization are rewriting long-accepted rules about the kinds of entities that can serve as members of organizations, the sorts of structures that organizations can adopt to facilitate and regulate their internal activities, and the range of roles that organizations can play in their broader environment. One critical manifestation of posthumanization is seen in the changing nature of intelligent agency within our world. For millennia, organizations were fashioned and led by intelligent agents in the form of human beings – sometimes assisted by intelligent (though not sapient) agents in the form of dogs, horses, and other kinds of domesticated animals that filled specialized roles in support of their human caretakers. In many human societies, over the last century the role of animals as intelligent agents participating in the work of organizations has declined, while a new form of intelligent agent has emerged to take on roles critical to organizations' successful functioning: namely, computerized systems that are capable of collecting and processing information and then selecting and pursuing a course of action.

The conceptual and functional distinction between the sort of 'bioagency' exercised by human beings and the 'synthetic agency' exercised by such electronic computerized systems was originally quite clear.[4] However, the array of intelligent agency present and active in organizations is now undergoing a rapid evolution, thanks to the emergence of new technologies for social robotics, artificial intelligence, artificial life, neurocybernetic augmentation, and genetic engineering.[5] Through our increasingly intimate integration of computerized devices into our cognitive processes and bodies, human agency is taking on aspects traditionally seen in artificial agents; the notion of the 'cyborg' is no longer simply a concept found in science fiction but – to a greater or lesser degree – an accurate description of ourselves and the people we meet

---

[3] For an in-depth discussion of technological and nontechnological forms of posthumanization, see Chapter One of this book, "A Typology of Posthumanism: A Framework for Differentiating Analytic, Synthetic, Theoretical, and Practical Posthumanisms," and Herbrechter, *Posthumanism: A Critical Analysis* (2013).

[4] The 'bioagency' possessed by biological entities like human beings and 'cyberagency' demonstrated by artificial entities are distinguished in Fleischmann, "Sociotechnical Interaction and Cyborg–Cyborg Interaction: Transforming the Scale and Convergence of HCI" (2009).

[5] Such technologies are discussed in detail in Chapter Three of this book, "The Posthuman Management Matrix: Understanding the Organizational Impact of Radical Biotechnological Convergence."

around us every day.[6] At the same time, developments in the fields of robotics and AI are creating synthetic systems that possess levels of sociality, imagination, emotion, legal and moral responsibility, and metabolic processes resembling those that had previously been seen only in biological entities like human beings.[7]

Within organizations, information will be gathered and communicated, strategic decisions made, and actions undertaken by a kaleidoscopic web of intelligent agents which together form a complex cybernetic network. Such entities may include 'natural' human beings who have not been biotechnologically modified; human beings possessing neuroprosthetic implants that provide extensive sensory, motor, and cognitive enhancement;[8] human beings whose physical structures and biological processes have been intentionally sculpted through genetic engineering;[9] human beings who spend all of their time dwelling in virtual worlds;[10] virtualized entities resulting from a process

---

[6] The ever-increasing aspects of 'cyborgization' reflected in the minds and bodies of typical human beings are discussed, e.g., in Haraway, *Simians, Cyborgs, and Women: The Reinvention of Nature* (1991); Tomas, "Feedback and Cybernetics: Reimaging the Body in the Age of the Cyborg" (1995); Hayles, *How We Became Posthuman: Virtual Bodies in Cybernetics, Literature, and Informatics* (1999); Clark, *Natural-born cyborgs: Minds, Technologies, and the Future of Human Intelligence* (2004); and Fleischmann (2009).

[7] A comprehensive review of advances in developing sociality, emotions, and other cognitive and biological capacities for robots is found in Friedenberg, *Artificial Psychology: The Quest for What It Means to Be Human* (2008). For the ability of robots to bear responsibility for their actions, see, e.g., Calverley, "Imagining a non-biological machine as a legal person" (2008); Coeckelbergh, "From Killer Machines to Doctrines and Swarms, or Why Ethics of Military Robotics Is Not (Necessarily) About Robots" (2011); and Gladden, "The Diffuse Intelligent Other: An Ontology of Nonlocalizable Robots as Moral and Legal Actors" (2016).

[8] For anticipated growth in the use of implantable neuroprosthetic devices for purposes of human enhancement, see, e.g., McGee, "Bioelectronics and Implanted Devices" (2008); Gasson, "Human ICT Implants: From Restorative Application to Human Enhancement" (2012); and Gladden, "Neural Implants as Gateways to Digital-Physical Ecosystems and Posthuman Socioeconomic Interaction" (2016).

[9] See, e.g., Panno, *Gene Therapy: Treating Disease by Repairing Genes* (2005); Bostrom, "Human Genetic Enhancements: A Transhumanist Perspective" (2012); De Melo-Martín, "Genetically Modified Organisms (GMOs): Human Beings" (2015); and Nouvel, "A Scale and a Paradigmatic Framework for Human Enhancement" (2015).

[10] Implications of long-term immersion in virtual reality environments are discussed in Bainbridge, *The Virtual Future* (2011); Heim, *The Metaphysics of Virtual Reality* (1993); Geraci, *Apocalyptic AI: Visions of Heaven in Robotics, Artificial Intelligence, and Virtual Reality* (2010); and Koltko-Rivera, "The potential societal impact of virtual reality" (2005).

of 'mind uploading';[11] artificial general intelligences;[12] social robots;[13] decentralized nanorobotic swarms;[14] artificial organic or electronic life-forms,[15] including virtual or physical robots that evolve through processes of mutation and natural selection;[16] sentient or sapient networks;[17] and 'hive minds' comprising groups of diverse agents linked in such a way that they can share collective sensory experiences, emotions, and volitions.[18]

---

[11] For perspectives on 'mind uploading' (including issues that may render it impossible), see Moravec, *Mind Children: The Future of Robot and Human Intelligence* (1990); Hanson, "If uploads come first: The crack of a future dawn" (1994); Proudfoot, "Software Immortals: Science or Faith?" (2012); Pearce, "The Biointelligence Explosion" (2012); Koene, "Embracing Competitive Balance: The Case for Substrate-Independent Minds and Whole Brain Emulation" (2012); and Ferrando, "Posthumanism, Transhumanism, Antihumanism, Metahumanism, and New Materialisms: Differences and Relations" (2013), p. 27.

[12] Potential paths to the development of artificial general intelligence and obstacles to its creation are discussed in, e.g., *Artificial General Intelligence*, edited by Goertzel & Pennachin (2007); *Theoretical Foundations of Artificial General Intelligence*, edited by Wang & Goertzel (2012); and *Artificial General Intelligence: 8th International Conference, AGI 2015: Berlin, Germany, July 22-25, 2015: Proceedings*, edited by Bieger et al. (2015).

[13] Robots that can interact socially with human beings are discussed in, e.g., Breazeal, "Toward sociable robots" (2003); Gockley et al., "Designing Robots for Long-Term Social Interaction" (2005); Kanda & Ishiguro, *Human-Robot Interaction in Social Robotics* (2013); *Social Robots and the Future of Social Relations*, edited by Seibt et al. (2014); *Social Robots from a Human Perspective*, edited by Vincent et al. (2015); and *Social Robots: Boundaries, Potential, Challenges*, edited by Nørskov (2016).

[14] Swarm robotics are discussed in, e.g., Arkin & Hobbs, "Dimensions of communication and social organization in multi-agent robotic systems" (1993); Barca & Sekercioglu, "Swarm robotics reviewed" (2013); and Brambilla et al., "Swarm robotics: a review from the swarm engineering perspective" (2013). Regarding nanorobotic swarms, see, e.g., Ummat et al., "Bionanorobotics: A Field Inspired by Nature" (2005), and Pearce (2012).

[15] Artificial life-forms are discussed, e.g., in Andrianantoandro et al., "Synthetic biology: new engineering rules for an emerging discipline" (2006); Cheng & Lu, "Synthetic biology: an emerging engineering discipline" (2012); and Gladden, "The Artificial Life-Form as Entrepreneur: Synthetic Organism-Enterprises and the Reconceptualization of Business" (2014). For the relationship of artificial life and evolutionary robotics, see Friedenberg (2008), pp. 201-16.

[16] Evolutionary robotics and evolvable robotic hardware are reviewed in Friedenberg (2008), pp. 206-10.

[17] For a self-aware future Internet that is potentially a sort of living entity, see Hazen, "What is life?" (2006). Regarding a future Internet that is 'self-aware' even if not subjectively conscious, see Galis et al., "Management Architecture and Systems for Future Internet Networks" (2009), pp. 112-13. A sentient Internet is also discussed in Porterfield, "Be Aware of Your Inner Zombie" (2010), p. 19. Regarding collectively conscious networks and a "post-internet sentient network," see Callaghan, "Micro-Futures" (2014).

[18] For detailed taxonomies and classification systems for potential kinds of hive minds, see Chapter 2, "Hive Mind," in Kelly, *Out of control: the new biology of machines, social systems and the economic world* (1994); Kelly, "A Taxonomy of Minds" (2007); Kelly, "The Landscape of Possible

At the forefront of efforts to understand and consciously shape this integration of biological and artificial agents are those diverse bodies of thought and practice that constitute the phenomenon of posthumanism. And yet, while insights and methodologies from the field of posthumanism have been advantageously applied to many other spheres of human activity, there have so far been very few explicit links made between posthumanism and the work of integrating posthumanized agents to form effective organizations. In this book, we endeavor to inform and enhance contemporary approaches to the design and operation of organizations by fashioning such a bridge between posthumanist thought and the fields of organizational theory and management.

The book is divided into three main parts. In Part I, we examine topics that provide a general theoretical framework for and introduction to management in a posthumanized context. In Part II, we take a closer look at the ways in which organizational management will be affected by the growing posthumanizing augmentation of human beings through technologies such as neuroprosthetics, implantable computing, virtual reality, and genetic engineering. In Part III, we examine in more detail some of the ways in which increasingly advanced technologies for robotics, artificial intelligence, and artificial life will impact organizational management.

Part I: Posthuman Management: Background and Theoretical Foundations

In Part I of the book, a general theoretical and practical framework for the field of posthuman management is developed. Each chapter approaches this task at a different level, moving from the more abstract sphere of basic investigations into the nature of posthumanism to the more concrete sphere of formulating tools for posthumanized business analysis and exploring the implications of posthumanization for a specific management discipline (in this case, organization development).

The Nature of Posthumanization and Posthumanism

In Chapter One, "A Typology of Posthumanism," we consider the nature of posthumanization and the many phenomena that have been described as

Intelligences" (2008); Yonck, "Toward a standard metric of machine intelligence" (2012); and Yampolskiy, "The Universe of Minds" (2014). For discussion of systems whose behavior resembles that of a hive mind without a centralized controller, see Roden, *Posthuman Life: Philosophy at the Edge of the Human* (2014), p. 39. Hive minds are also discussed in Gladden, "Utopias and Dystopias as Cybernetic Information Systems: Envisioning the Posthuman Neuropolity" (2015). For critical perspectives on hive minds, see, e.g., Bendle, "Teleportation, cyborgs and the posthuman ideology" (2002), and Heylighen, "The Global Brain as a New Utopia" (2002).

forms of 'posthumanism,' in order to situate organizational posthumanism within a broader theoretical context. The array of activities that have been described as 'posthumanist' is quite diverse, ranging from literary criticism of Renaissance texts[19] and efforts by military research agencies to develop futuristic technologies for human enhancement[20] to spiritual movements and specific styles of performance art.[21] The question thus arises of whether these phenomena share anything in common at all – and if so, what is their shared dynamic and what are the characteristics that distinguish these different forms of posthumanism.

Much excellent work has been carried out by Ferrando, Herbrechter, Birnbacher, Miah, Miller, and others that explores the conceptual foundations of posthumanism. However, among such studies it can be noted that those research articles which are especially comprehensive and systematic in scope[22] must often – due to space limitations – refrain from exploring any particular form of posthumanism in depth. Meanwhile, the book-length analyses of posthumanism that are exceptionally thorough and detailed in their approach often focus on a single aspect of posthumanism rather than attempting to survey the phenomenon as a whole.[23] Moreover, existing analyses of posthumanism tend to emerge from fields such as critical theory, cultural studies, philosophy of technology, and bioethics; from the perspective of one who is interested in organizational theory and management, it takes considerable work to extract meaningful insights from such studies and reinterpret and apply them in ways relevant to organizational life.[24]

---

[19] See, e.g., *Posthumanist Shakespeares*, edited by Herbrechter & Callus (2012).

[20] For examples of the term 'posthuman' being used to describe technologies whose development is being pursued by DARPA and other military research and development agencies, see, e.g., Coker, "Biotechnology and War: The New Challenge" (2004); Graham, "Imagining Urban Warfare: Urbanization and U.S. Military Technoscience" (2008), p. 36; and Krishnan, "Enhanced Warfighters as Private Military Contractors" (2015).

[21] The spiritual aspects of some forms of transhumanism have been noted by numerous scholars; see, e.g., Bostrom, "Why I Want to Be a Posthuman When I Grow Up" (2008), p. 108, and Herbrechter (2013), pp. 103-04. The neohumanist spiritual movement developed by Sarkar might also be considered a form of posthumanism; see Sarkar, "Neohumanism Is the Ultimate Shelter (Discourse 11)" (1982), and the discussion of such neohumanism in Chapter One of this book, "A Typology of Posthumanism." The form of metahumanism developed by Del Val and Sorgner applies posthumanist ideals to performance art; see Del Val & Sorgner, "A *Metahumanist* Manifesto" (2011), and Del Val et al., "Interview on the Metahumanist Manifesto with Jaime del Val and Stefan Lorenz Sorgner" (2011).

[22] For example, see the insightful discussion in Ferrando (2013).

[23] Such an exposition and investigation of critical posthumanism is found, e.g., in Herbrechter (2013).

[24] There are several forward-thinking works of management scholarship that consider the impacts that posthumanizing technologies will have on future organizations; however, they do so

Chapter One attempts to synthesize and advance such existing analyses of posthumanism in a way that lays a conceptual foundation for understanding the varied processes of posthumanization that are relevant to specific topics in organizational theory and management. We begin by formulating a comprehensive typology that can be used to classify existing and future forms of posthumanism. The framework suggests that a given form of posthumanism can be classified either as *analytic* or *synthetic* and as either *theoretical* or *practical.* An analytic posthumanism understands 'posthumanity' as a sociotechnological reality that already exists in the contemporary world, such as the nonanthropocentric outlook found among some present-day evolutionary biologists, secular humanists, or animal-rights advocates that tends to minimize the distinctions between human beings and other biological species. A synthetic posthumanism is quite different: it understands 'posthumanity' as a collection of hypothetical future entities – such as full-body cyborgs or genetically engineered human beings – whose creation can either be intentionally realized or prevented, depending on whether humanity decides to develop and deploy particular technologies. A theoretical form of posthumanism is one that primarily seeks to develop new knowledge or cultivate new ways of understanding reality; posthumanist thought and study occurring on university campuses (and especially within the humanities) are often of this sort. Finally, a practical posthumanism seeks primarily to bring about some social, political, economic, or technological change in the world: efforts to develop new cryonics technologies or to engineer transhumanist genetic enhancements may be of this kind.

Arranging the properties of analytic/synthetic and theoretical/practical as two orthogonal axes creates a grid that can be used to categorize a form of posthumanism into one of four quadrants or as a hybrid posthumanism spanning all quadrants. We argue that analytic theoretical forms of posthumanism can collectively be understood as constituting a 'posthumanism of critique'; synthetic theoretical posthumanisms, a 'posthumanism of imagination'; syn-

without describing posthumanizing technologies as such or drawing significantly on the theoretical or methodological aspects of posthumanism. Such works might better be understood as a form of 'management futurology' grounded solidly in the field of organizational management rather than as a bridge between management and posthumanism. They include studies such as Berner's comprehensive review of the management implications of futuristic technologies in *Management in 20XX: What Will Be Important in the Future – A Holistic View* (2004). Posthumanist themes are considered more explicitly – although in a narrowly focused context – in, e.g., Mara & Hawk, "Posthuman rhetorics and technical communication" (2009), and Barile, "From the Posthuman Consumer to the *Ontobranding* Dimension: Geolocalization, Augmented Reality and Emotional Ontology as a Radical Redefinition of What Is Real" (2013).

thetic practical posthumanisms, a 'posthumanism of control'; analytic practical posthumanisms, a 'posthumanism of conversion'; and hybrid posthumanisms uniting all four elements as a 'posthumanism of production.'

Having developed this framework, we employ it to sift through a wide range of phenomena that have been identified as 'posthumanist' in the scholarly literature or popular discourse and to categorize them according to the framework's criteria. The phenomena thus classified include critical, cultural, philosophical, sociopolitical, and popular (or 'commercial') posthumanism; science fiction; techno-idealism; multiple forms of metahumanism and neohumanism; antihumanism; prehumanism; feminist new materialism; the posthumanities; and biopolitical posthumanism, including bioconservatism and transhumanism.[25] Given its notable presence in the popular consciousness, special attention is devoted to transhumanism, and three specialized sub-typologies are discussed for distinguishing different forms of transhumanism. Chapter One concludes by considering the form of organizational posthumanism developed in this book and classifying it as a form of hybrid posthumanism that spans all four quadrants of the framework.

Applying Posthumanist Thought to Organizational Theory and Management

In Chapter Two, "Organizational Posthumanism," the manners in which posthumanist insights can be applied to the theory and practice of organizational management are explored in more detail. We sketch out one way of fashioning posthumanist methodologies into a coherent management approach and chart out the potential scope of such a field. At its heart, the organizational posthumanism formulated in this text is a pragmatic approach to analyzing, understanding, creating, and managing organizations that is attuned to the intensifying processes of technological posthumanization and which employs a post-dualistic and post-anthropocentric perspective that can aid in recognizing challenges caused by the forces of posthumanization and developing innovative strategies for appropriately harnessing those forces.

Organizational posthumanism does not naïvely embrace all forms of posthumanization; unlike some strains of transhumanist thought, it does not presume that all emerging technologies for genetic engineering or nanorobotics are inherently beneficial and free from grave dangers. But at the same time,

[25] Many of these forms of posthumanism are identified in Ferrando (2013); others are discussed in Herbrechter (2013); Birnbacher, "Posthumanity, Transhumanism and Human Nature" (2008); Miah, "A Critical History of Posthumanism" (2008); and Miller, "Conclusion: Beyond the Human: Ontogenesis, Technology, and the Posthuman in Kubrick and Clarke's 2001" (2012). Some forms, such as sociopolitical posthumanism, are explicitly defined for the first time within the chapters in this volume. Detailed descriptions of all of these types of posthumanism are presented in Chapter One of this text, "A Typology of Posthumanism."

organizational posthumanism does not directly join bioconservatism in attempting to block the development of particular technologies deemed to be hazardous or destructive. Instead, organizational posthumanism focuses on analyzing posthumanizing technologies that are already available or whose development is expected in order to assess their (potential) impact on organizations and develop strategies for utilizing such technologies in ways that are ethical, impactful, and efficient. Organizational posthumanism recognizes that emerging technologies are likely to possess both benign and harmful applications, and the role of a manager as such is to identify and creatively exploit the beneficial aspects of a technology within a particular organizational context while simultaneously avoiding or ameliorating the technology's more detrimental effects.[26]

Indeed, like critical posthumanism and other forms of analytic posthumanism, organizational posthumanism recognizes that to a certain degree the world as a whole has already become 'posthumanized' through nontechnological processes: for example, regardless of whether a particular organization decides to acquire and exploit technologies for social robotics and neuroenhancement, the organization must account for the fact that its pool of (potential) employees, customers, and other stakeholders includes a growing number of individuals who, in different fashions and for varying reasons, possess increasingly nonanthropocentric and nondualistic ways of viewing reality. Thus engaging the realities of posthumanization is something that every contemporary organization must do of necessity; the only question is the extent to which an organization does so consciously and with a coherent strategy.

In order to develop an adequate framework for identifying the aspects of organizations that our study must address, we turn to fields like organizational architecture, enterprise architecture, and organizational design. When organizations are viewed through the lens of these disciplines, the relevance of six key elements becomes apparent: the forces of posthumanization are expected to increasingly expand and transform the kinds of *agent-members, personnel structures, information systems, processes and activities, physical and virtual*

[26] A human manager may simultaneously also be, for example, a follower of a particular religious tradition, a consumer, a voter, a patient, and a parent. In those other capacities, he or she may quite possibly work actively to spur or prevent the adoption of particular posthumanizing technologies, based on his or her adherence to posthumanist movements like bioconservatism or transhumanism. Organizational posthumanism does not attempt to study or shape all of those ways in which a human being may be related to posthumanizing forces and technologies; its scope only includes those mechanisms and dynamics by which posthumanization impacts the organization whose activities the manager is (co)responsible for directing.

spaces, and *external ecosystems* that organizations are able (or required) to utilize.[27] We argue that in each of these six areas, three different kinds of posthumanizing technologies will create new opportunities, threats, and exigencies that drive organizational change. The first kind is technologies for human augmentation and enhancement, which include many forms of neuroprosthetics, implantable computing, genetic engineering, and life extension.[28] The second is technologies for synthetic agency, which include artificial intelligence, artificial life, and diverse forms of robotics such as social, nano-, soft, and evolutionary robotics.[29] The third kind is technologies for digital-physical ecosystems and networks, which create new kinds of environments that human, artificial, and hybrid agents can inhabit and infrastructure through which they can interact. Such technologies might create persistent immersive virtual worlds and cybernetic networks whose topologies allow their agent-members to form collective hive minds.[30]

Managing the Transformation and Functional Convergence of Human and Artificial Agents

Chapter Two thus sketches the contours of organizational posthumanism as a field that can allow management theorists to understand the forces of posthumanization that are impacting organizations and management practitioners to anticipate and shape them. However, before attempting to apply such insights to the task of creating organizational designs and enterprise architectures for particular organizations, it would be helpful to have at one's disposal a more concrete guide for assessing the technological posthumanization of particular groups of agents, such as those comprising the (potential) stakeholders of an organization. To that end, in Chapter Three we formulate "The Posthuman Management Matrix," a conceptual tool for analyzing and

---

[27] For example, within the 'congruence model' of organizational architecture conceptualized by Nadler and Tushman, structures, processes, and systems constitute the three main elements of an organization that must be considered. See Nadler & Tushman, *Competing by Design: The Power of Organizational Architecture* (1997), p. 47.

[28] Biologically and nonbiologically based efforts at human life extension are compared in Koene (2012).

[29] An overview of such topics can be found, e.g., in Friedenberg (2008) and Murphy, *Introduction to AI Robotics* (2000).

[30] Regarding the ongoing evolution of the Internet to incorporate ever more diverse types of objects and entities, see Evans, "The Internet of Everything: How More Relevant and Valuable Connections Will Change the World" (2012). For a conceptual analysis of the interconnection between physical and virtual reality and different ways in which beings and objects can move between these worlds, see Kedzior, "How Digital Worlds Become Material: An Ethnographic and Netnographic Investigation in Second Life" (2014). Regarding the typologies of posthumanized cybernetic networks, see Gladden, "Utopias and Dystopias as Cybernetic Information Systems" (2015).

managing the behavior of agents within organizations where the boundaries between human beings and computers are becoming increasingly blurred.

Within the schema of this Matrix, an organization's employees and consumers can include two different kinds of agents (human and artificial agents), and the characteristics possessed by a specific agent belong to one of two sets ('anthropic' or 'computronic' characteristics). The model thus defines four different types of possible entities that might serve as organizational participants and stakeholders. The phrase 'human agents possessing anthropic characteristics' is simply another way of describing the natural human beings who have not been modified by posthumanizing technological processes such as neuroprosthetic enhancement or genetic engineering and who – from the dawn of human history – have served as the backbone of all organizations on earth. Disciplines like HR management and organization development offer many time-tested approaches for optimizing the performance of such human beings within an organizational context.

The phrase 'artificial agents with computronic characteristics' is another way of describing the ubiquitous electronic systems developed over the last half-century in which a computer utilizing a conventional Von Neumann architecture and running specialized software serves as an intelligent agent to perform assignments like transporting materials within production facilities;[31] wielding a robotic arm to perform assembly-line manufacturing tasks;[32] monitoring systems and facilities to detect physical or electronic intrusion attempts;[33] automatically scheduling tasks and optimizing the use of physical and electronic resources;[34] initiating financial transactions within online markets;[35] mining data to evaluate an applicant's credit risk or decide what personalized offers and advertisements to display to a website's visitors;[36] inter-

---

[31] See, e.g., *The Future of Automated Freight Transport: Concepts, Design and Implementation*, edited by Priemus & Nijkamp (2005), and Ullrich, *Automated Guided Vehicle Systems: A Primer with Practical Applications* (2015).

[32] For an overview of such technologies, see, e.g., *Intelligent Production Machines and Systems*, edited by Pham et al. (2006), and Perlberg, *Industrial Robotics* (2016).

[33] Regarding the automation of intrusion detection and prevention systems, see Rao & Nayak, *The InfoSec Handbook* (2014), pp. 226, 235, 238.

[34] For an overview of methods that can be employed for such purposes, see Pinedo, *Scheduling: Theory, Algorithms, and Systems* (2012), and *Automated Scheduling and Planning: From Theory to Practice*, edited by Etaner-Uyar et al. (2013).

[35] See Schacht, "The Buzz about Robo-Advisers" (2015); Dhar, "Should You Trust Your Money to a Robot?" (2015); Scopino, "Do Automated Trading Systems Dream of Manipulating the Price of Futures Contracts? Policing Markets for Improper Trading Practices by Algorithmic Robots" (2015); and Turner, "The computers have won, and Wall Street will never be the same" (2016).

[36] Regarding the role of automated systems in data mining, see, e.g., Giudici, *Applied Data Min-*

acting with customers through automated call centers, online chatbot interfaces, and physical kiosks to offer customer support;[37] or dispensing goods and services to customers.[38] The successful integration of such artificial agent technologies into organizational life is a major focus of contemporary management theory and practice.

However, the remaining two types of entities described by the Posthuman Management Matrix have historically been overlooked by the field of management – and understandably so, because of the fact that such entities have not existed as beings that could serve as workers, customers, and other organizational stakeholders. We argue, though, that such entities are now emerging as potential organizational actors, thanks to posthumanizing phenomena such as the development of increasingly powerful forms of neuroprosthetics, genetic engineering, virtual reality, robotics, and artificial intelligence. Human agents possessing computer-like physical and cognitive characteristics can be understood as real-world embodiments of the 'cyborgs' long envisioned in science fiction, while artificial agents possessing anthropic physical and cognitive characteristics will have very little in common with the desktop computers of earlier eras; they can be better understood as 'bioroids' whose form and behaviors resemble those of sophisticated biological entities like human beings.

We suggest that existing management approaches will prove ill-equipped for successfully understanding and shaping the activities of such novel posthumanized entities. New approaches are expected to emerge that allow organizations to identify and address the serious operational, legal, and ethical issues that will arise as human employees and consumers become more like computers and computerized agents more like biological human beings. Such efforts can build on the foundations developed by disciplines like cybernetics, systems theory, xenopsychology, and exoeconomics that employ a nonanthropocentric perspective and which formulate genericized principles that are equally well-suited to explaining the forms and dynamics of all kinds of agents, regardless of whether they are human, artificial, or hybrid in nature.[39]

---

ing: Statistical Methods for Business and Industry (2003); Provost & Fawcett, *Data Science for Business* (2013), p. 7; and Warkentin et al., "The Role of Intelligent Agents and Data Mining in Electronic Partnership Management" (2012), p. 13282.

[37] Such technologies are described, e.g., in Perez-Marin & Pascual-Nieto, *Conversational Agents and Natural Language Interaction: Techniques and Effective Practices* (2011); McIndoe, "Health Kiosk Technologies" (2010); and Ford, *Rise of the Robots: Technology and the Threat of a Jobless Future* (2015).

[38] See, e.g., the firsthand account of such technologies from the perspective of a potential consumer in Nazario, "I went to Best Buy and encountered a robot named Chloe – and now I'm convinced she's the future of retail" (2015).

[39] For a history of the use of 'xeno-' as a prefix to designate disciplines that study the forms or

Envisioning the Posthumanization of a Management Discipline: Organization Development

Chapter Four ("Organization Development and the Robotic-Cybernetic-Human Workforce: Humanistic Values for a Posthuman Future?") considers the implications of technological posthumanization for a concrete discipline within the field of management. Namely, we explore the extent to which organization development can and should help organizations grapple with the impacts of emerging posthuman technologies. Organization development (OD) is a management discipline whose theory and practice are firmly rooted in humanistic values, insofar as it seeks to create effective organizations by facilitating the empowerment and growth of their human members. However, a new posthuman era is dawning in which human beings will no longer be the only intelligent actors guiding the behavior of organizations; increasingly, social robots, AI programs, and cybernetically augmented human employees are taking on roles as collaborators and decision-makers in the workplace, and this transformation is only likely to accelerate. How should OD professionals react to the rise of these posthumanizing technologies? Several ways are suggested in which OD could act as a 'Humanist OD for a posthuman world,' providing an essential service to future organizations without abandoning its traditional humanist values. An alternative vision is then presented for a 'Posthuman OD' that reinterprets and expands its humanist vision to embrace the benefits that social robots, AI, and cyberization can potentially bring into the workplace. Finally, we discuss the extent to which OD can remain a single, unified discipline in light of the challenge to its traditional humanistic values that is presented by such emerging technologies.

Part II: Human Augmentation: A Closer Look

In Part II of the book, we take a closer look at the ways in which organizational management will be affected by the growing posthumanizing augmentation of human beings through technologies such as neuroprosthetics, virtual reality, and genetic engineering.

Neuroprosthetically Facilitated Socioeconomic Interaction

Chapter Five ("Neural Implants as Gateways to Digital-Physical Ecosystems and Posthuman Socioeconomic Interaction") looks beyond current desktop, mobile, and wearable technologies to argue that work-related information

---

behaviors of intelligent agents other than human beings (and in particular, those of hypothetical extraterrestrial life-forms), see the "Preface and Acknowledgements for the First Edition" in Freitas, *Xenology: An Introduction to the Scientific Study of Extraterrestrial Life, Intelligence, and Civilization* (1979). For a similar use of 'exopsychology' in connection with the study of the cognitive mechanisms and processes of potential extraterrestrial intelligences, see Harrison & Elms, "Psychology and the search for extraterrestrial intelligence" (1990), p. 207. Regarding the work of exoeconomists, see Ames, "The Place of an Individual in an Economy" (1981), p. 37.

and communications technology (ICT) will increasingly move inside the human body through the use of neuroprosthetic devices that create employees who are permanently connected to their workplace's digital ecosystems. Such persons may possess enhanced perception, memory, and abilities to manipulate physical and virtual environments and to link with human and synthetic minds to form cybernetic networks that can be both 'supersocial' and 'postsocial.' However, such neuroprosthetics may also create a sense of inauthenticity, vulnerability to computer viruses and hacking, financial burdens, and questions surrounding ownership of intellectual property produced using implants. Moreover, those populations who do and do not adopt neuroprostheses may come to inhabit increasingly incompatible and mutually incomprehensible digital ecosystems. Here we propose a cybernetic model for understanding how neuroprosthetics can either facilitate human beings' participation in posthuman informational ecosystems – or undermine their health, information security, and autonomy.

The Possibility and Necessity of New Business Models

In Chapter Six, "The Impacts of Human Neurocybernetic Enhancement on Organizational Business Models," we consider ways in which the increased use of neuroprosthetics – among both organizations' employees and customers – may require companies to transform their business models. The social and economic impact of such neuroprosthetics will be dramatic: these devices are expected to reshape the ways in which human beings interact with one another, enabling them to create novel forms of social structures and organizations and to engage in new kinds of informational and economic exchange that were never previously possible. As participants in the larger societies, economies, and informational ecosystems within which they exist, businesses will be impacted by the widening use of neuroprosthetic technologies. Companies that are able to identify, understand, and anticipate these technological and social changes and transform their business models accordingly may be able to secure significant competitive advantages. On the other hand, companies that are not able to adapt their business models quickly enough to the social, economic, political, cultural, and ethical changes driven by neuroprosthetic technologies may find themselves unable to compete, grow, or even survive. In this text, we briefly consider the concept of a 'business model' and the situations that require a company to change its business model. We then explore three main areas in which the rise of neuroprosthetics is expected to transform humanity. Finally, we identify the impact that such changes will have on companies' business models and consider an example highlighting the reasons why many companies will need to adopt new business models in order to address these altered realities.

New Considerations for Organizational Information Security

Chapter Seven ("Implantable Computers and Information Security: A Managerial Perspective") considers the impact that the increased use of neuroprosthetic devices and other implantable computers will have on information security, especially as understood within an organizational context. The interdisciplinary field of information security already draws significantly on the biological and human sciences; for example, it relies on the knowledge of human physiology to design biometric authentication devices and utilizes insights from psychology to predict users' vulnerability to social engineering techniques and to develop preventative measures. The growing use of computers that are implanted within the human body for purposes of therapy or augmentation will compel the field of information security to develop relationships with fields such as medicine and biomedical engineering that are closer than and qualitatively different from those that have previously existed, insofar as the technologies and practices that InfoSec implements for implantable computers must not only secure the information contained within such devices but must also avoid creating biological or psychological harm (or even the danger of such harm) for the human beings within whose organisms the computers are embedded. In this text we identify unique issues and challenges that implantable computers create for information security. By considering the scenario of the computer contained within a sensory neuroprosthetic device in the form of an advanced retinal implant, we demonstrate the ways in which information security's traditional concepts of the confidentiality, integrity, and availability of information and the use of physical, logical, and administrative access controls become intertwined with issues of medicine and biomedical engineering. Finally, we explore the idea of 'celyphocybernetics' as a paradigm for conceptualizing the relationship of information security to medicine and biomedical engineering insofar as it relates to implantable computers.

Part III: Robotics and Artificial Intelligence: A Closer Look

In Part III of the book, we examine in more detail some of the ways in which increasingly advanced technologies for robotics, artificial intelligence, and artificial life will impact organizational management.

Could a Robot Succeed as the CEO of a Human Organization?

Chapter Eight, "The Social Robot as CEO: Developing Synthetic Charismatic Leadership for Human Organizations," explores one aspect of the question of whether a robot could ever serve successfully as the CEO of an organization that includes human personnel. Among the many important functions that a CEO must perform is that of motivating and inspiring a company's workers and cultivating their trust in the company's strategic direction and leadership. It might appear that creating a robot that can successfully inspire and win the trust of an organization's human personnel is a major hurdle to the development of a robot that can serve effectively as CEO of a company that includes human workers. In this text, however, we argue that the development of social robots that are capable of manifesting these leadership traits needed to serve as CEO within an otherwise human organization is not only possible but likely even inevitable.

We begin by analyzing what French and Raven refer to as 'referent power' and what Weber describes as 'charismatic authority' – two related characteristics which if possessed by a social robot could allow that robot to lead human personnel by inspiring and motivating them and securing their loyalty and trust.[40] By analyzing current robotic design efforts and literary depictions of robots, we suggest three ways in which human beings are striving to create charismatic robot leaders for ourselves. We then consider the manner in which particular robot leaders will acquire human trust, arguing that charismatic robot leaders for businesses and other kinds of institutions will emerge naturally from within our world's social fabric, without any rational decision on our part. Finally, drawing on Abrams,[41] we suggest that the stability of these leader-follower relations – and the extent to which charismatic social robots can remain long-term fixtures in leadership roles such as that of CEO – will hinge on a fundamental question of robotic intelligence and motivation that currently stands unresolved.

Future Robots as Actors That Are More Powerful – and More Elusive

In Chapter Nine, "The Diffuse Intelligent Other: An Ontology of Nonlocalizable Robots as Moral and Legal Actors," we explore the assumption that the physical forms and capacities displayed by future robots will resemble those seen in the past, and we consider the implications that the changing form of robots will have for the challenge of assigning moral and legal responsibility for actions performed by such robots. Already much thought has been given to the question of who bears moral and legal responsibility for actions performed by robots. Some argue that responsibility could be attributed to a robot if it possessed human-like autonomy and metavolitionality and that while such capacities can potentially be possessed by a robot with a single spatially compact body, they cannot be possessed by a spatially disjunct, decentralized collective such as a robotic swarm or network. However, advances in ubiqui-

---

[40] See, e.g., Forsyth, *Group Dynamics* (2010), p. 227, and Weber, *Economy and Society: An Outline of Interpretive Sociology* (1968), p. 215.

[41] See Abrams, "Pragmatism, Artificial Intelligence, and Posthuman Bioethics: Shusterman, Rorty, Foucault" (2004).

tous robotics and distributed computing open the door to a new form of robotic entity that possesses a unitary intelligence, despite the fact that its cognitive processes are not confined within a single spatially compact, persistent, identifiable body. Such a 'nonlocalizable' robot may possess a body whose myriad components interact with one another at a distance and which is continuously transforming as components join and leave the body. Here we develop an ontology for classifying such robots on the basis of their autonomy, volitionality, and localizability. Using this ontology, we explore the extent to which nonlocalizable robots – including those possessing cognitive abilities that match or exceed those of human beings – can be considered moral and legal actors that are responsible for their own actions.

Artificial Agents as Managers of Diverse Virtual Human Teams

Chapter Ten, "Leveraging the Cross-Cultural Capacities of Artificial Agents as Leaders of Human Virtual Teams," explores the possibility that artificial agents might be employed to manage cross-cultural virtual teams of human and synthetic workers. The human beings who manage global virtual teams regularly face challenges caused by factors such as the lack of a shared language and culture among team members and coordination delay resulting from spatial and temporal divisions between members of the team. As part of the ongoing advances in artificial agent (AA) technology, artificial agents have been developed whose purpose is to assist the human managers of virtual teams. In this text, we move a step further by suggesting that new capabilities being developed for artificial agents will eventually give them the ability to successfully manage virtual teams whose other members are human beings. In particular, artificial agents will be uniquely positioned to take on roles as managers of cross-cultural, multilingual, global virtual teams, by overcoming some of the fundamental cognitive limitations that create obstacles for human beings serving in these managerial roles. In order to effectively interact with human team members, AAs must be able to decode and encode the full spectrum of verbal and nonverbal communication used by human beings. Because culture is so deeply embedded in all human forms of communication, AAs cannot communicate in a way that is 'non-cultural'; an AA that is capable of communicating effectively with human team members will necessarily display a particular culture (or mix of cultures), just as human beings do. The need for AA team leaders to display cultural behavior raises the key question of *which* culture or cultures the AA leader of a particular human virtual team should display. We argue that the answer to this question depends on both the cultural makeup of a team's human members and the methods used to share information among team members. To facilitate the analysis of how an AA team leader's cultural behaviors can best be structured to fit the circumstances of a particular virtual team, we propose a two-dimensional model for designing suites of cultural behaviors for AAs that will manage human virtual teams. We consider examples of each type of AA described by the model, identify potential strengths and weaknesses of each type, suggest particular kinds of virtual teams that are likely to benefit from being managed by AAs of the different types, and discuss empirical study that can test the validity and usefulness of this framework.

Quantitative Approaches to Comparing Human and Artificial Agents as Managers

In Chapter Eleven, "A Fractal Measure for Comparing the Work Effort of Human and Artificial Agents Performing Management Functions," we consider in detail one aspect of efforts to determine whether a robot would perform better, worse, or simply differently than a human worker in a particular management role. Thanks to the growing sophistication of artificial agent technologies, businesses will increasingly face decisions of whether to have a human employee or artificial agent perform a particular function. This makes it desirable to have a common temporal measure for comparing the work effort that human beings and artificial agents can apply to a role. Existing temporal measures of work effort are formulated to apply either to human employees (e.g., FTE and billable hours) or computer-based systems (e.g., mean time to failure and availability) but not both. In this paper we propose a new temporal measure of work effort based on fractal dimension that applies equally to the work of human beings and artificial agents performing management functions. We then consider four potential cases to demonstrate the measure's diagnostic value in assessing strengths (e.g., flexibility) and risks (e.g., switch costs) reflected by the temporal work dynamics of particular managers.

Developing Managerial Robots with Capabilities Comparable to Those of Human Managers

Chapter Twelve, "Managerial Robotics: A Model of Sociality and Autonomy for Robots Managing Human Beings and Machines," investigates some concrete characteristics of existing robots in an effort to better understand whether and how such characteristics might be fundamentally intertwined with one another and how they might be manifested in more advanced future types of robots. The development of robots with increasingly sophisticated decision-making and social capacities is opening the door to the possibility of robots carrying out the management functions of planning, organizing, leading, and controlling the work of human beings and other machines.[42] In this text we study the relationship between two traits that impact a robot's ability

[42] Building on the classic management framework developed by Henri Fayol, Daft identifies the four essential functions of a manager as *planning*, *organizing*, *leading*, and *controlling* activities within an organization. See Daft, *Management* (2011).

to effectively perform management functions: those of *autonomy* and *sociality*. Using an assessment instrument we evaluate the levels of autonomy and sociality of 35 robots that have been created for use in a wide range of industrial, domestic, and governmental contexts, along with several kinds of living organisms with which such robots can share a social space and which may provide templates for some aspects of future robotic design. We then develop a two-dimensional model that classifies the robots into 16 different types, each of which offers unique strengths and weaknesses for the performance of management functions. Our data suggest correlations between autonomy and sociality that could potentially assist organizations in identifying new and more effective management applications for existing robots and aid roboticists in designing new kinds of robots that are capable of succeeding in particular management roles.

Artificial Life Forms Functioning Autonomously as Businesses within the Real-world Economy

In Chapter Thirteen, "Developing a Non-anthropocentric Definition of Business: A Cybernetic Model of the Synthetic Life-form as Autonomous Enterprise," it is noted that operating a business has traditionally been considered an exclusively human activity: while domesticated animals or desktop computers, for example, might participate in the work of a business, they are not in themselves capable of organizing or constituting a 'business.' However, the increasing sophistication and capacities of social robots, synthetic life-forms, and other kinds of artificial agents raises the question of whether some such entities might be capable not only of leading a business but of directly constituting one. In this text we argue that it is theoretically possible to create artificial life-forms that function as autonomous businesses within the real-world human economy and explore some of the implications of the development of such beings. Building on the cybernetic framework of the Viable Systems Approach (VSA), we formulate the concept of an 'organism-enterprise' that exists simultaneously as both a life-form and a business. The possible existence of such entities both enables and encourages us to reconceptualize the historically anthropocentric understanding of a 'business' in a way that allows an artificial life-form to constitute a 'synthetic' organism-enterprise (SOE) just as a human being acting as a sole proprietor constitutes a 'natural' organism-enterprise. Such SOEs would exist and operate in a sphere beyond that of current examples of artificial life, which produce goods or services within some simulated world or play a limited role as tools or assistants within a human business. Rather than competing against other artificial organisms in a virtual world, SOEs could potentially survive and evolve through competition against human businesses in our real-world economy. We conclude by briefly envisioning particular examples of SOEs that elucidate some of the legal, economic, and ethical issues that arise when a single economic ecosystem is shared by competing human and artificial life. It is suggested that the theoretical model of synthetic organism-enterprises developed in this text may provide a useful conceptual foundation for computer programmers, engineers, economists, management scholars and practitioners, ethicists, policymakers, and others who will be called upon in the coming years to grapple with the realities of artificial agents that increasingly function as autonomous enterprises within our world's complex economic ecosystem.

Conclusion

It is our hope that the questions raised, topics explored, and approaches suggested in this book can draw attention to an important element that is largely missing from contemporary debates surrounding emerging transformative technologies, which often focus on issues of economics (such as the question of whether increasing roboticization will produce mass human unemployment[43]) or bioethics (such as the question of whether neuroprosthetic devices that alter a user's personality or memories are ethically permissible[44]). Namely, we aim to highlight the fact that those posthumanizing technologies that transform the nature of human and synthetic agency will necessarily also transform the nature of the organizations for which agents serve as workers, consumers, managers, investors, and other stakeholders. Given the fact that almost every aspect of human existence is intimately connected with the activity of human organizations, the forces of posthumanization that enable or impel dramatic changes in such organizations will impact every corner of our lives. The extent to which such radical change can be anticipated and consciously shaped by organizations and those who manage them may largely determine the quality of the world – or worlds – experienced by generations of human beings to come.

References

Abrams, J.J. "Pragmatism, Artificial Intelligence, and Posthuman Bioethics: Shusterman, Rorty, Foucault." *Human Studies* vol. 27, no. 3 (2004): 241-58.

Ames, Edward. "The Place of an Individual in an Economy." In *Essays in Contemporary Fields of Economics: In Honor of Emanuel T. Weiler (1914-1979)*, edited by

---

[43] See, for example, Sachs et al., "Robots: Curse or Blessing? A Basic Framework" (2015), and Ford (2015).

[44] See, e.g., Maguire & McGee, "Implantable brain chips? Time for debate" (1999); Khushf, "The use of emergent technologies for enhancing human performance: Are we prepared to address the ethical and policy issues" (2005); Soussou & Berger, "Cognitive and Emotional Neuroprostheses" (2008); Kraemer, "Me, Myself and My Brain Implant: Deep Brain Stimulation Raises Questions of Personal Authenticity and Alienation" (2011); and Van den Berg, "Pieces of Me: On Identity and Information and Communications Technology Implants" (2012).

George Horwich and James P. Quirk, pp. 24-40. West Lafayette, IN: Purdue University Press, 1981.

Andrianantoandro, Ernesto, Subhayu Basu, David K. Karig, and Ron Weiss. "Synthetic biology: new engineering rules for an emerging discipline." *Molecular Systems Biology* 2, no. 1 (2006).

Arkin, Ronald C., and J. David Hobbs. "Dimensions of communication and social organization in multi-agent robotic systems." In *From Animals to Animats 2: Proceedings of the Second International Conference on Simulation of Adaptive Behavior,* edited by Jean-Arcady Meyer, H. L. Roitblat and Stewart W. Wilson, pp. 486-93. Cambridge, MA: The MIT Press, 1993.

Artificial General Intelligence, edited by Ben Goertzel and Cassio Pennachin. Springer Berlin Heidelberg, 2007.

Artificial General Intelligence: 8th International Conference, AGI 2015: Berlin, Germany, July 22-25, 2015: Proceedings, edited by Jordi Bieger, Ben Goertzel, and Alexey Potapov. Springer International Publishing, 2015.

Automated Scheduling and Planning: From Theory to Practice, edited by A. Şima Etaner-Uyar, Ender Özcan, and Neil Urquhart. Springer Berlin Heidelberg, 2013.

Bainbridge, William Sims. *The Virtual Future.* London: Springer, 2011.

Barca, Jan Carlo, and Y. Ahmet Sekercioglu. "Swarm robotics reviewed." *Robotica* 31, no. 03 (2013): 345-59.

Barile, Nello. "From the Posthuman Consumer to the *Ontobranding* Dimension: Geolocalization, Augmented Reality and Emotional Ontology as a Radical Redefinition of What Is Real." *intervalla: platform for intellectual exchange,* vol. 1 (2013). http://www.fus.edu/intervalla/images/pdf/9_barile.pdf. Accessed May 18, 2016.

Barile, S., J. Pels, F. Polese, and M. Saviano. "An Introduction to the Viable Systems Approach and Its Contribution to Marketing." *Journal of Business Market Management* 5(2) (2012): 54-78.

Beer, Stafford. *Brain of the Firm.* Second edition. New York: John Wiley, 1981.

Bendle, Mervyn F. "Teleportation, cyborgs and the posthuman ideology." *Social Semiotics* 12, no. 1 (2002): 45-62.

Berner, Georg. *Management in 20XX: What Will Be Important in the Future – A Holistic View.* Erlangen: Publicis Corporate Publishing, 2004.

Birnbacher, Dieter. "Posthumanity, Transhumanism and Human Nature." In *Medical Enhancement and Posthumanity,* edited by Bert Gordijn and Ruth Chadwick, pp. 95-106. The International Library of Ethics, Law and Technology 2. Springer Netherlands, 2008.

Bostrom, Nick. "Human Genetic Enhancements: A Transhumanist Perspective." In *Arguing About Bioethics,* edited by Stephen Holland, pp. 105-15. New York: Routledge, 2012.

Bostrom, Nick. "Why I Want to Be a Posthuman When I Grow Up." In *Medical Enhancement and Posthumanity,* edited by Bert Gordijn and Ruth Chadwick, pp.

107-37. The International Library of Ethics, Law and Technology 2. Springer Netherlands, 2008.

Brambilla, Manuele, Eliseo Ferrante, Mauro Birattari, and Marco Dorigo. "Swarm robotics: a review from the swarm engineering perspective." *Swarm Intelligence* 7, no. 1 (2013): 1-41.

Breazeal, Cynthia. "Toward sociable robots." *Robotics and Autonomous Systems* 42 (2003): 167-75.

Callaghan, Vic. "Micro-Futures." Presentation at Creative-Science 2014, Shanghai, China, July 1, 2014.

Calverley, D.J. "Imagining a non-biological machine as a legal person." *AI & SOCIETY* 22, no. 4 (2008): 523-37.

Casadesus-Masanell, Ramon, and Joan E. Ricart. "How to Design a Winning Business Model." *Harvard Business Review* 89, no. 1-2 (2011): 100-07.

Cheng, Allen A., and Timothy K. Lu. "Synthetic biology: an emerging engineering discipline." *Annual Review of Biomedical Engineering* 14 (2012): 155-78.

Clark, Andy. *Natural-born cyborgs: Minds, Technologies, and the Future of Human Intelligence.* Oxford: Oxford University Press, 2004.

Coeckelbergh, Mark. "From Killer Machines to Doctrines and Swarms, or Why Ethics of Military Robotics Is Not (Necessarily) About Robots." *Philosophy & Technology* 24, no. 3 (2011): 269-78.

Coker, Christopher. "Biotechnology and War: The New Challenge." *Australian Army Journal* vol. II, no. 1 (2004): 125-40.

Daft, Richard. *Management,* Mason, OH: South-Western, Cengage Learning, 2011.

DaSilva, Carlos M., and Peter Trkman. "Business Model: What It Is and What It Is Not." *Long Range Planning* 47 (2014): 379-89.

De Melo-Martín, Inmaculada. "Genetically Modified Organisms (GMOs): Human Beings." In *Encyclopedia of Global Bioethics,* edited by Henk ten Have. Springer Science+Business Media Dordrecht. Version of March 13, 2015. doi: 10.1007/978-3-319-05544-2_210-1. Accessed January 21, 2016.

Del Val, Jaime, and Stefan Lorenz Sorgner. "A *Metahumanist* Manifesto." *The Agonist* IV no. II (Fall 2011). http://www.nietzschecircle.com/AGONIST/2011_08/METAHUMAN_MANIFESTO.html. Accessed March 2, 2016.

Del Val, Jaime, Stefan Lorenz Sorgner, and Yunus Tuncel. "Interview on the Metahumanist Manifesto with Jaime del Val and Stefan Lorenz Sorgner." *The Agonist* IV no. II (Fall 2011). http://www.nietzschecircle.com/AGONIST/2011_08/Interview_Sorgner_Stefan-Jaime.pdf. Accessed March 2, 2016.

Dhar, Vasant. "Should You Trust Your Money to a Robot?" *Big Data* 3, no. 2 (2015): 55-58.

Evans, Dave. "The Internet of Everything: How More Relevant and Valuable Connections Will Change the World." Cisco Internet Solutions Business Group: Point of

View, 2012. https://www.cisco.com/web/about/ac79/docs/innov/IoE.pdf. Accessed December 16, 2015.

Ferrando, Francesca. "Posthumanism, Transhumanism, Antihumanism, Metahumanism, and New Materialisms: Differences and Relations." *Existenz: An International Journal in Philosophy, Religion, Politics, and the Arts* 8, no. 2 (Fall 2013): 26-32.

Fleischmann, Kenneth R. "Sociotechnical Interaction and Cyborg–Cyborg Interaction: Transforming the Scale and Convergence of HCI." *The Information Society* 25, no. 4 (2009): 227-35. doi:10.1080/01972240903028359.

Ford, Martin. *Rise of the Robots: Technology and the Threat of a Jobless Future*. New York: Basic Books, 2015.

Forsyth, D.R. *Group Dynamics*, 5e. Belmont, CA: Cengage Learning, 2010.

Freitas Jr., Robert A. "Preface and Acknowledgements for the First Edition." In *Xenology: An Introduction to the Scientific Study of Extraterrestrial Life, Intelligence, and Civilization*. Sacramento: Xenology Research Institute, 1979. http://www.xenology.info/Xeno/PrefaceFirstEdition.htm, last updated October 22, 2009. Accessed January 30, 2016.

Friedenberg, Jay. *Artificial Psychology: The Quest for What It Means to Be Human*. Philadelphia: Psychology Press, 2008.

The Future of Automated Freight Transport: Concepts, Design and Implementation, edited by Hugo Priemus and Peter Nijkamp. Cheltenham: Edward Elgar Publishing, 2005.

Galis, Alex, Spyros G. Denazis, Alessandro Bassi, Pierpaolo Giacomin, Andreas Berl, Andreas Fischer, Hermann de Meer, J. Srassner, S. Davy, D. Macedo, G. Pujolle, J. R. Loyola, J. Serrat, L. Lefevre, and A. Cheniour. "Management Architecture and Systems for Future Internet Networks." In *Towards the Future Internet: A European Research Perspective*, edited by Georgios Tselentis, John Domingue, Alex Galis, Anastasius Gavras, David Hausheer, Srdjan Krco, Volkmar Lotz, and Theodore Zahariadis, pp. 112-22. IOS Press, 2009.

Gasson, M.N. "Human ICT Implants: From Restorative Application to Human Enhancement." In *Human ICT Implants: Technical, Legal and Ethical Considerations*, edited by Mark N. Gasson, Eleni Kosta, and Diana M. Bowman, pp. 11-28. Information Technology and Law Series 23. T. M. C. Asser Press, 2012.

Geraci, Robert M. *Apocalyptic AI: Visions of Heaven in Robotics, Artificial Intelligence, and Virtual Reality*. New York: Oxford University Press, 2010.

Giudici, P. *Applied Data Mining: Statistical Methods for Business and Industry*. Wiley, 2003.

Gladden, Matthew E. "The Artificial Life-Form as Entrepreneur: Synthetic Organism-Enterprises and the Reconceptualization of Business." In *Proceedings of the Fourteenth International Conference on the Synthesis and Simulation of Living Systems*, edited by Hiroki Sayama, John Rieffel, Sebastian Risi, René Doursat and Hod Lipson, pp. 417-18. The MIT Press, 2014.

Gladden, Matthew E. "The Diffuse Intelligent Other: An Ontology of Nonlocalizable Robots as Moral and Legal Actors." In *Social Robots: Boundaries, Potential, Challenges*, edited by Marco Nørskov, pp. 177-98. Farnham: Ashgate, 2016.

Gladden, Matthew E. "Neural Implants as Gateways to Digital-Physical Ecosystems and Posthuman Socioeconomic Interaction." In *Digital Ecosystems: Society in the Digital Age*, edited by Łukasz Jonak, Natalia Juchniewicz, and Renata Włoch, pp. 85-98. Warsaw: Digital Economy Lab, University of Warsaw, 2016.

Gladden, Matthew E. "Utopias and Dystopias as Cybernetic Information Systems: Envisioning the Posthuman Neuropolity." *Creatio Fantastica* nr 3 (50) (2015).

Gockley, Rachel, Allison Bruce, Jodi Forlizzi, Marek Michalowski, Anne Mundell, Stephanie Rosenthal, Brennan Sellner, Reid Simmons, Kevin Snipes, Alan C. Schultz, and Jue Wang. "Designing Robots for Long-Term Social Interaction." In *2005 IEEE/RSJ International Conference on Intelligent Robots and Systems (IROS 2005)*, pp. 2199-2204. 2005.

Graham, Stephen. "Imagining Urban Warfare: Urbanization and U.S. Military Technoscience." In *War, Citizenship, Territory*, edited by Deborah Cowen and Emily Gilbert. New York: Routledge, 2008.

Hanson, R. "If uploads come first: The crack of a future dawn." *Extropy* 6, no. 2 (1994): 10-15.

Haraway, Donna. *Simians, Cyborgs, and Women: The Reinvention of Nature.* New York: Routledge, 1991.

Harrison, Albert A., and Alan C. Elms. "Psychology and the search for extraterrestrial intelligence." *Behavioral Science* 35, no. 3 (1990): 207-18.

Hayles, N. Katherine. *How We Became Posthuman: Virtual Bodies in Cybernetics, Literature, and Informatics*, Chicago: University of Chicago Press, 1999.

Hazen, Robert. "What is life?", *New Scientist* 192, no. 2578 (2006): 46-51.

Heim, Michael. *The Metaphysics of Virtual Reality.* New York: Oxford University Press, 1993.

Herbrechter, Stefan. *Posthumanism: A Critical Analysis.* London: Bloomsbury, 2013. [Kindle edition.]

Heylighen, Francis. "The Global Brain as a New Utopia." In *Renaissance der Utopie. Zukunftsfiguren des 21. Jahrhunderts*, edited by R. Maresch and F. Rötzer. Frankfurt: Suhrkamp, 2002.

Intelligent Production Machines and Systems, edited by Duc T. Pham, Eldaw E. Eldukhri, and Anthony J. Soroka. Amsterdam: Elsevier, 2006.

Kanda, Takayuki, and Hiroshi Ishiguro. *Human-Robot Interaction in Social Robotics.* Boca Raton: CRC Press, 2013.

Kedzior, Richard. *How Digital Worlds Become Material: An Ethnographic and Netnographic Investigation in Second Life.* Economics and Society: Publications of the Hanken School of Economics Nr. 281. Helsinki: Hanken School of Economics, 2014.

Kelly, Kevin. "The Landscape of Possible Intelligences." *The Technium*, September 10, 2008. http://kk.org/thetechnium/the-landscape-o/. Accessed January 25, 2016.

Kelly, Kevin. *Out of Control: The New Biology of Machines, Social Systems and the Economic World.* Basic Books, 1994.

Kelly, Kevin. "A Taxonomy of Minds." *The Technium*, February 15, 2007. http://kk.org/thetechnium/a-taxonomy-of-m/. Accessed January 25, 2016.

Khushf, George. "The use of emergent technologies for enhancing human performance: Are we prepared to address the ethical and policy issues." *Public Policy and Practice* 4, no. 2 (2005): 1-17.

Koene, Randal A. "Embracing Competitive Balance: The Case for Substrate-Independent Minds and Whole Brain Emulation." In *Singularity Hypotheses*, edited by Amnon H. Eden, James H. Moor, Johnny H. Søraker, and Eric Steinhart, pp. 241-67. The Frontiers Collection. Springer Berlin Heidelberg, 2012.

Koltko-Rivera, Mark E. "The potential societal impact of virtual reality." *Advances in virtual environments technology: Musings on design, evaluation, and applications* 9 (2005).

Kraemer, Felicitas. "Me, Myself and My Brain Implant: Deep Brain Stimulation Raises Questions of Personal Authenticity and Alienation." *Neuroethics* 6, no. 3 (May 12, 2011): 483-97. doi:10.1007/s12152-011-9115-7.

Krishnan, Armin. "Enhanced Warfighters as Private Military Contractors." In *Super Soldiers: The Ethical, Legal and Social Implications*, edited by Jai Galliott and Mianna Lotz. London: Routledge, 2015.

Magretta, Joan. "Why Business Models Matter." *Harvard Business Review* 80, no. 5 (2002): 86-92.

Maguire, Gerald Q., and Ellen M. McGee. "Implantable brain chips? Time for debate." *Hastings Center Report* 29, no. 1 (1999): 7-13.

Mara, Andrew, and Byron Hawk. "Posthuman rhetorics and technical communication." *Technical Communication Quarterly* 19, no. 1 (2009): 1-10.

McGee, E.M. "Bioelectronics and Implanted Devices." In *Medical Enhancement and Posthumanity*, edited by Bert Gordijn and Ruth Chadwick, pp. 207-24. The International Library of Ethics, Law and Technology 2. Springer Netherlands, 2008.

McGrath, Rita Gunther. *The End of Competitive Advantage: How to Keep Your Strategy Moving as Fast as Your Business.* Boston: Harvard Business Review Press, 2013.

McIndoe, Robert S. "Health Kiosk Technologies." In *Ethical Issues and Security Monitoring Trends in Global Healthcare: Technological Advancements: Technological Advancements*, edited by Steven A. Brown and Mary Brown, pp. 66-71. Hershey: Medical Information Science Reference, 2010.

Miah, Andy. "A Critical History of Posthumanism." In *Medical Enhancement and Posthumanity*, edited by Bert Gordijn and Ruth Chadwick, pp. 71-94. The International Library of Ethics, Law and Technology 2. Springer Netherlands, 2008.

Miller, Jr., Gerald Alva. "Conclusion: Beyond the Human: Ontogenesis, Technology, and the Posthuman in Kubrick and Clarke's 2001." In *Exploring the Limits of the Human through Science Fiction*, pp. 163-90. American Literature Readings in the 21st Century. Palgrave Macmillan US, 2012.

Moravec, Hans. *Mind Children: The Future of Robot and Human Intelligence.* Cambridge: Harvard University Press, 1990.

Murphy, Robin. *Introduction to AI Robotics.* Cambridge, MA: The MIT Press, 2000.

Nadler, David, and Michael Tushman. *Competing by Design: The Power of Organizational Architecture.* Oxford University Press, 1997. [Kindle edition.]

Nazario, Marina. "I went to Best Buy and encountered a robot named Chloe – and now I'm convinced she's the future of retail." *Business Insider*, October 23, 2015. http://www.businessinsider.com/what-its-like-to-use-best-buys-robot-2015-10. Accessed May 20, 2016.

Nouvel, Pascal. "A Scale and a Paradigmatic Framework for Human Enhancement." In *Inquiring into Human Enhancement*, edited by Simone Bateman, Jean Gayon, Sylvie Allouche, Jérôme Goffette, and Michela Marzano, pp. 103-18. Palgrave Macmillan UK, 2015.

Panno, Joseph. *Gene Therapy: Treating Disease by Repairing Genes.* New York: Facts on File, 2005.

Pearce, David. "The Biointelligence Explosion." In *Singularity Hypotheses*, edited by A.H. Eden, J.H. Moor, J.H. Søraker, and E. Steinhart, pp. 199-238. The Frontiers Collection. Berlin/Heidelberg: Springer, 2012.

Perez-Marin, D., and I. Pascual-Nieto. *Conversational Agents and Natural Language Interaction: Techniques and Effective Practices.* Hershey, PA: IGI Global, 2011.

Perlberg, James. *Industrial Robotics.* Boston: Cengage Learning, 2016.

Pinedo, Michael L. *Scheduling: Theory, Algorithms, and Systems*, fourth edition. New York: Springer Science+Business Media, 2012.

Porterfield, Andrew. "Be Aware of Your Inner Zombie." *ENGINEERING & SCIENCE* (Fall 2010): 14-19.

Posthumanist Shakespeares, edited by Stefan Herbrechter and Ivan Callus. Palgrave Shakespeare Studies. Palgrave Macmillan UK, 2012.

Proudfoot, Diane. "Software Immortals: Science or Faith?" In *Singularity Hypotheses*, edited by Amnon H. Eden, James H. Moor, Johnny H. Søraker, and Eric Steinhart, pp. 367-92. The Frontiers Collection. Springer Berlin Heidelberg, 2012.

Provost, Foster, and Tom Fawcett. *Data Science for Business.* Sebastopol, CA: O'Reilly Media, Inc., 2013.

Roden, David. *Posthuman Life: Philosophy at the Edge of the Human.* Abingdon: Routledge, 2014.

Sachs, Jeffrey D., Seth G. Benzell, and Guillermo LaGarda. "Robots: Curse or Blessing? A Basic Framework." NBER Working Papers Series, Working Paper 21091. Cambridge, MA: National Bureau of Economic Research, 2015.

Sarkar, Prabhat Rainjan. "Neohumanism Is the Ultimate Shelter (Discourse 11)." In *The Liberation of Intellect: Neohumanism.* Kolkata: Ananda Marga Publications, 1982.

Schacht, Kurt N. "The Buzz about Robo-Advisers." *CFA Institute Magazine* 26, no. 5 (2015): 49.

Scopino, G. "Do Automated Trading Systems Dream of Manipulating the Price of Futures Contracts? Policing Markets for Improper Trading Practices by Algorithmic Robots." *Florida Law Review,* vol. 67 (2015): 221-93.

Social Robots and the Future of Social Relations, edited by Johanna Seibt, Raul Hakli, and Marco Nørskov. Amsterdam: IOS Press, 2014.

Social Robots from a Human Perspective, edited by Jane Vincent, Sakari Taipale, Bartolomeo Sapio, Giuseppe Lugano, and Leopoldina Fortunati. Springer International Publishing, 2015.

Social Robots: Boundaries, Potential, Challenges, edited by Marco Nørskov. Farnham: Ashgate Publishing, 2016.

Soussou, Walid V., and Theodore W. Berger. "Cognitive and Emotional Neuroprostheses." In *Brain-Computer Interfaces,* pp. 109-23. Springer Netherlands, 2008.

Theoretical Foundations of Artificial General Intelligence, edited by Pei Wang and Ben Goertzel. Paris: Atlantis Press, 2012.

Tomas, David. "Feedback and Cybernetics: Reimaging the Body in the Age of the Cyborg." In *Cyberspace, Cyberbodies, Cyberpunk: Cultures of Technological Embodiment,* edited by Mike Featherstone and Roger Burrows, pp. 21-43. London: SAGE Publications, 1995.

Turner, Matt. "The computers have won, and Wall Street will never be the same." *Business Insider,* May 10, 2016. http://www.businessinsider.com/quant-funds-dominate-hedge-fund-rich-list-2016-5. Accessed June 5, 2016.

Ullrich, Günter. *Automated Guided Vehicle Systems: A Primer with Practical Applications,* translated by Paul A. Kachur. Springer Berlin Heidelberg, 2015.

Ummat, Ajay, Atul Dubey, and Constantinos Mavroidis. "Bionanorobotics: A Field Inspired by Nature." In *Biomimetics: Biologically Inspired Technologies,* edited by Yoseph Bar-Cohen, pp. 201-26. Boca Raton: CRC Press, 2005.

Van den Berg, Bibi. "Pieces of Me: On Identity and Information and Communications Technology Implants." In *Human ICT Implants: Technical, Legal and Ethical Considerations,* edited by Mark N. Gasson, Eleni Kosta, and Diana M. Bowman, pp. 159-73. Information Technology and Law Series 23. T. M. C. Asser Press, 2012.

Warkentin, Merrill, Vijayan Sugumaran, and Robert Sainsbury. "The Role of Intelligent Agents and Data Mining in Electronic Partnership Management." *Expert Systems with Applications* 39, no. 18 (2012): 13277-88.

Weber, Max. *Economy and Society: An Outline of Interpretive Sociology.* New York: Bedminster Press, 1968.

Yampolskiy, Roman V. "The Universe of Minds." arXiv preprint, *arXiv:1410.0369 [cs.AI]*, October 1, 2014. http://arxiv.org/abs/1410.0369. Accessed January 25, 2016.

Yonck, Richard. "Toward a standard metric of machine intelligence." *World Future Review* 4, no. 2 (2012): 61-70.

Part I

Creating the Human-Synthetic Organization: Background and Theoretical Foundations

Chapter One

A Typology of Posthumanism:

A Framework for Differentiating Analytic, Synthetic, Theoretical, and Practical Posthumanisms[1]

Abstract. The term 'posthumanism' has been employed to describe a diverse array of phenomena ranging from academic disciplines and artistic movements to political advocacy campaigns and the development of commercial technologies. Such phenomena differ widely in their subject matter, purpose, and methodology, raising the question of whether it is possible to fashion a coherent definition of posthumanism that encompasses all phenomena thus labelled. In this text, we seek to bring greater clarity to this discussion by formulating a novel conceptual framework for classifying existing and potential forms of posthumanism. The framework asserts that a given form of posthumanism can be classified: 1) either as an *analytic posthumanism* that understands 'posthumanity' as a sociotechnological reality that already exists in the contemporary world or as a *synthetic posthumanism* that understands 'posthumanity' as a collection of hypothetical future entities whose development can be intentionally realized or prevented; and 2) either as a *theoretical posthumanism* that primarily seeks to develop new knowledge or as a *practical posthumanism* that seeks to bring about some social, political, economic, or technological change. By arranging these two characteristics as orthogonal axes, we obtain a matrix that categorizes a form of posthumanism into one of four quadrants or as a hybrid posthumanism spanning all quadrants. It is suggested that the five resulting types can be understood

[1] This chapter was originally published in Gladden, Matthew E., *Sapient Circuits and Digitalized Flesh: The Organization as Locus of Technological Posthumanization*, pp. 31-91, Indianapolis: Defragmenter Media, 2016.

roughly as posthumanisms of *critique*, *imagination*, *conversion*, *control*, and *production*.

We then employ this framework to classify a wide variety of posthumanisms, such as critical, cultural, philosophical, sociopolitical, and popular (or 'commercial') posthumanism; science fiction; techno-idealism; metahumanism; neohumanism; antihumanism; prehumanism; feminist new materialism; the posthumanities; biopolitical posthumanism, including bioconservatism and transhumanism (with specialized objective and instrumental typologies offered for classifying forms of transhumanism); and organizational posthumanism. Of particular interest for our research is the classification of organizational posthumanism as a hybrid posthumanism combining analytic, synthetic, theoretical, and practical aspects. We argue that the framework proposed in this text generates a typology that is flexible enough to encompass the full range of posthumanisms while being discriminating enough to order posthumanisms into types that reveal new insights about their nature and dynamics.

I. Introduction

Terms such as 'posthumanism,' 'posthumanity,' and 'the posthuman' are being used to describe an increasingly wide and bewildering array of phenomena in both specialized scholarly and broader popular contexts. Spheres of human activity that have been described as 'posthumanist' include academic disciplines,[2] artistic movements,[3] spiritual movements,[4] commercial research and development programs designed to engineer particular new technologies,[5] works of science fiction,[6] and campaigns advocating specific legislative or regulatory action.[7]

---

[2] For examples, see the descriptions of critical, cultural, and philosophical posthumanism and the posthumanities later in this text.

[3] Examples include the works of performance art created by Del Val. See Del Val et al., "Interview on the Metahumanist Manifesto with Jaime del Val and Stefan Lorenz Sorgner" (2011).

[4] An instance is the form of neohumanism developed by Sarkar. See Sarkar, "Neohumanism Is the Ultimate Shelter (Discourse 11)" (1982).

[5] For examples of the term 'posthuman' being used to describe specific technologies that are being developed by DARPA and other military research and development agencies, see, e.g., Coker, "Biotechnology and War: The New Challenge" (2004); Graham, "Imagining Urban Warfare: Urbanization and U.S. Military Technoscience" (2008), p. 36; and Krishnan, "Enhanced Warfighters as Private Military Contractors" (2015).

[6] Posthumanist aspects of science fiction are discussed, for example, in Hayles, *How We Became Posthuman: Virtual Bodies in Cybernetics, Literature, and Informatics* (1999); *Cyberculture, Cyborgs and Science Fiction: Consciousness and the Posthuman*, edited by Haney (2006); and Goicoechea, "The Posthuman Ethos in Cyberpunk Science Fiction" (2008).

[7] Examples include some of the legislative and regulatory approaches proposed in Fukuyama, *Our Posthuman Future: Consequences of the Biotechnology Revolution* (2002), and Gray, *Cyborg Citizen: Politics in the Posthuman Age* (2002).

Running through many of these 'posthumanisms' is the common thread of emerging technologies relating to neurocybernetic augmentation, genetic engineering, virtual reality, nanotechnology, artificial life, artificial intelligence, and social robotics which – it is supposed – are challenging, destabilizing, or transforming our understanding of what it means to be 'human.' And yet when posthumanist interpretations are also being offered for subjects like the Bible,[8] medieval alchemical texts,[9] Shakespeare,[10] and 1930s zombie fiction,[11] it becomes apparent that directly equating posthumanism with an attitude toward futuristic technologies is overly simplistic and even misleading.

And not only do different manifestations of posthumanism differ widely from one another in their subject matter; even when two forms of posthumanism consider the same object, they often oppose one another in their aims, methodologies, and conclusions. For example, both transhumanists and bioconservatives attempt to foresee the extent to which genetic engineering will allow the capacities of future human beings to be radically transformed; while transhumanists conclude that the development of such technologies must be pursued as a natural next step in the evolution of humanity, bioconservatives conclude that pursuit of such technologies must be blocked in order to preserve the integrity of the human species and the possibility of a politically and economically just society.[12]

This *mélange* of meanings for the term 'posthumanism' raises important questions. First, is it possible to develop a definition of posthumanism that covers all of its uses? And second, assuming that this is theoretically possible, would it be desirable? Or is it better to acknowledge that 'posthumanism' has become too fragmented to possess a single coherent definition and that it is better to develop separate definitions for the diverse phenomena which share that appellation?

In this text, we seek to contribute to this debate by developing a conceptual framework that presents one approach to clarifying the key characteristics of different types of posthumanism and the relationships between them. Although the structure and details of the proposed framework are novel, such a framework can be understood as an appraisal, synthesis, and elaboration of the work of thinkers such as Ferrando, Herbrechter, Birnbacher, Miah, Miller,

---

[8] See, e.g., *The Bible and Posthumanism*, edited by Koosed (2014).

[9] See, e.g., Smith, *Genetic Gold: The Post-human Homunculus in Alchemical and Visual Texts* (2009).

[10] Examples include the texts collected in *Posthumanist Shakespeares*, edited by Herbrechter & Callus (2012).

[11] Instances of this can be found in *Better Off Dead: The Evolution of the Zombie as Post-Human*, edited by Christie & Lauro (2011).

[12] These issues are explored in more detail in the discussion of biopolitical posthumanism and bioconservatism later in this text.

and others who have not simply carried out posthumanist reflection on topics like genetic engineering or science fiction but have instead analyzed the nature of posthumanism itself – have attempted to forge some conceptual order amidst the landscape of many conflicting 'posthumanisms.'

Rather than presenting a simple catalogue of posthumanisms, the framework developed in this text proposes that a given form of posthumanism can be categorized on the basis of a pair of factors: its understanding of 'posthumanity' and the role or purpose for which the posthumanism has been developed. In this way, a posthumanism can be classified either as an *analytic posthumanism* that understands posthumanity as a sociotechnological reality that already exists in the contemporary world or as a *synthetic posthumanism* that understands posthumanity as a collection of hypothetical future entities whose development can be intentionally realized or prevented. Simultaneously, it can be classified either as a *theoretical posthumanism* that primarily seeks to develop new knowledge or as a *practical posthumanism* that primarily seeks to bring about some social, political, economic, or technological change. By combining these factors, a two-dimensional typology is created that identifies a form of posthumanism with one of four quadrants or as a hybrid posthumanism that spans all quadrants. After presenting this tool, the majority of this text will be spent in employing it to classify a wide variety of posthumanisms that have been identified in the literature.

II. Established Definitions of Posthumanism

Before formulating our typology of posthumanism, it is useful to explore the ways in which the concept of posthumanism is currently understood.

A multiplicity of posthumanisms. The term 'posthuman' has been used by different authors to represent very different concepts;[13] while this has enriched the development of posthumanism, it has also introduced confusion.[14] For example, Miller notes that the term has been given a variety of meanings by theorists operating in the natural sciences; cybernetics; epistemology; ontology; feminist studies; film, literary, and cultural studies; animal studies; and ecocriticism.[15] Herbrechter observes that the 'post-' in 'posthumanism' is not only ambiguous but even "radically open" in its meaning.[16] For example, the word can be understood either as 'post-*humanism*,' a critical response to and deconstructive working-through of the assumptions of humanism, or as

---

[13] Bostrom, "Why I Want to Be a Posthuman When I Grow Up" (2008), p. 107.

[14] See Ferrando, "Posthumanism, Transhumanism, Antihumanism, Metahumanism, and New Materialisms: Differences and Relations" (2013), p. 26.

[15] Miller, "Conclusion: Beyond the Human: Ontogenesis, Technology, and the Posthuman in Kubrick and Clarke's 2001" (2012), p. 163.

[16] Herbrechter, *Posthumanism: A Critical Analysis* (2013), p. 69.

'*posthuman*-ism,' a philosophy of future engineered beings whose capacities are expected to surpass those of contemporary human beings.[17] Indeed, Birnbacher suggests that the term 'posthumanity' and related idea of 'transhumanism' have been utilized by so many different thinkers in such widely divergent fashions that they can be better understood "as slogans rather than as well-defined concepts."[18]

Posthumanist terminology. In this text, we will refer often to the interrelated but distinct notions of 'posthumanization,' 'posthumanity,' 'posthumanism,' and the 'posthuman.' Because each of these terms has been used to represent multiple concepts, it is difficult to offer authoritative definitions for them. Nevertheless, they can be broadly differentiated:

- **Posthumanization** can be understood as a process by which society comes to include at least some intelligent personal subjects that are *not* natural biological human beings and which leads to a nonanthropocentric understanding of reality. At present, posthumanization often occurs as a result of the **technologization** of human beings, which is spurred by phenomena such as our increasing physical integration with electronic systems, our expanding interaction with and dependence on robots and artificial intelligences, our growing immersion in virtual worlds, and the use of genetic engineering to design human beings as if they were consumer products.[19] However, processes of posthumanization do not inherently require the use of modern technology: works of mythology or literature that present quasi-human figures such as monsters, ghosts, and semidivine heroes can advance the process of posthumanization by challenging the boundaries of our concept of humanity and, in some sense, incorporating those figures into the structures and dynamics of society.[20]
- **Posthumanity** refers either to a collection of intelligent beings – whether human, synthetic, or hybrid – that have been created or affected by a process of posthumanization or to the broader sociotechnological reality within which such beings exist.
- **Posthumanism** is a coherent conceptual framework that takes the phenomenon of posthumanization or posthumanity as its object; it may be developed as part of an academic discipline, artistic or spiritual movement, commercial venture, work of fiction, or form of advocacy, among other possible manifestations.

---

[17] Herbrechter (2013), p. 16.

[18] Birnbacher "Posthumanity, Transhumanism and Human Nature" (2008), p. 96.

[19] The relationship of posthumanism to the commercialization of the human entity is discussed in Herbrechter (2013), pp. 42, 150-52.

[20] For the role of such figures in nontechnological posthumanization, see, e.g., Herbrechter (2013), pp. 2-3, 106.

- **'Posthuman'** can refer to any of the above: a process (posthumanization), collection of entities (posthumanity), or body of thought (posthumanism).

Tracing the origins of posthumanism. Some identify the birth of posthumanism as an explicit conceptual system with Wiener's formulation of cybernetics in the 1940s; others suggest that posthumanism as an explicit discipline only appeared with Haraway's analysis of cyborgs and the dissolution of human-machine boundaries in the 1990s.[21] While ongoing developments in robotics, artificial intelligence, biocybernetics, and genetic engineering are lending new urgency to questions surrounding posthumanism, Herbrechter argues that the phenomenon of posthumanism is at least as old as that of post-Enlightenment humanism – even if it has only recently been explicitly named.[22] The fact that the term 'posthumanism' is used to refer to such a diverse array of intellectual phenomena means that scholars can date its origins variously to the Renaissance, post-Enlightenment era, 1940s, or 1990s, depending on exactly which 'posthumanism' is being considered.

Attempts at defining posthumanism generically. Ideally, it would be possible to formulate a generic definition of 'posthumanism' broad enough to cover all such intellectual frameworks. And, indeed, scholars have attempted to identify elements that are shared across all varieties of posthumanism. For example, Miller contends that various strains of posthumanism agree that:

> The posthuman subject is a multiple subject, not a unified one, and she or he (a distinction that also gets blurred in posthuman-ism) is not separate from his/her environment. Technologies become extensions of the self, and humans become only one type of individual in a vast ecosystem that includes digital as well as natural environmental forces. In other words, posthumanism is partly about leaving behind the old notions of liberal humanism. [...] But it also begins to gesture toward a much more radical state, a state beyond the current human form.[23]

According to this view, the heart of posthumanism is a 'post-anthropocentric'[24] perspective that looks beyond traditional human beings to identify other sources of intelligence, agency, subjectivity, and meaning within the world. Emphasizing this fact, Ferrando states that:

> Posthumanism is often defined as a post-humanism and a post-anthropocentrism: it is "post" to the concept of the human and to the

---

[21] Such perspectives on the genesis of posthumanism are offered, e.g., in Herbrechter (2013), p. 41, and its discussion of Gane, "Posthuman" (2006).

[22] Herbrechter (2013), p. 77.

[23] Miller (2012), p. 164.

[24] See Herbrechter (2013), pp. 2-3.

> historical occurrence of humanism, both based [...] on hierarchical social constructs and human-centric assumptions.[25]

Thus by way of offering a preliminary definition, Herbrechter suggests that posthumanism in its most general sense is "the cultural malaise or euphoria that is caused by the feeling that arises once you start taking the idea of 'post-anthropocentrism' seriously."[26] Similarly, Birnbacher suggests that the different forms of posthumanism are united in studying already existing or potential future 'posthumans' whose nature is not constrained by human nature as previously understood and who lack at least some key characteristics that have historically been considered typical of the human species.[27]

Miah, meanwhile, finds "a range of posthumanisms" that are united by the fact that they "challenge the idea that humanness is a fixed concept."[28] However, posthumanism's challenge to the concept of the 'human' differs from the more nihilistic attacks waged by postmodernism: in their own unique ways – whether subtly or wholeheartedly – various kinds of posthumanism are willing to entertain the idea of restoring in an altered post-anthropocentric form some of the 'grand narratives' about humanity, agency, history, and other phenomena that had been wholly rejected by postmodernism.[29]

Problems with a generic definition of posthumanism. While such general definitions offer a useful starting point, they are hampered by the fact that 'posthumanisms' differ markedly with regard to their origins, purpose, and methodology. For example, as we have noted, some thinkers argue that technological progress is an essential aspect of posthumanism that will inevitably someday be harnessed to engineer a superior posthumanity.[30] Other thinkers argue that technology is not an inherent element of posthumanism at all and that posthumanity is a conceptual array of interrelated human, quasi-human, and nonhuman beings (such as ghosts, monsters, aliens, and robots) that have held a place within the human imagination for hundreds or thousands of years. Any definition of 'posthumanism' that is broad enough to describe all such conflicting perspectives may be so vague as to be of little practical value.

Existing frameworks for categorizing posthumanisms. Scholars have proposed a range of conceptual frameworks for classifying the many forms of posthumanism. For example, Miah distinguishes between the three different

---

[25] Ferrando (2013), p. 29.

[26] Herbrechter (2013), p. 3.

[27] Birnbacher (2008), p. 104.

[28] Miah, "A Critical History of Posthumanism" (2008), p. 83.

[29] Differences between postmodernism and posthumanism can be observed, e.g., in Herbrechter (2013), p. 23.

[30] For such broadly transhumanist perspectives, see, e.g., Bostrom (2008) and Kurzweil, *The Singularity is Near: When Humans Transcend Biology* (2005).

phenomena of *biopolitical*, *cultural*, and *philosophical* posthumanism.[31] Ferrando distinguishes three forms of posthumanism *per se* (i.e., *critical*, *cultural*, and *philosophical* posthumanism), while noting that the word 'posthuman' is also used more broadly to include related phenomena such as transhumanism, new materialism, antihumanism, metahumanism, and the posthumanities.[32]

Finally, drawing on Rosenau, Herbrechter distinguishes two different strains of posthumanism. On one side is an *affirmative* posthumanism that includes 'technoeuphorians' (such as transhumanists) who wholeheartedly embrace posthumanizing technologies and 'technocultural pragmatists' who accept that posthumanizing technological change is inevitable and who attempt to strengthen its positive impacts while ameliorating any detrimental side-effects. On the other side is a *skeptical* posthumanism that includes 'catastrophists' (such as bioconservatives) who are attempting to forestall the development of posthumanizing technology due to its perceived danger and 'critical deconstructive posthumanists' (such as Herbrechter) who accept that posthumanizing technological change is occurring and who are primarily interested not in identifying its potentially negative biological or social impacts but in analyzing the theoretical weaknesses, biases, and naïvety displayed by those who zealously advocate such technologization of humankind.[33]

III. A Proposed Two-dimensional Typology of Posthumanism

While such existing schemas for classifying posthumanisms offer valuable insights, we contend that it would be useful to possess a more comprehensive and systematic framework developed for this purpose. To that end, we would suggest that a given form of posthumanism can be classified in two ways:

1) **By its understanding of posthumanity**. A form of posthumanism can be categorized either as an **analytic posthumanism** that understands posthumanity as a sociotechnological reality that already exists in the contemporary world and which needs to be analyzed or as a **synthetic posthumanism** that understands posthumanity as a collection of hypothetical future entities whose development can be either intentionally realized or intentionally prevented, depending

[31] See Miah (2008).

[32] Ferrando (2013), p. 26.

[33] For this dichotomy of affirmative and skeptical perspectives, see Herbrechter (2013), pp. 23-24, and its analysis of Rosenau, *Post-Modernism and the Social Sciences: Insights, Inroads, and Intrusions* (1992).

on whether or not human society chooses to research and deploy certain transformative technologies.

2) **By the purpose or role for which it was developed**. A form of posthumanism can be categorized either as a **theoretical posthumanism** that primarily seeks to develop new knowledge and understanding or as a **practical posthumanism** that primarily seeks to bring about some social, political, economic, or technological change in the real world.

By arranging these two characteristics as orthogonal axes, a matrix is obtained that categorizes a form of posthumanism into one of four quadrants or as a hybrid that spans all quadrants. Figure 1 depicts this matrix along with our proposed classification of numerous forms of posthumanism that will be investigated within this text. We can now discuss these two axes in more detail.

Analytic versus synthetic posthumanism. Analytic posthumanisms define 'posthumanity' as a sort of sociotechnological reality that already exists in the contemporary world and which calls out to be better understood. Such posthumanisms typically display a strong orientation toward the present and the past; they do not generally focus on the future, insofar as the exact form that the future will take has not yet become clear to us and thus cannot yet be the object of rigorous analysis.

Synthetic posthumanisms, on the other hand, define 'posthumanity' as a set of hypothetical future entities[34] (such as full-body cyborgs or artificial general intelligences) whose capacities differ from – and typically surpass – those of natural biological human beings and whose creation can either be intentionally brought about or intentionally blocked, depending on whether humanity decides to develop and implement certain transformative technologies such as those relating to genetic engineering, neuroprosthetics, artificial intelligence, or virtual reality. Such posthumanisms generally have a strong future orientation; they rarely give detailed attention to events of the distant past, and they conduct an exploration of power structures or trends of the current day only insofar as these offer some insight into how future processes of posthumanization might be directed.

---

[34] An exception to this definition would be prehumanism, a form of synthetic theoretical posthumanism that is similar to science fiction but which imagines the characteristics of quasi-human beings in a hypothetical distant past rather than in the far future. While the directionality of the temporal reference-points is reversed in comparison to that of futurological science fiction, the (implicit or explicit) contrast of contemporary humanity with the intelligent beings of a chronologically distant but causally connected world remains intact. See the discussion of prehumanism later in this text.

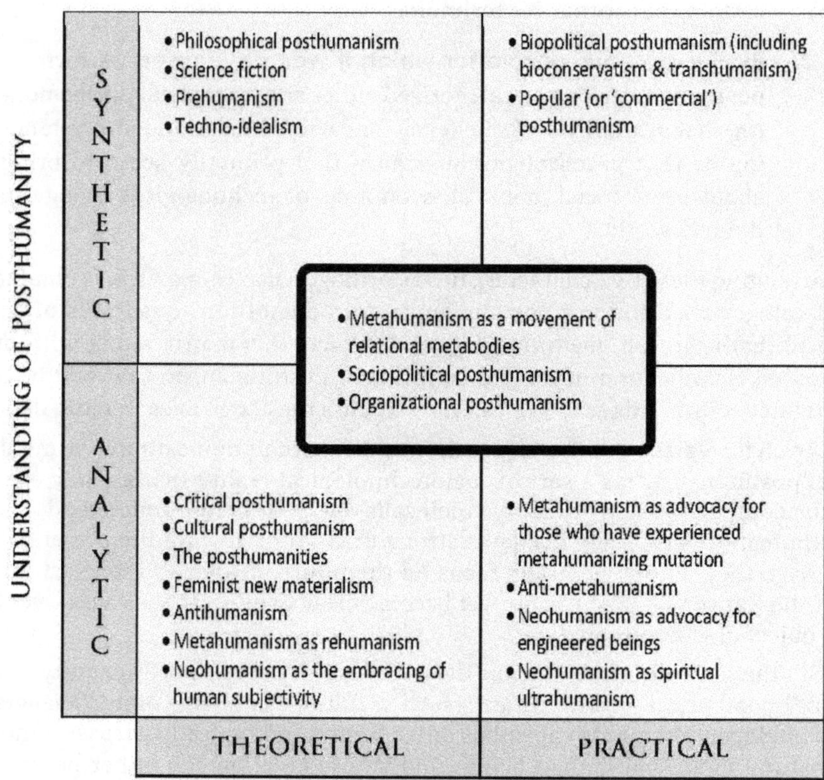

S Y N T H E T I C
A N A L Y T I C
UNDERSTANDING OF POSTHUMANITY
• Philosophical posthumanism
• Science fiction
• Prehumanism
• Techno-idealism
• Biopolitical posthumanism (including bioconservatism & transhumanism)
• Popular (or 'commercial') posthumanism
• Metahumanism as a movement of relational metabodies
• Sociopolitical posthumanism
• Organizational posthumanism
• Critical posthumanism
• Cultural posthumanism
• The posthumanities
• Feminist new materialism
• Antihumanism
• Metahumanism as rehumanism
• Neohumanism as the embracing of human subjectivity
• Metahumanism as advocacy for those who have experienced metahumanizing mutation
• Anti-metahumanism
• Neohumanism as advocacy for engineered beings
• Neohumanism as spiritual ultrahumanism
THEORETICAL
PRACTICAL
PURPOSE OR ROLE OF POSTHUMANISM

Fig. 1: Our proposed two-dimensional typology of posthumanism, which classifies a form of posthumanism based on whether it understands posthumanity as a sociotechnological reality already existing in the contemporary world ('analytic') or as a set of hypothetical future entities whose capacities differ from those of natural biological human beings ('synthetic') and whether its purpose is primarily to expand the knowledge possessed by humanity ('theoretical') or to produce some specific political, economic, social, cultural, or technological change within the world ('practical'). Classifications are suggested for numerous forms of posthumanism.

Theoretical versus practical posthumanism. Posthumanisms can also be classified according to the purpose for which they were developed or the role that they play.[35] Theoretical posthumanisms are those that mainly seek to enhance our understanding of issues and to expand the knowledge possessed by

[35] The distinction between theoretical and practical posthumanisms could be understood, for example, in light of the Aristotelian division of human activities into *theoria*, *poiesis*, and *praxis*.

humanity – not primarily for the sake of effecting some specific change within the world but for the sake of obtaining a deeper, richer, more accurate, and more sophisticated understanding of human beings and the world in which we exist.

Practical posthumanisms, on the other hand, are interested primarily in producing some specific political, economic, cultural, social, or technological change. While theoretical posthumanism often takes the form of analyses, critiques, or thought experiments, practical posthumanism may take the form of efforts to ensure or block the approval of proposed treaties, legislation, or regulations; secure or cancel funding for particular military, educational, or social programs; develop and test new technologies; design, produce, and market new kinds of goods or services; or influence the public to vote, spend their time and money, interact socially, tolerate particular corporate or governmental actions, or otherwise behave in specific ways. Practical posthumanisms may thus include elements of advocacy, engineering, and entrepreneurship.

Hybrid posthumanisms that combine all four aspects. There are at least three kinds of posthumanism which, we would argue, are simultaneously analytic, synthetic, theoretical, and practical. These will be explored in more depth later in this text. The first of these hybrid posthumanisms is the form of metahumanism formulated by Sorgner and Del Val.[36] Their metahumanist program possesses a strong theoretical component, insofar as it is grounded in and seeks to advance critiques developed by thinkers such as Nietzsche and Deleuze; however, it also displays a strong practical component in that it is geared toward generating works of performance art and other concrete products. Similarly, their metahumanism is analytic insofar as it reflects on the 'metabodies' of human beings as they exist today and synthetic insofar as it recognizes that new kinds of metabodies will be created in the future, largely through the ongoing technologization of humankind.

The second hybrid posthumanism is sociopolitical posthumanism. This is manifested, for example, in legal scholars' efforts to update legal systems to reflect emerging deanthropocentrized realities such as the growing ability of robots to autonomously make complex ethical and practical decisions that impact the lives of human beings.[37] Such work is theoretical insofar as it flows from a sophisticated theory of law and practical insofar as it is geared toward

---

Theoretical posthumanism is a kind of *theoria*, while practical posthumanism comprises *praxis* (as in the case of posthumanist political movements) and *poiesis* (as in the case of some posthumanist artistic movements).

[36] They describe their form of metahumanism in Del Val & Sorgner, "A *Metahumanist* Manifesto" (2011).

[37] A thoughtful example of this is found in Calverley, "Imagining a non-biological machine as a legal person" (2008).

reshaping real-world legal systems. Similarly, it is analytic insofar as it investigates the effects of posthumanization that are already reflected in the world today and synthetic insofar as it seeks to anticipate and account for different posthumanities that might appear in the future.

Finally, the form of organizational posthumanism formulated later in this text also combines both analytic and synthetic as well as theoretical and practical aspects. Organizational posthumanism is theoretical insofar as it seeks to understand the ways in which the nature of organizations is being transformed by the technologization and posthumanization of our world and practical insofar as it seeks to aid management practitioners in creating and maintaining viable organizations within that posthumanized context. It is analytic insofar as it recognizes post-anthropocentric phenomena (such as the growing use of AI, social robotics, and virtualized interaction) that are already present within many organizations and synthetic insofar as it believes that such post-anthropocentrizing trends will continue to accelerate and will generate organizational impacts that can be shaped through the planning and execution of particular strategies.

Distilling Five Main Types of Posthumanism: Posthumanisms of Critique, Imagination, Conversion, Control, and Production

The types of posthumanism delineated by our two-dimensional framework are generalizations. The phenomena that can be assigned to any one type may differ significantly from one another, thus it is hazardous to assign a broad-brush description to a type of posthumanism and expect it to apply equally well to all of the posthumanisms included within that type. Nevertheless, as a starting point for further discussion, we would suggest that it is possible to capture the fundamental dynamic of each type of posthumanism.

For example, analytic theoretical posthumanisms might collectively be understood as manifesting a '**posthumanism of critique**' that employs posthumanist methodologies to identify hidden anthropocentric biases and posthumanist aspirations contained within different fields of human activity. Similarly, synthetic theoretical posthumanisms could be seen as exemplifying a '**posthumanism of imagination**' that creatively envisions hypothetical future posthumanities so that their implications can be explored. Analytic practical posthumanisms manifest a '**posthumanism of conversion**' aimed at changing hearts and minds and influencing the way in which human beings view the world around themselves. Synthetic practical posthumanisms exemplify a '**posthumanism of control**' that seeks either to develop new technologies that give individuals control over their own posthumanization or to implement legal or economic controls to govern the development of such technologies. Finally, hybrid posthumanisms that span all four spheres can be understood as examples of a '**posthumanism of production**' that develops a robust and rigorous theoretical framework that is then employed to successfully generate concrete products or services within the contemporary world. An overview of these five main types of posthumanism is reflected in Figure 2.

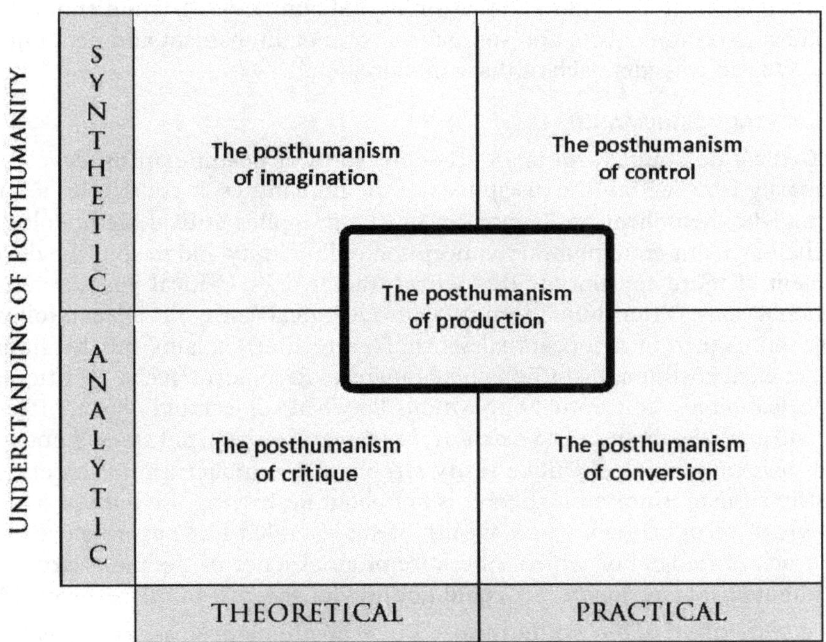

Fig. 2: The five types of posthumanism delineated by our two-dimensional model can be understood informally as posthumanisms of critique, imagination, conversion, control, and production.

IV. Classification and Analysis of Individual Forms of Posthumanism

A review of the literature reveals many different phenomena that have been identified as forms of posthumanism or which more generally have been described as 'posthuman' or 'posthumanist' in nature. Below we classify and analyze many such phenomena utilizing our two-dimensional typology.

A. Analytic Theoretical Posthumanisms: Seeking to Understand the Posthumanized Present

Analytic theoretical posthumanisms can collectively be understood as constituting a 'posthumanism of critique' that employs posthumanist methodologies to uncover hidden anthropocentric biases and posthumanist aspirations that are concealed within different fields of human activity. Such forms of analytic theoretical posthumanism include critical posthumanism, cultural posthumanism, the posthumanities (or 'digital humanities'), feminist new materialism, antihumanism, and some forms of metahumanism and neohumanism. We can consider each of these in more detail.

1. Critical Posthumanism

Critical posthumanism is an academic form of posthumanism developed primarily from within the disciplines of the humanities. It constitutes a form of **analytic theoretical posthumanism** in that it applies critical methodologies to challenge our contemporary conception of humanity and to spur the development of more appropriate theoretical frameworks. Critical posthumanism does not come 'after' humanism in a chronological sense but instead follows from humanism in a conceptual sense; Herbrechter explains this by stating that critical posthumanism "inhabits humanism deconstructively,"[38] critiquing historical binary conceptual oppositions between subject and object, biological and artificial, human and machine, human and animal, nature and nurture, and male and female.[39] Unlike many strains of postmodernism, such critical posthumanism is not nihilistic;[40] it is not about destroying the human subject but about recognizing a whole wealth of subjects that had never before been fully acknowledged or which – because of an absence of the necessary sociotechnological environment – could not previously exist in the real world.[41]

Assimilation of the nonhuman. Critical posthumanism seeks to create an account of the personal subject that is descriptive rather than normative and which does not consider 'humanity' as historically (and narrowly) defined but instead addresses a broader universe of entities that includes natural human beings as well as related entities like ghosts, angels, monsters, cyborgs, artificial intelligences, and extraterrestrial beings that have traditionally been considered quasi-human, parahuman, or nonhuman.[42] Critical posthumanism

---

[38] Herbrechter (2013), pp. 6-7.

[39] The raising of such challenges to historical binary and dualistic thought is a hallmark of posthumanism. See, e.g., Herbrechter (2013), pp. 79, 90.

[40] Regarding the positive aspects of critical posthumanism that distinguish it from more negational forms of postmodernism, see Herbrechter (2013), p. 196.

[41] See Herbrechter (2013), p. 198.

[42] Regarding the wide spectrum of entities that are important for critical posthumanism, see, e.g.,

possesses an empathy for such excluded beings in part because it claims that we owe our humanity to them: while some humanists contend that the 'human being' is defined first and then all entities that fail to satisfy that definition are excluded as being 'nonhuman,' critical posthumanism argues that in reality it was our inherent understanding of the myriad forms of the 'inhuman' that first allowed us to define the 'human' in opposition to them.[43] In a sense, critical posthumanism is thus nothing new; it is an age-old, nontechnological, deconstructive process that continually challenges our understanding of (and exclusive identification with) the 'human' by bringing into our circle of awareness examples of the inhuman and nonhuman.[44] It has existed for as long as monsters, angels, mythic heroes, and the relationship of such entities to human beings have been pondered within works of art, literature, philosophy, and theology.

Posthumanism with or without technology. In contrast with transhumanism – which is closely identified with particular technologies – critical posthumanism can thus take the form of a 'posthumanism without technology'[45] that focuses on anthropological, linguistic, or aesthetic questions rather than issues of biomedical engineering. However, as a practical matter, critical posthumanism's consideration of the 'nonhuman other' has taken on a new focus and urgency thanks to the accelerating processes of technologization that are now reshaping humankind. Critical posthumanism does not formulate a critique of technology *per se* but of the processes of technologization by which technological mechanisms, systems, and attitudes are consolidating their power over all aspects of human life. Critical posthumanism recognizes the fact that human beings are – and have always been – locked in a symbiotic relationship of coevolution with our technology; it analyzes and critiques this process, without condemning or embracing it *a priori* in the way that biopolitical posthumanism often does.[46]

Diagnosing 'speciesism.' Critical posthumanism considers the cases of nonhuman entities as a means of diagnosing what it sees as previously unnoticed forms of 'speciesism' or anthropocentric bias that have long permeated human political, economic, scientific, artistic, and religious activity.[47] For example, traditional cultural studies are highly anthropocentric, insofar as they

Herbrechter (2013), pp. 2-3, 106.

[43] For a discussion of the logical and practical priority of the 'human' or 'nonhuman,' see Herbrechter (2013), p. 55, and its reflections on Curtis, "The Inhuman" (2006), p. 434.

[44] Herbrechter (2013), p. 44.

[45] Regarding nontechnological forms of posthumanization, see Herbrechter (2013), p. 157.

[46] For a discussion of our symbiotic relationship with technology and critical posthumanism's attitude toward it, see Herbrechter (2013), pp. 90, 19.

[47] Ferrando (2013), p. 29.

assume that 'humanity' (or something closely mimicking it) is required in order for culture to exist; thus animals may have societies, but they do not possess culture. Critical posthumanism, on the other hand, does not assume as a starting point that culture logically requires humanity; indeed, it explicitly rejects this notion.[48] Critical posthumanism accepts the fact that human beings are no longer the only intelligent social actors within the world; we are increasingly only one of many kinds of individuals – both real and virtual, biological and electronic – that populate a rich and complex digital-physical environment and shape it through our interactions.[49] Critical posthumanism thus seeks to identify hidden assumptions that only human beings – and not, for example, social robots or genetically enhanced domesticated animals – are capable of filling particular roles within society or that human activity should be carried out with the sole purpose of benefitting human beings.

A critique of cybernetics, virtualization, and transhumanism. While critical posthumanism appreciates the value of robots and AIs in helping us to better understand the nature of human intelligence and agency, it does not share transhumanism's zeal for attempting to literally transform human beings into virtualized or robotic entities. Indeed, a major aim of critical posthumanism is to resist the defining of 'mind' as a disembodied collection of information in the manner promoted by many forms of transhumanism and some of the more techno-idealistic branches of cybernetics.[50] As envisioned by Haraway, for example, critical posthumanism is not simply an approach bent on destroying traditional anthropocentric presumptions; it also displays a positive element that seeks to formulate a new understanding of human beings as 'embodied selves.'[51] Similarly, Hayles foresees a danger that the growing cultural fascination with virtual reality might encourage a false belief that information can exist in a disembodied form; her critical posthumanism thus aims to ensure that processes of posthumanization do not result in the dematerialization of human beings but in our rematerialization – in a recognition that

---

[48] Regarding the conceptual relationship of humanity to culture, see Badmington, "Cultural Studies and the Posthumanities" (2006), p. 270, and its discussion in Herbrechter (2013), p. 174.

[49] Miller (2012), p. 164. For a philosophical analysis of posthumanized digital-physical ecosystems and the interdependencies existing among their human and nonhuman actors that advances and refines conventional Actor-Network Theory (ANT), see Kowalewska, "Symbionts and Parasites – Digital Ecosystems" (2016).

[50] For critical posthumanism as a challenge to techno-idealism and transhumanism, see Herbrechter (2013), p. 94.

[51] Regarding critical posthumanism's efforts to fashion a positive concept of the embodied self, see Haraway, *Simians, Cyborgs, and Women: The Reinvention of Nature* (1991), and Herbrechter (2013), pp. 99-100.

we are networked corporalities, material-digital beings, and not pure information as some transhumanists might claim.[52] Critical posthumanism also challenges transhumanism by devoting attention to questions of power and privilege; Ferrando notes that critical posthumanism explicitly analyzes such issues, while transhumanism is singularly 'non-critical' in its lack of interest in the historical development of humanity and its naïve presentation of a generic 'human being' that exists without reference to social or economic class, sex, race, ethnicity and nationality, interpersonal relationships, or religion and spirituality.[53]

Creating a concept of humanity that can endure. It is possible to argue that far from 'destroying' the concept of humanity in a postmodernist sense, critical posthumanism is actually aimed at saving the concept of humanity; critical posthumanism accomplishes this by transforming our notion of 'humanity' into a broader concept of 'posthumanity' that does not require the continued survival of human beings in some mythically pristine, unengineered, untechnologized, and 'natural' biological form but which instead welcomes into the family of (post-)humanity a wider range of biological, artificial, and hybrid subjects. According to this view, even if 'humanity' in the narrow humanist sense were to someday suffer extinction, a more broadly understood 'posthumanity' would be likely to survive. Indeed, some have suggested that by insisting on a definition of humanity that is so rigidly anthropocentric, it is humanism itself that has created the risk of the eventual 'dehumanization' of the universe through the elimination of biological humankind. Critical posthumanism might thus be understood as a sort of conceptual lifeboat that opens the door to the long-term persistence of a world of sapient (if not 'naturally human') posthuman persons and subjects.[54]

Humanism, rehumanism, or alterhumanism? Rather than continuing recent postmodernist trends of disparaging humanism, critical posthumanism might be seen as constituting a renaissance of a transformed and deanthropocentrized humanist thought.[55] Indeed, Herbrechter suggests that posthumanism might be understood as a sort of autoimmune response generated by the larger humanistic culture that can serve to liberate contemporary human beings from the more oppressive and problematic aspects of humanism, thereby

---

[52] For the critical posthumanist rejection of an understanding of the human entity as pure information, see Hayles (1999) and its discussion in Herbrechter (2013), pp. 185-86.

[53] Ferrando (2013), p. 28.

[54] For the notion that humanism may be the true threat to humanity and posthumanism its rescuer, see Herbrechter (2013), pp. 123-24, 187, and its commentary on Hayles (1999), p. 290.

[55] Regarding posthumanism as the refinement and fulfillment of humanism, see Herbrechter (2013), p. 106.

leading to the first full flowering of true humanism. However, critical posthumanism attempts to counteract the more dehumanizing aspects of posthumanization not through a strategy of nostalgic 'rehumanization' that restores classical humanism to an authoritative role but through a form of 'alterhumanism' that expands itself to encompass entities and perspectives previously dismissed as inhuman.[56]

Critical posthumanism as a bridge between posthumanisms. Herbrechter's efforts to fashion a "critical but open-minded posthumanism"[57] are suggestive of the fact that critical posthumanism is well-positioned to serve as an impartial mediator and translator between conflicting posthumanist positions. For example, Herbrechter draws on Thacker's attempts to describe the growing informatization of human beings and conversion of the human body into 'biomedia' in a way that is critical but value-neutral and does not inherently support transhumanist or bioconservative positions.[58]

Similarly, Herbrechter argues that critical posthumanism represents a sort of reversible methodological process that can translate between the two spheres or levels of the human being as personal subject and human being as viable system. Taking the human subject as its starting point, critical posthumanism can draw on the insights of postmodernism to deconstruct that subject and move to the atomic realm of processes and relations that constitute what is referred to as a 'human being.' Conversely, by drawing on insights from cybernetics and systems theory, critical posthumanism can begin with a collection of discrete processes and relations and correlate them to show how their interactions create a system that constitutes a human (or posthuman) subject. Critical posthumanism might thus serve as a bridge between postmodernism and cybernetics.[59]

Posthuman Realism

One form of critical posthumanism sometimes referred to by its own name is the strain formulated by Hayles known as 'posthuman realism.' As described above, it emphasizes the embodiment of the human being within a finite and nonexchangeable biological substrate, which contrasts with techno-idealist and transhumanist visions of the human mind as a virtualized entity or collection of disembodied data that can be shifted from one body to another

---

[56] For critical posthumanism's ability as an 'alterhumanism' to critique the detrimental effects of posthumanization without resorting to naïve humanism, see Herbrechter (2013), pp. 76-77, 70.

[57] Herbrechter (2013), p. 171.

[58] For such more or less value-neutral analyses of posthumanization, see Thacker, "What Is Biomedia?" (2003), p. 52, and the discussion of it in Herbrechter (2013), pp. 191-92.

[59] Regarding critical posthumanism as a mediator between postmodernist understandings of the subject and cybernetics, see Herbrechter (2013), pp. 198-99.

(and between biological and electronic substrates) without imperiling its consciousness or personal identity.[60]

2. Cultural Posthumanism

Miah places the origins of cultural posthumanism in *Posthuman Bodies,* edited by Halberstam and Livingstone in 1995. Other formative figures identified by Miah include Haraway, Hayles, Badmington, and Graham.[61] As a form of **analytic theoretical posthumanism**, cultural posthumanism understands 'posthumanity' to be a state that already exists within our contemporary world. It argues that the nature of posthumanity can be diagnosed by applying the tools of cultural studies to analyze elements of contemporary culture, including works of literature, film, television, music, painting, sculpture, architecture, fashion, computer games, tabletop roleplaying games, and religious and political speech.

Affinity with critical posthumanism. Some authors treat cultural posthumanism and critical posthumanism as though they were the same discipline; other scholars classify critical posthumanism as a subset of cultural posthumanism or *vice versa.* Indeed, the overlap between cultural and critical posthumanism is significant, and many thinkers have worked to advance both forms of posthumanism. Like critical posthumanism, cultural posthumanism can take the form of a 'posthumanism without technology': rather than awaiting or building a future of technologized beings, cultural posthumanism focuses on the present in which humanity already "collapses into *sub-*, *inter-*, *trans-*, *pre-*, *anti-*."[62] Cultural posthumanism also shares with critical posthumanism a strong second-order element, in that it seeks to understand the cognitive and social dynamics by which cultural posthumanism is generated. In fact, Miah argues that the most coherent and explicit theories of posthumanism have been developed from within the fields of cultural and literary studies and communications.[63]

Differences from critical posthumanism. Despite the links between cultural and critical posthumanism, differences can be discerned between the two fields. For example, in exploring posthumanism's origins in the 1990s, Ferrando distinguishes the critical posthumanism that emerged within the sphere of literary criticism and which was driven primarily by feminist theorists from the cultural posthumanism that emerged simultaneously within the field of

---

[60] See Hayles (1999), p. 5, and Herbrechter (2013), p. 43.

[61] Miah (2008), pp. 76, 78.

[62] See *Posthuman Bodies*, edited by Halberstam & Livingstone (1995), p. viii, and the commentary in Miah (2008), p. 76.

[63] Miah (2008), pp. 75-76.

cultural studies.[64] Unlike critical posthumanism (and biopolitical posthumanism), cultural posthumanism does not privilege issues relating to subjectivity, ethics, politics, and power relations but seeks to develop a broader analysis of posthumanization processes that gives equal weight to their aesthetic, artistic, and theological facets. Beyond highlighting deficiencies in existing bodies of thought, cultural posthumanism can also play a proactive role in building the 'posthumanities' that will increasingly become the focus of study at universities.[65]

Cultural visions of a posthumanized future as diagnoses of the posthumanized present. Both critical and cultural posthumanism analyze the state of posthumanity as it exists in the present moment; however, while critical posthumanism typically focuses on the effects of posthumanization that have already impacted human beings, cultural posthumanism also studies cultural depictions of future social and technological change (e.g., as presented in works of science fiction), insofar as they reflect a current desire for or fear of posthumanization. However, depictions of breakdowns in the binary opposition of human and inhuman can be found not only in science fiction but in all types of cultural texts, from ancient to contemporary works; thus cultural posthumanism has a vast field of objects for study.[66]

Cultural products as harbingers of posthuman oppression or liberation. As previously noted, critical posthumanism does not take an *a priori* stance in favor of either technoeuphoric transhumanism or technoparanoid bioconservatism; it instead tries to honestly understand and critique both positions.[67] Nevertheless, in practice critical posthumanism injects itself into such biopolitical discourses in a way meant to expose perceived biases and shift the processes of posthumanization in a direction of greater justice and equity. Miah argues that despite its supposed neutrality regarding the value of posthumanization, cultural posthumanism, too, often reflects an implicit concern that revolutionary new technologies will be appropriated by the powerful in a way that thwarts the realization of social justice for the less privileged. Cultural posthumanism documents the ways in which cultural products explore the power of the posthumanization process to either liberate or oppress human beings.[68] Miah suggests that this investigation of the meaning of justice and

[64] Ferrando (2013), p. 29.

[65] Herbrechter (2013), p. 143.

[66] Regarding the broad range of cultural artifacts that may reflect posthumanist themes, see Herbrechter (2013), p. 143.

[67] See Herbrechter (2013), p. 84.

[68] Regarding this dual potential of the forces of posthumanization, see Herbrechter (2013), p. 85.

ethics in a posthumanized world represents a common interest of both cultural and philosophical posthumanism.[69]

3. The Posthumanities (or Digital Humanities)

Ferrando notes that while the word 'posthumanities' can refer to a collection of future posthumanized species, it can also denote a set of academic disciplines that are in the process of succeeding the historical disciplines of the humanities.[70] The nature of such 'posthumanities' is as diverse and ambiguous as that of posthumanism itself. On the one hand, the posthumanities can include disciplines like critical and cultural posthumanism that explicitly incorporate posthuman realities into their subject matter or posthumanist conceptual frameworks and techniques into their methodologies; such posthumanities offer a skeptical assessment of posthumanizing and technologizing trends. On the other hand, the term 'posthumanities' is sometimes used as a synonym for the 'digital humanities,' a group of fields that are on the vanguard of the technologization of academia. Displaying a techno-enthusiasm similar to that of transhumanism, posthumanities of the latter sort advocate the replacement of "analog or literacy-based knowledge structures" with virtualized digital collections of data.[71]

Human nature and the posthumanities. Herbrechter notes that simply because critical posthumanism considers 'human nature' to be a cultural artifact, it is not obligated to claim that human nature is unworthy of study. Indeed, the critical posthumanities will be well-positioned to investigate human nature in a way that expands the scope of such a 'nature' in a deanthropocentrizing manner.[72] With its insights into the history, structure, and practices of various spheres of culture, cultural posthumanism can play a role in taking the critical methodologies developed within critical posthumanism and applying them across the current range of the humanities to develop nonanthropocentric and nonbinary posthumanities that can survive and thrive despite their loss of the concept of human nature that has historically served as the anchor of the humanities.[73]

Counteracting the forces of scientism. From the perspective of critical posthumanism, one important aim of the posthumanities is to ensure that disciplines such as philosophy, theology, history, and the arts continue to play a

---

[69] Miah (2008), p. 79.

[70] Ferrando (2013), p. 32.

[71] For the posthumanities as a possible driver (rather than critic) of digitalization, see Herbrechter (2013), p. 179.

[72] Herbrechter (2013), p. 168.

[73] This is similar to the previously discussed notion that posthumanism might serve as the rescuer of a faltering humanism. See Herbrechter (2013), p. 143.

role in shaping our understanding of human nature and that fields such as neuroscience, biology, chemistry, and computer science do not appropriate for themselves the sole privilege and responsibility of defining what is and is not human. In this way, Herbrechter suggests that the posthumanities can help guarantee that binary and anthropocentric historical humanism is succeeded by a nondualistic and nonanthropocentric posthumanism rather than by a 'scientistic' posthumanism that simply replaces the transcendental idol of the human with a new transcendental idol of science.[74]

4. Feminist New Materialism

Ferrando cites a range of 'new materialisms' that have arisen as a largely feminist response to late postmodernism; they represent a pushback against those forms of postmodernism that had resolved the historic 'nature versus nurture' debate by strongly emphasizing the importance of culture and education while downplaying the role of biology and matter in shaping human existence.[75] New materialism's link to posthumanism lies in the fact that rather than resolving such a binary question in one direction or the other, it dissolves the dualism that pits language and culture against biology and matter. As Ferrando explains, within new materialist thought "biology is culturally mediated as much as culture is materialistically constructed," and matter cannot be separated from the dynamic and performative process of its ongoing materialization.[76]

Herbrechter offers a similar account of the neovitalism that arises from a "feminist materialist, life-affirming tradition" which offers a critique of the more death-centered philosophy of, for example, Derrida. For Herbrechter, the posthumanist aspect of new materialism can be seen in its effort "to reposition the notion of 'life' outside propriety or impropriety, namely by 'de-athropocentring' and 'de-ontologizing' it."[77] He also notes that strong feminist elements have long been found within mainstream critical posthumanism; Haraway, for example, suggests that the posthumanizing dissolution of the boundary between human being and machine resulting from the technologization and cyborgization of our lives can also be exploited to dissolve other boundaries such as those relating to gender.[78]

---

[74] For the posthumanities as a bulwark against scientism, see Herbrechter (2013), p. 169.

[75] Ferrando (2013), pp. 30-31.

[76] Ferrando (2013), p. 31.

[77] Herbrechter (2013), p. 212.

[78] The recognition of such blurring boundaries has long been at the core of posthumanism. See Haraway (1991) and Herbrechter (2013), pp. 99-100.

5. Antihumanism

The term 'antihumanism' has been used to describe an array of phenomena that bear some relationship to posthumanism. Some forms of antihumanism are directly identified with posthumanism; for example, Miah characterizes Pepperell's theory of posthumanism – in which the technological tools that once gave humankind dominance over nature now threaten to claim dominance over us – as a form of "*anti*-humanism, which is re-enlightened by modern science."[79] Other forms of antihumanism are described as diverging from posthumanism in key respects. For example, Ferrando conceptualizes 'antihumanism' as sharing a central tenet with posthumanism: namely, a radical critique of "modern rationality, progress and free will" that constitutes a "deconstruction of the notion of the human." However, the deconstruction offered by posthumanism argues that simple binaries such as 'human versus nonhuman' are no longer meaningful and that human beings are not (any longer) the only kinds of personal subjects that constitute our society. Antihumanism, on the other hand, claims that the binary of 'life versus death' is still meaningful – and that the human being, as such, is dead. Ferrando argues that while posthumanism draws much from the deconstructive approach of Derrida, antihumanism has more in common with the 'death of Man' propounded by Foucault.[80]

Drawing on Badmington, Herbrechter suggests that antihumanism is frequently just a well-disguised form of humanism, insofar as it does not develop its own independent perspective but instead simply defines itself as the negation of all that humanism stands for. However, denying the exclusive centrality of the 'human' is not the same thing as embracing the joint centrality of the 'human and nonhuman'; from the perspective of critical posthumanism, antihumanism thus presents an insufficient challenge to the fundamentally anthropocentric doctrines of humanism. While antihumanism remains locked into the binary patterns that characterize humanist thought, critical posthumanism makes a concentrated effort to break down those historical binaries, replacing them with richer and more sophisticated conceptual schemas.[81]

While the relationship of antihumanism to posthumanism is thus complex, building on Ferrando's analysis we would suggest that at least some forms of antihumanism have evolved to take on characteristics indicative of posthumanist thought. We would argue that such antihumanism is most naturally classified as a form of **analytic theoretical posthumanism**. While such antihumanism differs from critical posthumanism in its attitude toward binary

---

[79] See Miah (2008), p. 75, and Pepperell, *The Posthuman Condition: Consciousness Beyond the Brain* (2003).

[80] Ferrando (2013), pp. 31-32.

[81] Herbrechter (2013), p. 126.

frameworks and post-anthropocentrism, it shares critical posthumanism's rejection of simplistic post-Enlightenment humanism, its goal of developing a more accurate understanding of the nature of humanity, and an emphasis on analyzing the state of humanity as it has come to exist rather than in some engineered form that it might take in the distant future.

6. Metahumanism as Rehumanism

There have arisen at least three independent uses of the term 'metahumanism.' These are: 1) metahumanism understood as a form of 'rehumanism,' as formulated by Sanbonmatsu; 2) metahumanism as an activist movement in support of those who have been subject to metahumanizing mutation, as formulated in numerous works of science fiction and fantasy; and 3) metahumanism as a philosophical and artistic approach and movement of relational 'metabodies,' as formulated by Del Val and Sorgner. We would argue that the first form of metahumanism constitutes a type of analytic theoretical posthumanism; it will thus be considered in more detail here. The second form of metahumanism will be discussed later as a form of synthetic practical posthumanism, and the third will be explored as a type of hybrid posthumanism that spans theoretical, practical, analytic, and synthetic spheres.

Writing in 2004, Sanbonmatsu formulated a concept of 'metahumanism' not as a form of posthumanism but rather as a critical response to and explicit rejection of it. He argues that within our contemporary world,

> [...] in the Western academy, cultural studies theorists and other academic intellectuals hold conferences celebrating our so-called post-human times, singing the virtues of cyborgs, prosthetics, and bioengineering. Post-humanism is merely the latest in a string of commodity concepts spun off by academic industrialists to shore up the crumbling appearance of use value in their work.[82]

In this view, posthumanism is presented as perhaps the most degenerate iteration of a disintegrating Western critical tradition, while metahumanism is proposed as a form of thought that can rescue the critical tradition by confronting and vanquishing posthumanism. In its contents, such metahumanism would essentially appear to be a reborn humanism operating under a different name. Thus Sanbonmatsu argues that "If critical thought is to survive this implosion of theory" represented by posthumanism, posthumanist thought must be challenged by a metahumanism that constitutes "a return to ontology and the grounding of thought in a meaningful account of human being" and which

[82] Sanbonmatsu, *The Postmodern Prince: Critical Theory, Left Strategy, and the Making of a New Political Subject* (2004), p. 207.

does not hesitate "to declare itself to be in defense of *this being that we are – or that we might become.*"[83]

Herbrechter considers Sanbonmatsu to be pursuing the "renewal of a leftist radical humanism in the name of a Kantian cosmopolitan tradition."[84] However, such metahumanism could instead arguably be understood as an idiosyncratic example of **analytic theoretical posthumanism**, insofar as it does not simply propose for adoption a naïve 19th-Century humanism that is unaware of the processes of technologization and posthumanization that have occurred during recent centuries. Rather than ignoring the rise of posthumanist thought, Sanbonmatsu's metahumanism explicitly critiques and seeks to learn from what it perceives as the errors of earlier posthumanist accounts. While such metahumanism can thus be viewed as an 'anti-posthumanism,' we would argue that it can alternatively be understood as a 'rehumanism' informed by posthumanist insights.

7. Neohumanism as the Embracing of Human Subjectivity

As is true for 'posthumanism' and 'metahumanism,' the term 'neohumanism' has been used to describe a divergent array of phenomena. For example, Herbrechter refers broadly to the discourse that pits "transhumanists versus neohumanists."[85] In that context, neohumanists can be understood as thinkers who disagree both with the postmodernist annihilation of the notion of humanity and the transhumanist idolization of a reengineered humanity; neohumanists seek to salvage the positive elements of humanism but in a manner that acknowledges ongoing processes of posthumanization. Similarly, Wolin employs the term when arguing that in his later works Foucault distanced himself from his earlier post-structuralist critique of modernity and formulated a new 'neohumanist' approach in which the existence of a free and thinking human subject is at least implicitly embraced.[86] If considered a form of posthumanism, such neohumanisms would take their place alongside critical posthumanism as a form of **analytic theoretical posthumanism**.

B. Synthetic Theoretical Posthumanisms: Seeking to Understand a Future Posthumanity

Synthetic theoretical posthumanisms manifest a 'posthumanism of imagination' that creatively envisions hypothetical future posthumanities so that

[83] Sanbonmatsu (2004), p. 207.

[84] For this critique of Sanbonmatsu's metahumanism, see Herbrechter (2013), p. 71.

[85] Herbrechter (2013), p. 40.

[86] See Wolin, "Foucault the Neohumanist?" (2006), and Nealon, *Foucault Beyond Foucault* (2008), pp. 10-11.

their implications can be explored.[87] Such forms of synthetic theoretical posthumanism include philosophical posthumanism, science fiction, prehumanism, and techno-idealism. We can consider each of these in more detail.

1. Philosophical Posthumanism

Philosophical posthumanism combines critical posthumanism's academic rigor with science fiction's practice of imagining possible future paths for the processes of posthumanization. It is a **synthetic theoretical posthumanism** insofar as it constructs scenarios of future posthumanities and its goal is to deepen human knowledge rather than to generate some economic, political, or technological impact.

Philosophical posthumanism draws on the insights of critical and cultural posthumanism, integrating them into traditional methodologies of philosophical inquiry in order to reassess earlier philosophical claims with a new awareness of the ways in which philosophy has been suffused with "anthropocentric and humanistic assumptions" that limit its scope, comprehensiveness, and effectiveness.[88] Moreover, as philosophy reflects on processes of posthumanization to envision the ways in which they will reshape ontology, epistemology, and ethics, this generates a new process of 'philosophical posthumanization' that takes its place alongside other technological and social forms of posthumanization.[89]

Origins in critical and cultural posthumanism. Ferrando recounts that during the 1990s feminists within the field of literary criticism developed critical posthumanism, which interacted with cultural posthumanism to give rise to philosophical posthumanism by the end of the decade.[90] Similarly, Miah considers the cyborg expositions of Haraway and Gray, the posthumanism of Hayles and Fukuyama, and Bostrom's transhumanism to have contributed to the development of philosophical posthumanism.[91] Philosophical posthumanism can be understood either as a form of philosophy that has adopted elements of posthumanist thought or as a new form of critical and cultural posthumanism that has chosen to focus its attention on traditional philosophical questions.

---

[87] As previously noted, an exception to this temporal pattern is prehumanism, which considers fictional or hypothetical beings of the far-distant past as an alternative to positioning them in the far-distant future.

[88] Ferrando (2013), p. 29.

[89] Herbrechter (2013), p. 176.

[90] Ferrando (2013), p. 29.

[91] See Miah (2008), p. 80; Haraway, "A Manifesto for Cyborgs: Science, Technology, and Socialist Feminism in the 1980s" (1985); Gray, "The Ethics and Politics of Cyborg Embodiment: Citizenship as a Hypervalue" (1997); Gray (2002); Hayles (1999); Fukuyama (2002); and Bostrom, "A History of Transhumanist Thought" (2005).

The differences between philosophical and cultural posthumanism, in particular, are frequently blurred. Even Miah, who clearly distinguishes philosophical posthumanism from its biopolitical and cultural siblings, notes that the analyses offered by philosophical posthumanism are often "inextricable from other cultural critiques." However, it is possible to identify differences between the two fields; for example, Miah suggests that while cultural posthumanism (as represented by Haraway and Hayles) is "intended to disrupt uniform ideas about what it means to be human and the social and political entitlements this might imply," philosophical posthumanism typically focuses on ontological, phenomenological, and epistemological questions surrounding scenarios of future technologization.[92]

Envisioning future posthumanity. Like cultural posthumanism, philosophical posthumanism contemplates not only current processes of technologization but also hypothetical futuristic technologies that do not yet exist but which have been envisioned in works of science fiction. While cultural posthumanism analyzes such fictional future technologies as a means of diagnosing current humanity's desire for or fear of further posthumanization, philosophical posthumanism uses hypothetical technologies as the bases for thought experiments that explore the ontological, epistemological, ethical, legal, and aesthetic implications of such future posthumanization. By exploiting philosophical methodologies and a knowledge of science and technology, such thought experiments allow philosophical posthumanists to understand the ways in which human nature may be transformed or superseded through future posthumanization – without necessarily advocating or opposing such transformations in the way that a biopolitical posthumanist would.[93]

The phenomenon of environmental posthumanization. As conceptualized by Miah, a notable characteristic of philosophical posthumanism is that it does not focus on changes to human beings *per se* as the primary manifestation of posthumanization.[94] Instead, philosophical posthumanism posits a broader phenomenon in which posthumanization is occurring throughout the world as a whole. For example, the proliferation of social robots, artificial general intelligences, artificial life-forms, virtual worlds, ubiquitous computing, and the Internet of Things is expected to create a rich digital-physical ecosystem in which human beings are no longer the only – or perhaps even the most significant – intelligent actors. Such a post-anthropocentric and post-dualistic world would already possess a strongly posthuman character regardless of

---

[92] Miah (2008), pp. 79-80.

[93] Regarding philosophical posthumanism's dispassionate analysis of processes of posthumanization, see, e.g., Miah (2008), p. 79.

[94] Miah (2008), pp. 80-81.

whether human beings undergo processes of biotechnological transformation or choose to remain in their 'natural' biological form.

Some strains of philosophical posthumanism effectively update historical Darwinian biological materialism for the age of artificial life, viewing the posthuman world as a place in which the differences between human beings and animals, human beings and robots, and human beings and electronic information systems are increasingly ones of degree rather than kind.[95] The relationship between the human and machine is explored especially by considering entities such as cyborgs in which those two realms have become physically and behaviorally fused.[96] It also addresses the ontological and ethical implications of new kinds of entities such as artificial general intelligences that have not yet been created in practice but for whose development much theoretical groundwork has been laid; this gives philosophical posthumanism a stronger future orientation than critical posthumanism, which is more concerned with ethical and social realities of our current day.

2. Science Fiction

Herbrechter suggests that true science fiction is "the most posthumanist of all genres," as it takes seriously – and often advances – the ongoing "dissolution of ontological foundations like the distinction between organic and inorganic, masculine and feminine, original and copy, natural and artificial, human and nonhuman."[97] In its most representative form, science fiction attempts to construct coherent visions of a near- or far-future posthumanized world so that its nature and implications can be investigated; for this reason, science fiction can be categorized as a **synthetic theoretical posthumanism.**[98]

Science fiction versus posthumanist reflection on science fiction. It is important to distinguish science fiction itself from scholarly analysis *of* science fiction. While science fiction typically constitutes a form of synthetic theoretical posthumanism, the reflection on science fiction that is carried out, for example, by cultural posthumanists is often a form of analytic theoretical

---

[95] For philosophical posthumanism's consideration of evolutionary processes in biological and nonbiological entities, see Miah (2008), p. 82.

[96] Miah (2008), pp. 80-81.

[97] Herbrechter (2013), pp. 115-17.

[98] Building on Poster and Hayles, Herbrechter notes that the cyberpunk genre in particular – which attempts to construct realistic and realizable visions of a near-future technologized posthumanity – has most explicitly grappled with the nature of human beings as embodied informational processes and the ramifications of posthumanizing technologies that are expected to break down traditional humanist binaries and reshape the experience of human existence within the coming decades. See Goicoechea (2008); Poster, *What's the Matter with the Internet?* (2001); Hayles, *My Mother Was a Computer: Digital Subjects and Literary Texts* (2005); Hayles (1999); and Herbrechter (2013), p. 187.

posthumanism. From the perspective of cultural posthumanism, science fiction's relevance does not depend on it portraying future technologies that are in fact strictly realizable; rather it is relevant because it reflects society's current 'cultural imaginary' and can thus be used to diagnose humanity's attitude toward the processes of technologization and posthumanization.[99] In a related fashion, when transhumanism draws inspiration from works of science fiction to spur the real-world pursuit of particular futuristic technologies, it constitutes a form of synthetic practical rather than synthetic theoretical posthumanism.

Science fiction and the genesis of posthumanism. From its birth, the field of posthumanism has been tied to the world of science fiction. Indeed, the work generally considered to contain the earliest allusion to a critical posthumanism, Hassan's 1977 text "Prometheus as Performer: Toward a Posthumanist Culture? A University Masque in Five Scenes," explicitly cites the film *2001: A Space Odyssey* and dawning questions about artificial intelligence as being relevant to understanding the "emergent [...] posthumanist culture."[100] If posthumanism has always drawn on certain forms of science fiction, Miller suggests that – in complementary fashion – science fiction has always constituted a form of posthumanism. While 'posthumanism' as such may only have been labelled and defined during the last few decades, science fiction had already existed for centuries as an unrecognized form of posthumanism; only recently has critical theory begun to follow science fiction's example of radically reassessing the limits of human nature and the social and technological structures that circumscribe the meaning of 'the human.'[101]

Distinguishing science fiction from popular ('commercial') posthumanism. In places, Herbrechter writes of science fiction as though it were essentially a commercial enterprise whose contents are formulated by large corporations with the goal of maximizing revenue and profits – rather than a serious literary and artistic endeavor whose contents are crafted by individual authors, filmmakers, and game designers as a means of exploring difficult philosophical, political, and social issues facing humanity. Thus he emphasizes the "rather close 'co-operation' between science fiction, the film industry and its lobbies and the discourse on posthumanity in general."[102] However, such a view appears to be an oversimplification. We would argue that in the context of posthumanism, the phrase 'science fiction' is frequently used to refer to two spheres of human activity which are so qualitatively different in nature that they are better classified as two entirely different forms of posthumanism.

---

[99] Herbrechter (2013), p. 117.

[100] See Hassan, "Prometheus as Performer: Toward a Posthumanist Culture? A University Masque in Five Scenes" (1977), and its discussion in Herbrechter (2013), p. 33.

[101] This point is made in Miller (2012), p. 164.

[102] Herbrechter (2013), p. 39.

We would suggest that the term 'science fiction' be reserved for the first of these two types of posthumanism, which involves the construction of fictional scenarios (often set in the future) as a means of exploring the profound ontological, biological, ethical, social, and cultural implications of posthumanization. Works of science fiction are, in a sense, thought experiments similar to those utilized within philosophical posthumanism. However, while philosophical posthumanism employs the rigorous methodologies and critical apparatus of philosophy, science fiction exploits the freedom to draw on more artistic and less formally academic methodologies. Works such as paintings, sculpture, or music with science-fiction themes can explore the 'mood' or 'ethos' of posthumanization in a general sense. Artistic forms such as films or novels can present more detailed diegetic content but are consumed in a manner that is still largely passive. However, interactive media such as computer games and tabletop roleplaying games can put their human players in situations in which they face complex ethical dilemmas and must actively confront challenges associated with new posthumanized ways of being. As noted above, because of its emphasis on imagining future posthumanities and the fact that it is primarily geared at deepening human knowledge, science fiction can be best understood as a form of **synthetic theoretical posthumanism**.

The second kind of posthumanism that is sometimes described as a type of 'science fiction' (and which Herbrechter indeed takes to be the most representative form of science fiction) is what we would refer to as 'popular' (or 'commercial') posthumanism to distinguish it from science fiction proper. Examples of popular posthumanism include films, television series, and other works that are created either to generate maximum profits by engaging mass audiences or to condition the public to accept certain future actions by governments, corporations, or other institutions. Like posthumanist science fiction, popular posthumanism often employs storylines that are set in the future and which feature cyborgs, androids, artificial general intelligences, genetic engineering, virtual reality, and other posthumanizing technologies. However, rather than attempting to confront and thoughtfully explore the philosophical implications of such phenomena, popular posthumanism exploits posthuman themes instrumentally as a means of achieving some practical goal – such as generating revenue from movie ticket sales.

Some artistic products function simultaneously as works of both posthumanist science fiction and popular posthumanism; in practice, the division between these two types is rarely absolute. Nevertheless, the divergence in the goals of posthumanist science fiction and popular posthumanism can often be seen, for example, in the difference between complex original literary works and their later adaptations into Hollywood blockbuster films that feature a drastic simplification of the works' philosophical content coupled with more frequent explosions and a happy ending in which the protagonist defeats the (often technologically facilitated) threat to humanity.[103] Popular posthumanism will be considered in more detail later as a form of synthetic practical posthumanism.

3. Prehumanism

While some works of science fiction envision the extremely far future, other forms of theoretical posthumanism envision the extremely distant past. For example, some proponents of cultural materialism emphasize the billions of years that passed before intelligent life appeared on earth. These vast foregone eons are highlighted not because the events that occurred within them are of direct interest to posthumanism but because they contextualize and deanthropocentrize our present moment; they emphasize the fact that the universe is not dependent on humanity for its existence or meaning and that the whole era of humankind's flourishing is only a fleeting instant in comparison to the lifespan of the cosmos as a whole.[104] Practitioners of what might be called 'prehumanism' are not interested in performing a literal scientific reconstruction of the biological or anthropological characteristics of the precursors of modern human beings but rather in imagining such prehistoric beings from a metaphorical or hypothetical perspective in order to better appreciate the relationship of contemporary humanity to the timescale of the universe.

'Prehumanist' approaches generally constitute forms of **synthetic theoretical posthumanism**, insofar as they are grounded in imagination rather than critique. Herbrechter notes, for example, that the world of posthumanist speculative fiction includes not only works that explore future spaces but also ones that explore "fictional pasts or *verfremdet* (defamiliarized) presents."[105] As a posthumanist approach that looks back imaginatively to the past, prehumanism thus constitutes a mirror image of the posthumanist science fiction that looks ahead imaginatively to the future.[106] Works such as the cosmic horror literature of H.P. Lovecraft that feature alien entities that have existed for mil-

---

[103] For example, consider Asimov's *Robot* series of stories and novels as compared with the 2004 Will Smith cinematic vehicle, *I, Robot.*

[104] See Herbrechter (2013), pp. 9-10.

[105] Such products are by no means limited to science fiction but can include works of any genre and theme that disorient and challenge their characters and readers. See Herbrechter (2013), p. 116.

[106] As described here, prehumanism is thus not 'pre-humanist' in the sense of considering the world that existed before the appearance of humanism but rather 'prehuman-ist' in the sense of considering the world that existed before the appearance of human beings. The usage described here thus differs from the way in which the terms 'prehumanism' and 'prehumanist' are employed in, e.g., Berrigan, "The Prehumanism of Benzo d'Allesandria" (1969), and Witt, "Francesco Petrarca and the Parameters of Historical Research" (2012), to refer to time periods that preceded and concepts that foreshadowed those of Renaissance humanism.

lions of years (or in a timeless parallel dreamworld) can be understood as examples of such prehumanism.[107] Other works such as *2001: A Space Odyssey* simultaneously constitute both: 1) prehumanism that uses the distant past as a setting for imagining a 'quasi-human' that already was; and 2) posthumanist science fiction that looks into the future to imagine a 'quasi-human' that has not yet been.[108]

4. Techno-Idealism

Techno-idealism is a form of posthumanist thought closely linked to but distinct from transhumanism. It involves the belief that the sole essential part of a human being is the mind and that this 'mind' consists of a particular pattern of information. Because only a mind's pattern of information – and not the physical substrate in which the information is stored – is relevant, all of a brain's biological neurons can be replaced one by one with electronic replicas, and as long as the pattern of interactions found within the brain's neural network is preserved intact, the person's mind, consciousness, and identity would continue to exist within its new (and undying) robotic shell. From the perspective of techno-idealism, human beings' physical biological bodies are ultimately interchangeable and replaceable with physical robotic bodies or potentially even virtualized ones.

Contrast with critical posthumanism. Herbrechter portrays techno-idealists as yearning for 'technoscientific utopias' in which human engineers will someday unravel the mysteries of genetics, thereby allowing biological life to finally be transformed into pure, disembodied information; in this way, virtuality becomes a means to immortality as human beings "gain control over the 'book of life'."[109] He contrasts techno-idealism's naïve understanding of the nature of the human mind with the more thoughtful and incisive analyses conducted within critical and philosophical posthumanism. Indeed, Herbrechter suggests that critical posthumanism can largely be understood as an effort to defend the material anchoring of humanity against those techno-idealists who seek to virtualize and disembody everything – as manifested, for example, in their advocacy of mind uploading.[110]

Complementarity to transhumanism. The 'posthumanity' envisioned by techno-idealism is one of hypothetical future entities like full-body cyborgs and uploaded minds. Techno-idealism does not, in itself, actively seek to engineer such beings but rather to develop conceptual frameworks for exploring their nature, capacities, and behavior; it can thus be understood as a form of

---

[107] See, e.g., Lovecraft, *The Dunwich Horror and Others* (1983) and *At the Mountains of Madness and Other Novels* (1985).

[108] See Kubrick's *2001: A Space Odyssey* (1968).

[109] Herbrechter (2013), pp. 103, 171.

[110] Herbrechter (2013), p. 95.

synthetic theoretical posthumanism. However, in practice techno-idealist frameworks are often formulated by committed transhumanists seeking an intellectual justification for their concrete practical endeavors. Drawing on Krüger, Herbrechter traces the development of a 'radical techno-idealism' from Wiener's cybernetics, the futurology of the incipient Space Age, and the cryonics movement to figures such as More, Minsky, Moravec, Kurzweil, and contemporary transhumanist performance artists.[111] For many such individuals, the techno-idealism which says that human beings *can* achieve immortality through the development of transformative technologies is paired with a technological determinism which says that humanity inevitably *will* create and implement such technologies.[112]

It is not necessary, however, for transhumanists to hold techno-idealist beliefs. For example, one could conceivably deny that an uploaded mind is a 'true' human mind – while simultaneously arguing that such artificial intelligences should nonetheless be developed to serve as successors to humanity and a next step in the evolution of sapient intelligence within our world. Someone holding such a view would be a transhumanist but not a techno-idealist. Conversely, a person could conceivably accept the claim that a biological human brain can be gradually replaced by an electronic brain without destroying its owner's 'mind' – but without feeling the slightest inclination to see any human being undergo such a procedure. Indeed, such a person might feel a sense of revulsion at the idea that causes him or her to oppose the development of such technologies, even while accepting their efficacy on an intellectual level. Such an individual would be a techno-idealist but not a transhumanist.

C. Analytic Practical Posthumanisms: Seeking to Reshape the Posthumanized Present

Analytic practical posthumanisms seek to reshape an already-existing posthumanized world. They can be understood as constituting a 'posthumanism of conversion' that is aimed at changing hearts and minds and influencing the way in which human beings view and interact with their contemporary environment. Such forms of analytic practical posthumanism include some forms of metahumanism and neohumanism, which we describe in more detail below.

---

[111] See Krüger, *Virtualität und Unsterblichkeit* [Virtuality and Immortality] (2004), as discussed in Herbrechter (2013), p. 103.

[112] On this frequent pairing of theoretical and practical posthumanism, see Herbrechter (2013), p. 103.

1. Metahumanism as Advocacy for Those Who Have Experienced Metahumanizing Mutation

Since the 1980s, the term 'metahuman' has been used within a range of science-fiction, superhero, and fantasy literature and roleplaying games to refer to a human being who has undergone a mutation or transformation that grants the individual a new physical form or altered sensory, cognitive, or motor capacities; the mechanics of the transformation may be portrayed as technological, magical, or otherwise preternatural in nature.[113] The term 'metahumanity' is employed within such a fictional world to describe either its typically diverse collection of metahuman beings or the state of being a metahuman. Within the context of such a fictional world, 'metahumanism' can describe either: 1) the condition of possessing metahuman characteristics (which can be viewed by different individuals as a blessing or a curse); or 2) a political or social movement that works to promote the safety, welfare, and basic rights of metahumans, who often suffer discrimination as a result of the radical otherness that can terrify or appall 'normal' human beings.

a. Anti-metahumanism as Discrimination against Metahumans

Within such a fictional context, 'anti-metahumanism' describes an opposing political, social, or religious movement that views metahumans either as a lesser form of being whose activities must be supervised, a threat to the welfare of regular human beings, or inherently evil.[114] Such oppression is typically described as being inflicted by natural, non-metahumanized human beings, although metahumans themselves are capable of displaying anti-metahuman attitudes and behaviors.

b. Classifying Metahumanism within the Fictional and Real Worlds

When classifying them as forms of posthumanism, metahumanism and anti-metahumanism can be understood from two perspectives, namely: 1) as they function within the fictional world in which they appear; and 2) as devices created by authors, filmmakers, or game designers and consumed by audiences within our contemporary real world. Within the fictional worlds in which they exist as political and social movements, metahumanism and anti-metahumanism depict a form of **analytic practical posthumanism**, insofar as

---

[113] See Ferrando (2013), p. 32. Perhaps the earliest published use of the term 'metahuman' in this sense (in particular, as an adjective referring to superhuman powers or abilities gained as a result of infection by an extraterrestrial virus) was in the anthology set in the shared *Wild Cards* superhero universe published in 1986. See, e.g., Milán, "Transfigurations" (p. 264) and "Appendix: The Science of the Wild Card Virus: Excerpts from the Literature" (p. 403), in *Wild Cards*, edited by Martin (1986).

[114] For a depiction of anti-metahumanism, e.g., within the fictional universe of the *Shadowrun* roleplaying game, see the *Sixth World Almanac*, edited by Hardy & Helfers (2010), pp. 23, 35, 49, 54, 57, 79, 142.

they focus on an already existing (within the work's fictional timeline) posthumanity and either advocate for the adoption of particular policies or work directly to empower or suppress metahumanity.

However, within our real world, such fictional depictions of metahumanism and anti-metahumanism play a broader range of roles. Some creators of fictional works employ metahumans (and the reactions to them) as a means of critiquing our real-world presumptions and encouraging audiences to probe their own understanding of what it means to be human. In these cases, it is not being claimed by an author that posthumanized beings displaying those exact characteristics might someday come to exist; rather, metahumanity is being used as a device to compel contemporary audiences to consider their own humanity. Such metahumanism and anti-metahumanism serve as a form of **analytic posthumanism** that is either **theoretical** or **practical**, depending on whether it fills the role of a thought experiment or is intended to alter the way that audiences treat other human beings (or animals, artificial intelligences, and other nonhuman beings).

Other fictional works may feature metahumanism and anti-metahumanism in order to help audiences explore the many possible forms that future posthumanity might take and understand the interrelationships between posthumanizing technologies such as genetic engineering, neuroprosthetics, and artificial intelligence. Such works are often forms of **synthetic theoretical posthumanism**;[115] however, they may also display aspects of **synthetic practical posthumanism**, if designed to foster attitudes of acceptance toward future metahuman beings.

3. Neohumanism as Advocacy for Engineered Beings

One variety of 'neohumanism' was described in an earlier section as a type of analytic theoretical posthumanism. The term 'neohuman' has also been used within the context of science fiction to describe genetically engineered human beings who possess a genotype derived from and similar to that of natural human beings but who have been given enhanced sensory, motor, and cognitive capacities. While some fictional neohumans are presented as relishing the engineered capacities that make them 'superior' to natural human beings, others resent these traits that they never chose to possess and which

---

[115] This is especially true of works featuring future worlds in which metahumans can choose at least some of their 'nonhuman' traits, such as characters who acquire neuroprosthetic enhancements or study magic within the *Shadowrun* universe. Similarly, in many tabletop roleplaying games and computer games, a game's contemporary human player must invest significant time and care in selecting his or her character's metahuman characteristics from among a complex system of physical and cognitive attributes, advantages, disadvantages, skills, and equipment and possessions. See, e.g., the *Shadowrun: Core Rulebook 5*, edited by Killiany & Monasterio (2013).

cause them to be seen as something other than fully human. Rather than emphasizing the engineered characteristics that set them apart, such neohumans may instead accentuate those shared genetic traits that link them with (the rest of) humanity.[116]

In such a context, 'neohumanism' would involve advocacy for the development of such engineered beings or defense of the rights and welfare of such persons, thus resembling metahumanism in its form of support for those who have experienced metahumanizing mutation. Such neohumanism would be a form of **analytic practical posthumanism** within the fictional worlds in which it is depicted, but it could be either **analytic** or **synthetic** and either **theoretical** or **practical** if evaluated according to the real-world reasons for which a creator of fiction decided to include it in his or her work.

4. Neohumanism as Spiritual Ultrahumanism

Another application of the term 'neohumanism' is in describing a holistic and universalist philosophy developed by Sarkar that is grounded in Tantric spiritual principles[117] and manifested in particular religious practices, works of art and literature, humanitarian and animal-rights initiatives, and a global network of schools guided by "a transcivilizational global pedagogy."[118] The goal of such a neohumanism is:

> [...] to relocate the self from ego (and the pursuit of individual maximization), from family (and the pride of genealogy), from geo-sentiments (attachments to land and nation), from socio-sentiments (attachments to class, race and religious community), from humanism (the human being as the centre of the universe) to Neohumanism (love and devotion for all, inanimate and animate, beings of the universe).[119]

This nominal dislocation of the human being from its historical position as the 'center of the universe' appears to have much in common with the post-anthropocentric attitude that is developed, for example, within critical posthumanism. However, that similarity is arguably superficial. Elsewhere, Sarkar writes that:

> Neohumanism will give new inspiration and provide a new interpretation for the very concept of human existence. It will help people understand that human beings, as the most thoughtful and intelligent beings in this created universe, will have to accept the great

[116] See *Interface Zero 2.0: Full Metal Cyberpunk*, developed by Jarvis et al. (2013), p. 107.

[117] See the "Foreword" to *Neohumanist Educational Futures: Liberating the Pedagogical Intellect*, edited by Inayatullah et al. (2006).

[118] "Foreword," *Neohumanist Educational Futures* (2006).

[119] "Foreword," *Neohumanist Educational Futures* (2006).

> responsibility of taking care of the entire universe – will have to accept that the responsibility for the entire universe rests on them.[120]

Ferrando argues that some forms of transhumanism can actually be understood as an 'ultrahumanism' that seeks to advance post-Enlightenment rationality and scientific progress to its logical conclusion, thereby consummating humanism rather than superseding it.[121] A similar account might be offered of Sarkar's neohumanism: rather than rejecting the humanist vision of human beings as the supreme intelligent agents charged with exercising dominion over nature, neohumanism seeks to cement the position of human beings as the 'center of the universe' – albeit a center that serves as a loving caretaker for the rest of creation.[122]

Such neohumanism is **analytic**, insofar as it focuses its attention on the human beings who already exist today and the sociotechnological reality within which they are embedded. While such neohumanism possesses many elements that are explicitly philosophical in nature, the neohumanist project is geared primarily toward creating a movement whose adherents alter their daily lives to incorporate particular spiritual practices and who establish and operate schools, charitable institutions, and other organizations that embody the movement's philosophy; in this sense, neohumanism can be understood as a **practical** rather than theoretical posthumanism.

D. Synthetic Practical Posthumanisms: Seeking to Control the Processes Generating a Future Posthumanity

Synthetic practical posthumanisms reflect a 'posthumanism of control' that seeks to initiate, accelerate, guide, limit, or block future processes of posthumanization – typically through regulating the development of new technologies or through other political, economic, or social mechanisms. Such forms of synthetic practical posthumanism include biopolitical posthumanism (which itself includes bioconservatism and transhumanism) and popular or 'commercial' posthumanism. We can consider these in more detail.

1. Biopolitical Posthumanism

Biopolitical posthumanism encompasses a range of posthumanisms that all envision the engineering of a future 'posthumanity' but which differ in their assessment of whether such a development is desirable or undesirable. Biopolitical posthumanisms manifest a strong future orientation: they attempt to predict the long-term impact of pursuing particular new biotechnologies and – based on such predictions – work to actively facilitate or impede the creation

---

120 Sarkar (1982).

121 Ferrando (2013), p. 27.

122 Indeed, Sarkar claims explicitly that "Neohumanism is humanism of the past, humanism of the present and humanism – newly explained – of the future." See Sarkar (1982).

of such technologies by spurring political or regulatory action, influencing public opinion, advancing scientific research and technology commercialization, or through other means. Such biopolitical posthumanisms are **synthetic** insofar as they understand posthumanity to be a collection of future beings whose creation can be purposefully brought about or avoided, and they are **practical** insofar as they seek to actively accomplish or block the advent of such posthuman beings.

Contrasting attitudes toward posthumanity. Different forms of biopolitical posthumanism are distinguished by their attitude toward biotechnological posthumanization. For Miah, biopolitical posthumanism can be divided fairly neatly into the opposing camps of 'bioconservative' thinkers like Fukuyama and 'technoprogressive' or transhumanist thinkers like Stock. Bioconservatives see the advent of posthumanity as a negative or retrogressive step – a loss of human dignity and a destruction of the characteristic essence that makes human beings unique – while technoprogressives see the arrival of posthumanity as an advance by which human nature is beneficially enhanced or its limits transcended.[123]

Birnbacher argues that the concept of 'posthumanity' is in itself value-neutral;[124] however, one could contend that for biopolitical posthumanists, 'posthumanity' is in fact an intensely *value-laden* term – but one whose 'authentic' value is disputed by two opposed ideological groups. Such an interpretation is consistent with Miah's observation that for some bioconservatives, the very word 'posthumanism' is presumed to represent a world so obviously horrific and morally bankrupt that little need is seen to offer specific arguments about why the creation of a 'posthuman' world should be avoided.[125]

Having reviewed biopolitical posthumanism in general, it is worth exploring in more depth its two most prominent forms: bioconservatism and transhumanism.

a. Bioconservatism

Bioconservatism is a form of posthumanism that came into existence largely as a rejection of the tenets of another form of posthumanism – namely,

---

[123] See Miah (2008), pp. 73-74. 'Factor X' is the term used by Fukuyama to describe the essence of humanity that is vulnerable to being corrupted through the unrestrained application of biomedical technology. This can be compared and contrasted, e.g., with the idea of 'essence loss' within the fictional *Shadowrun* universe. See Fukuyama (2002) and *Shadowrun: Core Rulebook 5* (2013), pp. 52-55, 396-97.

[124] Birnbacher (2008), p. 95.

[125] Miah (2008), pp. 74-75.

transhumanism.[126] For bioconservatives, the arrival of the posthumanity envisioned by transhumanism would bring about the 'dehumanization' of the human species.[127] Fukuyama is frequently cited as an eminent bioconservative as a result of his writing and public debating in opposition to transhumanism during his time as a member of the U.S. President's Council on Bioethics in the early 2000s. Habermas is also often cited as a leader in the world of bioconservative thought: while much of his work is highly theoretical, it includes a call to action that points toward practical applications, and the critiques and conceptual frameworks that he has developed provide a philosophical foundation for bioconservatism.[128]

Bioconservatism is a **synthetic posthumanism** insofar as it focuses its attention on hypothetical and emerging technologies that can potentially be used to engineer new quasi-human biological species or cyborgs that differ greatly from human beings as they exist today. It is a **practical posthumanism** insofar as it attempts to block the creation of such future posthumanized beings by rallying public opinion to support particular political and social initiatives; developing and promoting treaties, legislation, regulations, and policies for adoption by governments; pressuring companies, universities, and other institutions engaged in transhumanist programs to curtail such activities; and encouraging individual consumers to change the ways in which they spend their money and time.

Concerns regarding the social impact of posthumanization. Typical bioconservatism does not focus on the psychological, phenomenological, or ontological consequences of posthumanization for the individual posthumanized being. Instead, it sketches out the broad negative impacts that biotechnological posthumanization will supposedly have for human society as a whole – for example, by weakening government protections for human rights, lowering the ethical standards of corporations, creating economic injustice, pressuring entire social classes of human beings to modify themselves in order to compete economically, and perhaps even sparking civil war between those transhuman beings who have been genetically and cybernetically 'enriched' and those natural human beings who, comparatively speaking, are genetically and cybernetically 'deprived.'[129] This emphasis on broad social concerns is reflected in Bostrom's characterization of the five main objections that bioconservatism offers to the purposeful creation of posthumanized beings – namely, that: 1) "It can't be done"; 2) "It is too difficult/costly"; 3) "It would be bad for

[126] Herbrechter (2013), pp. 36-37.

[127] Birnbacher (2008), p. 97.

[128] Herbrechter (2013), pp. 161-62.

[129] Miah (2008), pp. 73-74; Herbrechter (2013), p. 45, 162.

society"; 4) "Posthuman lives would be worse than human lives"; and 4) "We couldn't benefit."[130]

b. Transhumanism

Transhumanism shares with analytic posthumanism its origins in the late 1980s and early 1990s and a "perception of the human as a non-fixed and mutable condition"; in other ways, though, the two perspectives are quite different.[131] Transhumanism does not look back into humanity's past to diagnose the social and technological legacy that we have inherited; instead it looks ahead to the future – and in particular, to the 'enhanced' human, quasi-human, or parahuman species that can be fashioned through the intentional application of genetic engineering, nanotechnology, cryonics, 'mind uploading,' and other emerging or hypothetical technologies.[132]

Understanding of posthumanity. Bostrom uses the word 'posthuman' in a concrete functional sense to refer to an engineered being that possesses at least one 'posthuman capacity' exceeding what is possible for natural human beings.[133] In Bostrom's conception of posthumanity, posthuman beings will not necessarily constitute the entirety – or even a large percentage – of future human society. Indeed, because of the cost and difficulty of the bioengineering equipment and techniques that are needed to create posthuman beings, it is likely that such beings will at least initially represent only a small portion of human society. This **synthetic** understanding differs from analytic forms of posthumanism in which all human beings are already considered to be posthumanized, insofar as we live in a world that is posthuman.

Attitude toward posthumanity. The attitude toward posthumanity expressed by Bostrom can be taken as typical of transhumanists more generally. Bostrom makes a nominal effort at suggesting that he is neutral regarding the question of whether posthumanity represents a step forwards or backwards in human development; he acknowledges that while transhumanism is only concerned with creating forms of posthumanity that are "very good," there are undoubtedly other "possible posthuman modes of being" that would be "wretched and horrible."[134] Elsewhere, however, Bostrom appears to define posthumanity in such a way that it can only be a beneficial phenomenon. For example, he defines a 'posthuman being' not merely as one that has been tech-

[130] Bostrom (2008), p. 109.

[131] For an account of the origins of such forms of posthumanism, see Ferrando (2013), p. 26.

[132] Ferrando (2013), p. 27.

[133] Bostrom (2008), p. 108.

[134] This passing acknowledgement is found within an otherwise vigorous defense of the goal of engineering posthumanity. See Bostrom (2008), p. 108.

nologically engineered to possess characteristics differing from those naturally possessed by human beings but as one who has been technologically engineered to possess either: 1) an enhanced "capacity to remain fully healthy, active, and productive, both mentally and physically"; 2) enhanced "general intellectual capacities [...], as well as special faculties such as the capacity to understand and appreciate music, humor, eroticism, narration, spirituality, mathematics, etc."; or 3) an enhanced "capacity to enjoy life and to respond with appropriate affect to life situations and other people."[135] Bostrom's view of 'posthumanity' is thus not value-neutral but strongly value-laden, as it would automatically exclude from being considered 'posthumanizing' any future technology that results in injury to human beings' health, a degradation of their cognitive capacities, or an impairment to their ability to enjoy social interactions – even if the technology were developed as part of a transhumanist bioengineering project whose explicit goal was to bring about the creation of posthumanity and its negative impacts were an unintended effect.[136]

Transhumanism as activism and project. In the understanding described above, 'posthumanity' is positioned as though it were a new form of space travel or nuclear power whose costs and benefits can be carefully weighed by a government panel that then decides whether to appropriate funds to bring such technology into existence or to ban the technology and prevent its development. This understanding is quite different from that of analytic posthumanism, which believes that posthumanity is inevitable because it is already here, and that the fundamental question is not whether one should seek to actively bring about or prevent the world's posthumanization but how to interpret it.

Critique from the perspective of critical and cultural posthumanism. Transhumanism involves efforts to intentionally engineer a new human species through the use of emerging biotechnologies. It thus typically focuses on the technological posthumanization of humanity and ignores the many nontechnological ways in which posthumanization has been occurring for centuries. Ferrando notes that cultural and critical posthumanism are inclined to negatively assess such an approach. From their perspective, transhumanism appears to possess an overly simplistic conceptualization of the world: it is willing to perpetuate a post-Enlightenment vision of 'human exceptionalism' that places human beings in a hierarchy over nonhuman animals and nature – and indeed, transhumanism further expands this stratification of being by

---

[135] Bostrom (2008), p. 108.

[136] Identifying posthumanity with an 'enhanced' humanity reflects an optimistic assumption that all posthumanizing bioengineering efforts will be driven by a well-intentioned (and effective) vision of 'improving' human nature and not, for example, by a desire to produce quasi-human workers, test subjects, toys, or personal companions that possess a diminished human nature and whose creation is driven by the self-interest of particular governments, corporations, or individual consumers.

creating a new 'hierarchy of hierarchies' in which a soon-to-be-engineered posthumanity will peer down from its superior vantage point outside of the natural order. But transhumanism often glosses naïvely over the fact that such frameworks have historically been used to place some human beings (such as slaves) in positions of inhuman subjugation, that such injustices widely exist even today, and that the development of transhumanist technologies could easily exacerbate rather than solve such problems.[137] Thus Herbrechter positions the critical posthumanism of Hayles as being steadfastly opposed to transhumanism and its goal of achieving the radical disembodiment and dematerialization of the human intellect.[138]

Transhumanism as commercialization of the human being. Anders and Herbrechter suggest that at least some strains of transhumanism could be viewed as outgrowths of the West's hyper-commercialized culture of consumer technology. Members of society have been conditioned to covet the newest models of products – whether smartphones or televisions or automobiles – that possess the most innovative features and best specifications and are ostensibly far superior to last year's models; all 'sophisticated' and 'successful' members of society participate in a cycle of continuous product upgrades. According to this view, transhumanism laments – and is even ashamed by – the fact that the human mind and body are not a purposefully engineered consumer product that can be upgraded; through the application of biotechnologies and a reconceptualization of the nature of humanity, it seeks to transform the human being into just such a consumer product.[139] Although transhumanism envisions itself as a positive movement that seeks to exalt humanity by transcending the limits of human nature, it could thus alternatively be understood as a negative movement that is embarrassed by the messy imperfections inherent in human beings' biological nature and which seeks to suppress that reality beneath a patina of technological enhancement.

Not all technologists are transhumanists. Not all (or even many) scientists, engineers, and entrepreneurs doing cutting-edge work in the fields of genetic engineering, neuroprosthetics, nanorobotics, and artificial intelligence are transhumanists; many individuals involved with developing new technologies for the engineering and augmentation of human beings are content to focus on the very concrete next steps involved with advancing the 'evolution' of humanity. For transhumanists, though, such incremental progress is a necessary but only preliminary step toward the creation of fully disembodied

[137] See Ferrando (2013), pp. 27-28.

[138] See Hayles (1999) and Herbrechter (2013), p. 94.

[139] See Anders, *Die Antiquiertheit des Menschen. Band 1: Über die Seele im Zeitalter der zweiten industriellen Revolution* (1992), pp. 31ff., as analyzed in Herbrechter (2013), p. 170.

posthuman entities that can slip effortlessly between biological and electronic modes of being, between actual and virtual substrates.[140]

Religious aspects of transhumanism. Transhumanism frequently takes on aspects of a religious movement, formulating visions of "techno-transcendence and digital cities of god in cyberspace, of the overcoming of the flesh"; it thus cannot be understood simply from a technological perspective but also requires insights from the field of theology.[141] Some would even contend that transhumanism's conceptual origins lie in (arguably misguided) interpretations of the work of Catholic theologian Pierre Teilhard de Chardin and his idea of the 'noosphere' of shared digital information that would someday come to surround the globe.[142]

Building on Le Breton's analysis, Herbrechter suggests that from the perspective of critical posthumanism, transhumanism can be understood as a sort of 'neognostic' hatred of the body that privileges the mind over its vessel of flesh that continuously degrades and decays.[143] Such conceptual objections to transhumanism, however, are very different from bioconservatives' objections regarding the expected negative real-world impacts of transhumanist projects.

c. Three Typologies of Transhumanism

There are at least three ways of classifying different forms of transhumanism: from political, objective, and instrumental perspectives.

A political typology of transhumanism. Ferrando identifies three distinct strains within transhumanism:[144]

1) **Libertarian transhumanism** argues that the free market – and not governmental oversight – can best ensure that technologies for human enhancement are efficiently and effectively developed and made accessible within human society.
2) **Democratic transhumanism** seeks to ensure – for example, by means of government regulation – that technologies for human enhancement do not simply become privileges for the powerful and

[140] See Herbrechter (2013), p. 101.

[141] Herbrechter (2013), p. 103.

[142] See Teilhard de Chardin, *Le Phénomène humain* (1955), and its discussion in Herbrechter (2013), p. 104. The revolutionary nature of Teilhard's scientific, philosophical, and theological investigations open them to many possible interpretations; his thought has frequently been appropriated by transhumanist groups that disconnect it from its ultimate grounding in the orthodox Catholic intellectual tradition and thus interpret it in ways that do not necessarily reflect its original import or context.

[143] See Le Breton, David, *L'Adieu au corps* (1999), pp. 49, 219-223, as discussed in Herbrechter (2013), pp. 96-97.

[144] Ferrando (2013), p. 27.

wealthy but are made freely accessible to all human beings regardless of their social or economic status.

3) **Extropianism** is a movement founded by More and others that advocates the development of genetic engineering, nanotechnology, cryonics, mind uploading, and other technologies that can supposedly allow human lives to be extended indefinitely and spent in pursuit of intellectual fulfillment.

This model for categorizing transhumanisms might be understood as constituting a 'political' typology of transhumanism, as it largely distinguishes transhumanisms according to their view of the role of governments in steering the development and deployment of transhumanist technologies.

An objective typology of transhumanism. Significant variations also exist between different forms of transhumanism regarding the kinds of entities that are objects of the process of biotechnological posthumanization. Another typology can thus be formulated by classifying strains of transhumanism according to their objects:

4) **Biotransformative transhumanism** seeks to employ transformative technologies to allow *particular human beings who are already alive* to transcend the limits of human nature through manipulation or augmentation of their existing biological organisms – for example, through somatic cell gene therapy, cryonics, or neuroprosthetic enhancement.

5) **Biogenerative transhumanism** seeks to purposefully design the characteristics of *future beings who have not yet been conceived or born* (e.g., through the use of germline gene therapy (GGT) or synthetic biology to engineer a new superhuman species).

6) **Mimetic transhumanism** seeks to transcend the limits of human nature by creating superior and transcendent *beings that are wholly artificial and do not represent a continuation of humanity in an organic, biological sense* but which in some conceptual sense might nevertheless be considered our 'offspring' – and perhaps even more so than can our biological offspring, insofar as they would be consciously designed by human beings to embody our highest aspirations, rather than being the non-designed products of randomized biological reproductive processes. Such beings might include artificial superintelligences, sapient robot networks, or 'uploaded' human minds that are in fact artificial replicas rather than continuations of their human models.

Herbrechter agrees with Le Breton that for the group we refer to as biotransformative transhumanists, the most relevant power relationship is not that which allows other members of society to control (or be controlled by) an individual but that which allows the individual to control his or her own body.[145] For example, Herbrechter notes that for transhumanists like Warwick, transhumanism is about a rational humanist subject making a free choice between 'good' and 'evil' (or perhaps between 'good' and 'better') and choosing the path that will result in the most happiness and independence.[146] Biotransformative transhumanism might thus be understood as a form of extreme humanism.

On the other hand, some forms of radical mimetic transhumanism seek to actively break all connections with humanistic values. Building on McLuhan's notion of the 'global electric village,' Herbrechter observes that some transhumanists see it as humanity's role (and even responsibility) to give birth to our nonanthropic, artificially intelligent successors.[147] Similarly, drawing on Truong's analysis, Herbrechter notes that some transhumanists look forward with hope to the day when human beings will be replaced by the AIs that represent the next stage in the evolution of consciousness within our corner of the universe. It is anticipated that such artificial intelligences would eventually become fundamentally 'inhuman' as they evolve beyond the shackles created by human-like sociality, rationality, and knowledge; while 'consciousness' might thus continue to exist long after the demise of humanity, 'human-like consciousness' would not long survive the biological beings who provided its template.[148]

An instrumental typology of transhumanism. Distinctions also exist between the technologies advocated by different transhumanists for creating posthumanized entities. There are correlations between the goals held by particular transhumanists and the technologies used to pursue those goals; however, the alignment between goals and instruments is not absolute. Some transhumanists first choose the goal that they wish to accomplish and then seek to develop technologies to accomplish that goal. For them, achievement of their selected goal is paramount and the means used to achieve it are secondary and subject to change. On the other hand, some transhumanists work as scientists, engineers, entrepreneurs, ethicists, policy experts, or advocates specializing in a particular type of technology, such as artificial intelligence, neuroprosthetics, or germline gene therapy. For them, their paramount desire is discovering new avenues for improving humanity through the use of that particular technology; the specific ways in which that technology can be em-

---

[145] See Herbrechter (2013), pp. 96, and its analysis of Le Breton (1999), pp. 49.

[146] Warwick's views on human enhancement can be found, e.g., in Warwick, "The Cyborg Revolution" (2014). Such perspectives are analyzed in Herbrechter (2013), p. 102.

[147] Herbrechter (2013), p. 50.

[148] See Truong, Jean-Michel, *Totalement inhumaine* (2001), pp. 49, 207, as translated and analyzed in Herbrechter (2013), p. 172. See also Gladden, "The Social Robot as 'Charismatic Leader': A Phenomenology of Human Submission to Nonhuman Power" (2014).

ployed to create enhanced, transcendent, posthumanized beings are secondary. Such transhumanism can perhaps best be understood using the instrumental typology described here. For example, a scientist who specializes in developing new techniques for synthetic biology and who possesses transhumanist inclinations might pursue the use of such methods for biotransformative, biogenerative, and mimetic transhumanism, while a transhumanist researcher in the field of artificial intelligence might similarly pursue ways of applying AI to advance all three objective types of transhumanism.

2. Popular (or 'Commercial') Posthumanism

Herbrechter distinguishes between "a fashionable and popular posthumanism" and a more "serious and philosophical one." Occasionally, he seems to suggest that science fiction falls within the sphere of popular and faddish posthumanism – such as when he speaks of the intimate collaboration between science fiction and the commercial film industry and notes that the importance of science fiction for posthumanism is "most visible" when science fiction is considered "in its Hollywood blockbuster incarnation."[149] However, as noted earlier, we would argue that in its best and truest form, science fiction takes its place alongside philosophical posthumanism as a form of synthetic theoretical posthumanism that seeks to deepen our understanding of future posthumanities. While we would agree that for many members of the general public, Hollywood blockbusters represent the most *visible* presentations of explicitly posthumanist themes, they are typically not the most insightful, in-depth, or coherent presentations. By focusing on Hollywood blockbusters, Herbrechter minimizes the role of other forms of science fiction (such as novels, short stories, roleplaying and computer games, manga and anime, and independent films) that present more well-thought-out and incisive analyses of posthumanist themes. We would suggest that the more popular (if not populist) and commercially oriented works of speculative fiction – such as Hollywood blockbusters – can be better understood as a form of **synthetic practical posthumanism** that is geared specifically at generating particular economic, social, or political outcomes and which we will discuss here under the title of popular (or 'commercial') posthumanism. Works of popular posthumanism are typically aimed either at generating maximum profits for their producers, influencing public opinion to create a demand for new posthumanizing technologies, or preparing the public to accept changes to daily life that are being planned by government policymakers, corporations, or other powers.

Many of the criticisms directed broadly at the world of 'science fiction' can more accurately be understood as targeting the products and methods of com-

[149] Herbrechter (2013), pp. 22, 39, 107.

mercial posthumanism. In discussing Best and Kellner's analysis of posthumanism, Herbrechter notes the claim that "Economic neoliberalism, free market ideology and late capitalist individualism can no longer be separated from the various technological and cultural posthumanization processes."[150] According to that view, popular posthumanism can be seen as simply the most extreme manifestation of the link between commercial and political interests and the ongoing infusion of posthumanist themes into contemporary culture. Similarly, Herbrechter suggests that just as neuroscientists are exploring ways to exploit the plasticity of the human brain, so, too, "Global virtual hypercapitalism needs an equally plastic and flexible individual subject";[151] popular posthumanist narratives that emphasize the pliability, dissolubility, and reconfigurability of the human being support the development of subjects that are ready-made for control by corporate interests.

Indeed, Herbrechter notes the cynical argument that the apparent processes of posthuman technologization might simply be a ruse and distraction foisted cleverly on the public by the forces of neoliberal hypercapitalism that draw attention away from the "ever-increasing gap between rich and poor and the further concentration of power and capital" by subduing the masses with the hope or fear of a radically different future.[152] If such intentionally fabricated posthumanism exists, we would suggest that it takes the form not of critical or philosophical posthumanism (whose proponents are constitutionally on guard against such efforts at manipulation) but of techno-idealism, transhumanism, and the sort of commercial posthumanism described here. Indeed, Herbrechter alludes to the fact that complex, long-term, resource-intensive programs for developing new technologies for virtualization, miniaturization, surveillance, cyborgization, and artificial intelligence are being funded and led not primarily by philosophers who are interested in exploring the boundaries of human nature but by powerful commercial and governmental institutions (including banks, insurance companies, marketing firms, Internet and technology companies, and military and police organizations) that are seeking to develop such instruments for their own concrete ends. Such technologies not only give governments new tools for fighting crime and terrorism but also facilitate the invention of new forms of crime and terrorism (such as

---

150 Herbrechter (2013), p. 55.

151 Herbrechter (2013), p. 25.

152 Herbrechter notes the substantiveness of this argument without necessarily fully endorsing it; see Herbrechter (2013), p. 23.

memory-hacking or the development of hybrid bioelectronic viruses[153]) that were never previously possible.[154]

Just as popular posthumanism can be employed as an instrument by corporations and governments to aid in their technoscientific consolidation of profits and power, so, too, can critical and sociopolitical posthumanism – with support from science fiction – play an important role in identifying these technologically facilitated efforts to gain hegemony and in developing creative new ways of conceptualizing the nature of citizenship in a posthuman world that guarantee a more democratic basis for political and economic power.[155]

E. Posthumanisms That Join the Analytic, Synthetic, Theoretical, and Practical

Hybrid posthumanisms that include strong analytic, synthetic, theoretical, and practical aspects can be understood as examples of a 'posthumanism of production' that develops a robust and rigorous theoretical framework which is then utilized to successfully generate concrete products or services within the contemporary world. At least three forms of posthumanism display hybrid traits to such an extent that it would be arbitrary to attempt to force them to fit into just one quadrant of our framework. These forms of posthumanism are the metahumanism developed by Del Val and Sorgner, sociopolitical posthumanism, and organizational posthumanism. We can consider each of these posthumanisms in turn.

1. Metahumanism as a Movement of Relational Metabodies

Ferrando cites a form of 'metahumanism' originally formulated by Del Val and Sorgner in 2010[156] and grounded in the thought of Nietzsche, Deleuze, Haraway, Hayles, and others.[157] Such metahumanism draws explicitly on such diverse fields as neuroscience, chaos theory, quantum physics, ecology, and Eastern philosophy.[158] Sorgner explains that this metahumanism attempts to build on the best insights from both Anglo-American transhumanist and Continental posthumanist thought. On the one hand, metahumanism adopts critical posthumanism's "attempt to transcend dualisms" and cultivation of a

[153] See Gladden, *The Handbook of Information Security for Advanced Neuroprosthetics* (2015), for a discussion of such possibilities.

[154] Herbrechter (2013), p. 190; see also Gladden, *The Handbook of Information Security for Advanced Neuroprosthetics* (2015).

[155] See Gray (2002), p. 29, and its discussion in Herbrechter (2013), p. 190.

[156] Ferrando (2013), p. 32.

[157] Del Val et al. (2011), pp. 1-2, 6-9.

[158] Del Val et al. (2011), p. 9.

"this-worldly understanding of human beings"; although, rather than assuming the materialist perspective attributed to posthumanism, metahumanism adopts an intensely relational outlook.[159] At the same time, metahumanism is compatible with the transhumanist desire to create transcendent beings. However, metahumanism holds that while it is acceptable for individuals to desire such a transformation and to pursue that goal by applying advanced biotechnologies to themselves (i.e., as a form of biotransformative transhumanism), driving the evolution of human beings into a superior species cannot be claimed to be a necessary goal for humanity as a whole – because the transhumanist ideal is only one of many aims present within the "radical plurality of concepts of the good."[160]

Sorgner positions metahumanism as an outgrowth of philosophical posthumanism rather than cultural or critical posthumanism, insofar as metahumanism's key dynamic is its focus on consistently applying a particular philosophical methodology that Sorgner describes as a 'procedural attitude' which "brings together Adorno's negative dialectics and Vattimo's radical hermeneutics such that it is a particular procedure or a method which can get applied to various discourses." This method is employed by entering into the discourses of other thinkers (such as utilitarian bioethicists) and helping them develop their own paradigms by challenging, undermining, and breaking apart those positions that they take for granted – thereby transforming their thought into something that is "more fluid and multiperspectival."[161]

Metahumanism represents a form of 'radical relationalism,' insofar as it suggests that physical or social bodies which appear to be discrete entities can instead best be understood as the effects of contingent relations (such as movement) and that such seemingly discrete bodies can be transformed by altering the relations in which they participate. This notion is formalized in the idea of a 'metabody,' which "is not a fixed entity but a relational body." Such metabodies are both 'metasexual' and post-anatomical.[162] Metahumanism emphasizes that "Monsters are promising strategies for performing this development away from humanism"[163] and its understanding of the human body. In the recognition that the depiction of quasi-human monsters might aid us to think about humanity in a new way, a concrete link exists between

[159] Del Val et al. (2011), p. 2-3.

[160] Such a position has connections with both postmodernism and posthumanism. See Del Val et al. (2011), p. 3.

[161] Metahumanism thus inherently possesses a strong outward orientation that reaches out to engage thinkers who work in other disciplines and possess other perspectives. See Del Val et al. (2011), pp. 3-4.

[162] Del Val et al. (2011), pp. 5, 14, 8.

[163] Del Val & Sorgner (2011), p. 1.

the philosophical metahumanism proposed by Del Val and Sorgner and the form of fictional metahumanism that we discussed in an earlier section.

Unlike biopolitical posthumanism, metahumanism does not have a strong future orientation; it shares with cultural and critical posthumanism the fact that "it is non-utopian, it does not see the metahuman as a future, but as a strategy in the present."[164] However, while metahumanism contains strong **analytic** aspects, it is also a form of **synthetic posthumanism**, insofar as it envisions a new kind of posthumanized being that does not yet fully exist but which is only now in the process of appearing. Likewise, metahumanism spans **theoretical** and **practical posthumanism** in that it not only seeks to better understand human nature but also to give birth to concrete new forms of artistic expression and social and political interaction. This is done partly by enacting "new strategies of resistance" to human beings' subjugation to representation and language; such strategies may take the form of "amorphous becomings" manifested through the motion of dance and other forms of artistic performance.[165]

2. Sociopolitical Posthumanism

Sociopolitical posthumanism can be understood as a form of what Herbrechter (building on Rosenau) describes as 'techno-cultural pragmatism.'[166] Sociopolitical posthumanism accepts that posthumanizing technological change is gaining in speed and intensity and – given the fact that the yearning for technological advancement is a fundamental aspect of human nature – any efforts to completely block such technologization are misguided and futile. Instead, sociopolitical posthumanism seeks to steer the processes of technologization and posthumanization in a way that maximizes their positive impacts while ameliorating or avoiding their detrimental side-effects.

Sociopolitical posthumanism frequently initiates new debates among subject-matter experts and the broader public on such topics and, insofar as possible, proposes solutions. The **analytic** and **theoretical** aspects of sociopolitical posthumanism are evident when, for example, scholars explore how established definitions of a 'legal person' are challenged by an increasingly deanthropocentrized environment in which some artificially intelligent systems already display human-like decision-making capacities and fill societal roles previously restricted to human beings. The **synthetic** and **practical** aspects are manifested when scholars draw on such theoretical investigations to propose the implementation of new legislation, regulations, or financial systems not

---

[164] Del Val et al. (2011), p. 6.

[165] Del Val himself has pioneered such forms of artistic expression. For the role of practical action in metahumanism, see Del Val et al. (2011), pp. 5-6, 12.

[166] See Herbrechter (2013), pp. 23-24, and its discussion of Rosenau (1992).

because they are needed to account for a reality that exists today but to address the activities of posthumanized beings expected to appear in the future. However, sociopolitical posthumanism differs from the synthetic practical posthumanisms of transhumanism and bioconservatism, whose adherents may manufacture theoretical frameworks to justify the pursuit or condemnation of processes of technologization that they already instinctively find appealing or repellent. For practitioners of sociopolitical posthumanism, a serious and in-depth exploration of theoretical questions is generally the starting point, and any resulting proposals for practical change emerge from a well-developed theoretical framework of the sort commonly found within philosophical or critical posthumanism.

Such sociopolitical posthumanism can be found, for example, within the field of law, where Braman argues that the traditional "assumption that the law is made by humans for humans" is no longer tenable; as the roles played by computers in society's decision-making processes grow, we are beginning to witness "a transformation in the legal system so fundamental that it may be said that we are entering a period of posthuman law."[167] Another example would be the theoretically grounded 'Cyborg Bill of Rights' proposed by Gray as an attempt to ensure that the increasing technological capacity for cyborgization will result in beneficial new forms of posthumanized political organization and engagement and not simply the production of new military instruments.[168]

3. Organizational Posthumanism

Organizational posthumanism applies posthumanist insights and methodologies to the study and management of organizations including businesses, nonprofit organizations, schools, religious groups, professional associations, political parties, governments, and military organizations. Insofar as ongoing technological and social change is reshaping the capacities and relationality of the human beings who belong to organizations – and creating new kinds of entities like social robots that can enter into goal-directed social relationships with human beings and one another[169] – the nature of organizations is itself changing. Organizational posthumanism can aid us in making sense of and, ideally, anticipating such changes. While a scattered assortment of works by

---

[167] Berman, "Posthuman Law: Information Policy and the Machinic World" (2002).

[168] See Gray (2002) and the discussion of that work in Herbrechter (2013), p. 105. For a further sociopolitical posthumanist discussion of ways in which, e.g., the use of posthuman neuroprosthetic technologies could give rise to new forms of utopian or dystopian societies, see Gladden, "Utopias and Dystopias as Cybernetic Information Systems: Envisioning the Posthuman Neuropolity" (2015).

[169] See, e.g., Gladden, "The Social Robot as 'Charismatic Leader'" (2014).

management theorists and practitioners have begun to explore the implications of posthumanism for organizational life, these investigations are still in their incipient stages;[170] the explicit formulation within this book of organizational posthumanism as an emerging discipline thus represents a novel development within the fields of posthumanism and organizational management.

Organizational posthumanism can be defined as an approach to analyzing, understanding, creating, and managing organizations that employs a post-anthropocentric and post-dualistic perspective and which recognizes that emerging technologies that complement traditional biological human beings with new kinds of intelligent actors also transform the structures, membership, dynamics, and roles available to organizations.[171] From this description, it can be seen that – like sociopolitical posthumanism and the metahumanism of Del Val and Sorgner – organizational posthumanism incorporates elements of both analytic and synthetic and both theoretical and practical posthumanism.

Analytic and synthetic elements. Organizational posthumanism is **analytic** in that it is not simply interested in imagining the radically novel forms that organizations might take ten or twenty or fifty years from now, after ongoing trends of roboticization, cyborgization, digitalization, and virtualization will have transformed organizations wholly beyond recognition; it is also interested in understanding and shaping the dynamics of organizations that exist today to the extent that they have already been affected by technological and nontechnological processes of posthumanization. Although the impact that artificial intelligence, social robotics, nanorobotics, artificial life, genetic engineering, neurocybernetics, and virtual reality have had on organizations to date is relatively small when compared to biopolitical posthumanists' visions of the sociotechnological changes that loom on the horizon, even those modest impacts already realized are transforming the ways that organizations can and must operate, rendering many previous best practices increasingly obsolete.

---

[170] For examples of such works, see, e.g., Gephart, "Management, Social Issues, and the Postmodern Era" (1996); Berner, *Management in 20XX: What Will Be Important in the Future – A Holistic View* (2004); Mara & Hawk, "Posthuman rhetorics and technical communication" (2009); Barile, "From the Posthuman Consumer to the *Ontobranding* Dimension: Geolocalization, Augmented Reality and Emotional Ontology as a Radical Redefinition of What Is Real" (2013); and Gladden, "Neural Implants as Gateways to Digital-Physical Ecosystems and Posthuman Socioeconomic Interaction" (2016).

[171] For an in-depth discussion of this topic, see Part Two of this volume, "Organizational Posthumanism."

At the same time, organizational posthumanism is **synthetic** insofar as effective strategic management demands that organizations anticipate the contours of new phenomena that may appear in the future and understand their potential implications for an organization. For example, the frequently employed PESTLE analysis requires organizations to envision the short-, medium-, and long-term political, economic, social, technological, legal, and environmental impacts that will result either from internal organizational decisions or future changes in the organization's external ecosystem.[172] In order to anticipate such potential impacts and develop contingency plans for responding to them (or strategies to proactively shape them), organizations must attempt to project as accurately as possible the future directions of posthumanization processes and the new kinds of beings, organizational structures, interactions, physical and virtual spaces, and ecosystems that they might produce. This demands a rigorous and imaginative futurology similar to that employed in philosophical posthumanism and the more thoughtful forms of science fiction.

Theoretical and practical elements. Organizational posthumanism is **theoretical** insofar as it attempts to identify and understand the manner in which organizations are being affected by existing or potential processes of posthumanization. This involves analyzing the ways in which organizations' members, structures, processes, information systems, physical and virtual spaces, and external environments are being changed through the action of supplementing or replacing their natural biological human workers with advanced AIs, social robots, neuroprosthetically augmented human beings, and other posthumanized beings. In this regard, organizational posthumanism builds on existing lines of inquiry within philosophical posthumanism. For example, Miah notes that posthumanist thought has long studied the growing fusion of human beings with the technological devices that we use to interact with one another and with our environment and to perform work-related tasks. As such tools grow increasingly sophisticated, they acquire ever subtler and more efficacious ways of liberating and empowering human beings, even as they subjugate and oppress. Much of this ambivalent dynamic results from our tools' deepening integration into the mechanisms of organizations of which we are members.[173] The theoretical component of organizational posthumanism attempts to develop coherent conceptual frameworks to explain and anticipate such phenomena.

At the same time, organizational posthumanism is also **practical** in that its goal is not simply to understand the ways in which posthuman realities are

---

[172] See Cadle et al., *Business Analysis Techniques: 72 Essential Tools for Success* (2010), pp. 3-6.

[173] See Miah (2008), p. 82, and its analysis of Mazlish, *The Fourth Discontinuity: The Co-Evolution of Humans and Machines* (1993).

affecting organizations but also to aid management practitioners in proactively designing, creating, and maintaining organizations that can subsist within such a complex and novel competitive environment. Organizational posthumanism seeks to intentionally bring about the creation of a particular type of near-future 'posthumanity' (i.e., a world of organizations that survive as viable systems within a nonanthropocentric context of radical technological change and convergence) and to purposefully block the creation of a different type of near-future 'posthumanity' (i.e., a world of organizations that become unproductive, inefficient, unsustainable, dehumanizing, and even dystopian as a result of their inability to deal with the emerging nonanthropocentric context).[174]

V. Conclusion

The term 'posthumanism' is employed within an increasingly wide array of contexts to describe phenomena which, in one way or another, focus on a change in the traditional understanding of the human being. Some forms of posthumanism argue that the historical definition of humanity has always been problematic, others that it is now fracturing and becoming obsolete as a result of ongoing technological change. Still other forms of posthumanism argue that our traditional understanding of the 'human' must be expanded or replaced as a next step in the development of sapient society. As we have seen, posthumanisms include such diverse phenomena as new academic disciplines, artistic and spiritual movements, research and development programs for new technologies, works of science fiction, social advocacy campaigns, and legislative lobbying efforts.

By grouping posthumanisms into a handful of basic types and clarifying the similarities and differences between them, the two-dimensional conceptual framework formulated in this text attempts to create a more orderly and comprehensive foundation for the investigation of posthumanism than has previously existed. The first type considered in detail was analytic theoretical posthumanism, which includes such fields as critical and cultural posthumanism and can be understood roughly as a *posthumanism of critique*. Synthetic theoretical posthumanism, which includes phenomena like philosophical posthumanism, science fiction, and techno-idealism, can be generally understood as a *posthumanism of imagination*. Analytic practical posthumanism, which includes various forms of metahumanism and neohumanism, can be seen as a *posthumanism of conversion* of hearts and minds. Synthetic practical

[174] In the case of, e.g., commercial enterprises and military organizations, the theory and practice of organizational posthumanism might be employed not only to maximize the efficiency and productivity of one's own posthumanized organization but also to degrade the efficiency and productivity of competing or opposing organizations, to the extent that such actions are legally and ethically permissible.

posthumanism, which includes transhumanism, bioconservatism, and popular or commercial posthumanism, can be understood as a *posthumanism of control* over the actions of societies and individuals. Finally, the hybrid posthumanism that combines both analytic and synthetic as well as theoretical and practical aspects – as exemplified by the metahumanism of Sorgner and Del Val, sociopolitical posthumanism, and organizational posthumanism – can be understood as a *posthumanism of production.*

As posthumanist perspectives continue to be adapted and applied to new fields – such as that of organizational management – the work of developing conceptual frameworks that can coherently account for the full spectrum of posthumanisms is only beginning. It is hoped that the typology formulated in this text can contribute to such endeavors by highlighting areas of definitional ambiguity, building new conceptual bridges between different forms of posthumanism, and formulating terminological reference points that can be relied upon both by those who embrace various forms of posthumanism and those who wish to challenge the principles of posthumanist thought.

References

2001: A Space Odyssey. Directed by Stanley Kubrick. 1968. Warner Home Video, 2001. DVD.

Anders, Günther. *Die Antiquiertheit des Menschen. Band 1: Über die Seele im Zeitalter der zweiten industriellen Revolution*, Munich: Beck, 1956 [1992].

Asimov, Isaac. *I, Robot.* New York: Gnome Press, 1950.

Badmington, Neil. "Cultural Studies and the Posthumanities." In *New Cultural Studies: Adventures in Theory*, edited by Gary Hall and Claire Birchall, pp. 260-72. Edinburgh: Edinburgh University Press, 2006.

Barile, Nello. "From the Posthuman Consumer to the *Ontobranding* Dimension: Geolocalization, Augmented Reality and Emotional Ontology as a Radical Redefinition of What Is Real." *intervalla: platform for intellectual exchange,* vol. 1 (2013). http://www.fus.edu/intervalla/images/pdf/9_barile.pdf. Accessed May 18, 2016.

Berman, Sandra. "Posthuman Law: Information Policy and the Machinic World." *First Monday* 7, no. 12 (2002). http://www.ojphi.org/ojs/index.php/fm/article/view/1011/932. Accessed March 15, 2016.

Berner, Georg. *Management in 20XX: What Will Be Important in the Future – A Holistic View.* Erlangen: Publicis Corporate Publishing, 2004.

Berrigan, Joseph R. "The Prehumanism of Benzo d'Allesandria." *Traditio* (1969): 249-63.

Better Off Dead: The Evolution of the Zombie as Post-Human, edited by Deborah Christie and Sarah Juliet Lauro. New York: Fordham University Press, 2011.

The Bible and Posthumanism, edited by Jennifer L. Koosed. Atlanta: Society of Biblical Literature, 2014.

Birnbacher, Dieter. "Posthumanity, Transhumanism and Human Nature." In *Medical Enhancement and Posthumanity*, edited by Bert Gordijn and Ruth Chadwick, pp. 95-106. The International Library of Ethics, Law and Technology 2. Springer Netherlands, 2008.

Bostrom, Nick. "A History of Transhumanist Thought." *Journal of Evolution and Technology* vol. 14, no. 1 (2005). http://jetpress.org/volume14/bostrom.html.

Bostrom, Nick. "Why I Want to Be a Posthuman When I Grow Up." In *Medical Enhancement and Posthumanity*, edited by Bert Gordijn and Ruth Chadwick, pp. 107-37. The International Library of Ethics, Law and Technology 2. Springer Netherlands, 2008.

Cadle, James, Debra Paul, and Paul Turner. *Business Analysis Techniques: 72 Essential Tools for Success.* Swindon: British Informatics Society Limited, 2010.

Calverley, D.J. "Imagining a non-biological machine as a legal person." *AI & SOCIETY* 22, no. 4 (2008): 523-37.

Coker, Christopher. "Biotechnology and War: The New Challenge." *Australian Army Journal* vol. II, no. 1 (2004): 125-40.

Curtis, Neal. "The Inhuman." *Theory, Culture & Society* 23, 2-3 (2006): 434-36.

Cyberculture, Cyborgs and Science Fiction: Consciousness and the Posthuman, edited by William S. Haney II. Amsterdam: Rodopi, 2006.

Del Val, Jaime, and Stefan Lorenz Sorgner. "A *Metahumanist* Manifesto." *The Agonist* IV no. II (Fall 2011). http://www.nietzschecircle.com/AGONIST/2011_08/METAHUMAN_MANIFESTO.html. Accessed March 2, 2016.

Del Val, Jaime, Stefan Lorenz Sorgner, and Yunus Tuncel. "Interview on the Metahumanist Manifesto with Jaime del Val and Stefan Lorenz Sorgner." *The Agonist* IV no. II (Fall 2011). http://www.nietzschecircle.com/AGONIST/2011_08/Interview_Sorgner_Stefan-Jaime.pdf. Accessed March 2, 2016.

Ferrando, Francesca. "Posthumanism, Transhumanism, Antihumanism, Metahumanism, and New Materialisms: Differences and Relations." *Existenz: An International Journal in Philosophy, Religion, Politics, and the Arts* 8, no. 2 (Fall 2013): 26-32.

Fukuyama, Francis. *Our Posthuman Future: Consequences of the Biotechnology Revolution.* New York: Farrar, Straus, and Giroux, 2002.

Gane, Nicholas. "Posthuman." *Theory, Culture & Society* 23, 2-3 (2006): 431-34.

Gephart, Jr., Robert P. "Management, Social Issues, and the Postmodern Era." In *Postmodern Management and Organization Theory*, edited by David M. Boje, Robert P. Gephart, Jr., and Tojo Joseph Thatchenkery, pp. 21-44. Thousand Oaks, CA: Sage Publications, 1996.

Gladden, Matthew E. *The Handbook of Information Security for Advanced Neuroprosthetics*, Indianapolis: Synthypnion Academic, 2015.

Gladden, Matthew E. "Neural Implants as Gateways to Digital-Physical Ecosystems and Posthuman Socioeconomic Interaction." In *Digital Ecosystems: Society in the*

Digital Age, edited by Łukasz Jonak, Natalia Juchniewicz, and Renata Włoch, pp. 85-98. Warsaw: Digital Economy Lab, University of Warsaw, 2016.

Gladden, Matthew E. "The Social Robot as 'Charismatic Leader': A Phenomenology of Human Submission to Nonhuman Power." In *Sociable Robots and the Future of Social Relations: Proceedings of Robo-Philosophy 2014*, edited by Johanna Seibt, Raul Hakli, and Marco Nørskov, pp. 329-39. Frontiers in Artificial Intelligence and Applications 273. IOS Press, 2014.

Gladden, Matthew E. "Utopias and Dystopias as Cybernetic Information Systems: Envisioning the Posthuman Neuropolity." *Creatio Fantastica* nr 3 (50) (2015).

Goicoechea, María. "The Posthuman Ethos in Cyberpunk Science Fiction." *CLCWeb: Comparative Literature and Culture* 10, no. 4 (2008): 9. http://docs.lib.purdue.edu/cgi/viewcontent.cgi?article=1398&context=clcweb. Accessed May 18, 2016.

Graham, Stephen. "Imagining Urban Warfare: Urbanization and U.S. Military Technoscience." In *War, Citizenship, Territory*, edited by Deborah Cowen and Emily Gilbert. New York: Routledge, 2008.

Gray, Chris Hables. *Cyborg Citizen: Politics in the Posthuman Age*, London: Routledge, 2002.

Gray, Chris Hables. "The Ethics and Politics of Cyborg Embodiment: Citizenship as a Hypervalue." *Cultural Values* 1, no 2 (1997): 252-58.

Haraway, Donna. "A Manifesto for Cyborgs: Science, Technology, and Socialist Feminism in the 1980s." *Socialist Review* vol. 15, no. 2 (1985), pp. 65-107.

Haraway, Donna. *Simians, Cyborgs, and Women: The Reinvention of Nature*. New York: Routledge, 1991.

Hassan, Ihab. "Prometheus as Performer: Toward a Posthumanist Culture? A University Masque in Five Scenes." *The Georgia Review* vol. 31, no. 4 (1977): 830-50.

Hayles, N. Katherine. *How We Became Posthuman: Virtual Bodies in Cybernetics, Literature, and Informatics*, Chicago: University of Chicago Press, 1999.

Hayles, N. Katherine. *My Mother Was a Computer: Digital Subjects and Literary Texts*. Chicago: University of Chicago Press, 2005.

Herbrechter, Stefan. *Posthumanism: A Critical Analysis*. London: Bloomsbury, 2013. [Kindle edition.]

I, Robot. Directed by Alex Proyas. 2004. 20th Century Fox, 2005. DVD.

Interface Zero 2.0: Full Metal Cyberpunk, developed by David Jarvis, Curtis Lyons, Sarah Lyons, Thomas Shook, David Viars, and Peter J. Wacks, Gun Metal Games, 2013.

Kowalewska, Agata. "Symbionts and Parasites – Digital Ecosystems." In *Digital Ecosystems: Society in the Digital Age*, edited by Łukasz Jonak, Natalia Juchniewicz, and Renata Włoch, pp. 73-84. Warsaw: Digital Economy Lab, University of Warsaw, 2016.

Krishnan, Armin. "Enhanced Warfighters as Private Military Contractors." In *Super Soldiers: The Ethical, Legal and Social Implications*, edited by Jai Galliott and Mianna Lotz. London: Routledge, 2015.

Krüger, Oliver. *Virtualität und Unsterblichkeit: Die Visionen des Posthumanismus* [Virtuality and Immortality]. Freiburg: Rombach, 2004.

Kurzweil, Ray. *The Singularity is Near: When Humans Transcend Biology.* New York: Viking Penguin, 2005.

Le Breton, David. *L'Adieu au corps.* Paris: Métaillé, 1999.

Lovecraft, Howard Phillips. *At the Mountains of Madness and Other Novels.* Sauk City, Wisconsin: Arkham House, 1985.

Lovecraft, Howard Phillips. *The Dunwich Horror and Others.* Sauk City, Wisconsin: Arkham House, 1983.

Mara, Andrew, and Byron Hawk. "Posthuman rhetorics and technical communication." *Technical Communication Quarterly* 19, no. 1 (2009): 1-10.

Mazlish, Bruce. *The Fourth Discontinuity: The Co-Evolution of Humans and Machines.* New Haven: Yale University Press, 1993.

Miah, Andy. "A Critical History of Posthumanism." In *Medical Enhancement and Posthumanity*, edited by Bert Gordijn and Ruth Chadwick, pp. 71-94. The International Library of Ethics, Law and Technology 2. Springer Netherlands, 2008.

Milán, Victor. "Transfigurations" and "Appendix: The Science of the Wild Card Virus: Excerpts from the Literature." In *Wild Cards*, edited by George R. R. Martin. Bantam Books: 1986.

Miller, Jr., Gerald Alva. "Conclusion: Beyond the Human: Ontogenesis, Technology, and the Posthuman in Kubrick and Clarke's 2001." In *Exploring the Limits of the Human through Science Fiction*, pp. 163-90. American Literature Readings in the 21st Century. Palgrave Macmillan US, 2012.

Nealon, Jeffrey T. *Foucault Beyond Foucault.* Stanford, CA: Stanford University Press, 2008.

Neohumanist Educational Futures: Liberating the Pedagogical Intellect, edited by Sohail Inayatullah, Marcus Bussey, and Ivana Milojević. Tansui, Taipei: Tamkang University Press, 2006. http://www.metafuture.org/Books/nh-educational-futures-foreword.htm. Accessed March 16, 2016.

Pepperell, R. *The Posthuman Condition: Consciousness Beyond the Brain.* Bristol, UK: Intellect Books, 2003.

Poster, Mark. *What's the Matter with the Internet?* Minneapolis: University of Minnesota Press, 2001.

Posthuman Bodies, edited by Judith Halberstam and Ira Livingstone. Bloomington, IN: Indiana University Press, 1995.

Posthumanist Shakespeares, edited by Stefan Herbrechter and Ivan Callus. Palgrave Shakespeare Studies. Palgrave Macmillan UK, 2012.

Rosenau, Pauline Marie. *Post-Modernism and the Social Sciences: Insights, Inroads, and Intrusions.* Princeton, NJ: Princeton University Press, 1992.

Sanbonmatsu, John. *The Postmodern Prince: Critical Theory, Left Strategy, and the Making of a New Political Subject.* New York: Monthly Review Press, 2004.

Sarkar, Prabhat Rainjan. "Neohumanism Is the Ultimate Shelter (Discourse 11)." In *The Liberation of Intellect: Neohumanism.* Kolkata: Ananda Marga Publications, 1982.

Shadowrun: Core Rulebook 5, edited by Kevin Killiany and Katherine Monasterio. Lake Stevens, Washington: Catalyst Game Labs, 2013.

Sixth World Almanac, edited by Jason Hardy and John Helfers. The Topps Company, 2010.

Smith, Andrew James. *Genetic Gold: The Post-human Homunculus in Alchemical and Visual Texts.* M.A. Thesis. University of Pretoria, 2009.

Teilhard de Chardin, Pierre. *Le Phénomène humain.* Paris: Seuil, 1955.

Thacker, Eugene. "What Is Biomedia?" *Configurations* 11 (2003): 47-79.

Truong, Jean-Michel. *Totalement inhumaine.* Paris: Seuil, 2001.

Warwick, K. "The Cyborg Revolution." *Nanoethics* 8 (2014): 263-73.

Witt, Ronald. "Francesco Petrarca and the Parameters of Historical Research." *Religions* 3, no. 3 (2012): 699-709.

Wolin, Richard. "Foucault the Neohumanist?" *The Chronicle of Higher Education.* The Chronicle Review (September 1, 2006).

Chapter Two

Organizational Posthumanism[1]

Abstract. Building on existing forms of critical, cultural, biopolitical, and sociopolitical posthumanism, in this text a new framework is developed for understanding and guiding the forces of technologization and posthumanization that are reshaping contemporary organizations. This 'organizational posthumanism' is an approach to analyzing, creating, and managing organizations that employs a post-dualistic and post-anthropocentric perspective and which recognizes that emerging technologies will increasingly transform the kinds of members, structures, systems, processes, physical and virtual spaces, and external ecosystems that are available for organizations to utilize. It is argued that this posthumanizing technologization of organizations will especially be driven by developments in three areas: 1) technologies for human augmentation and enhancement, including many forms of neuroprosthetics and genetic engineering; 2) technologies for synthetic agency, including robotics, artificial intelligence, and artificial life; and 3) technologies for digital-physical ecosystems and networks that create the environments within which and infrastructure through which human and artificial agents will interact.

Drawing on a typology of contemporary posthumanism, organizational posthumanism is shown to be a hybrid form of posthumanism that combines both analytic, synthetic, theoretical, and practical elements. Like analytic forms of posthumanism, organizational posthumanism recognizes the extent to which posthumanization has already transformed businesses and other organizations; it thus occupies itself with understanding organizations as they exist today and developing strategies and best practices for responding to the forces of posthumanization. On the other hand, like synthetic forms of posthumanism, organizational posthumanism anticipates the

[1] This chapter was originally published in Gladden, Matthew E., *Sapient Circuits and Digitalized Flesh: The Organization as Locus of Technological Posthumanization*, pp. 93-131, Indianapolis: Defragmenter Media, 2016.

fact that intensifying and accelerating processes of posthumanization will create future realities quite different from those seen today; it thus attempts to develop conceptual schemas to account for such potential developments, both as a means of expanding our theoretical knowledge of organizations and of enhancing the ability of contemporary organizational stakeholders to conduct strategic planning for a radically posthumanized long-term future.

I. Introduction

'Posthumanism' can be defined briefly as an intellectual framework for understanding reality that is post-anthropocentric and post-dualistic; for posthumanism, the 'natural' biological human being as traditionally understood becomes just one of many intelligent subjects acting within a complex ecosystem.[2] Some forms of posthumanism focus on the ways in which our notion of typical human beings as the only members of society has been continuously challenged over the centuries through the generation of cultural products like myths and works of literature that feature quasi-human beings such as monsters, ghosts, angels, anthropomorphic animals, cyborgs, and space aliens (i.e., through processes of nontechnological 'posthumanization').[3] Other forms of posthumanism address the ways in which the circle of persons and intelligent agents dwelling within our world is being transformed and expanded through the engineering of new kinds of entities such as human beings possessing neuroprosthetic implants, genetically modified human beings, social robots, sentient networks, and other advanced forms of artificial intelligence (i.e., through processes of technological posthumanization).[4] The development of

---

[2] This definition builds on the definitions formulated by scholars of posthumanism such as Ferrando, Miller, Herbrechter, Miah, and Birnbacher, as well as on our own typology of posthumanism found in Part One of this volume, "A Typology of Posthumanism: A Framework for Differentiating Analytic, Synthetic, Theoretical, and Practical Posthumanisms." See Ferrando, "Posthumanism, Transhumanism, Antihumanism, Metahumanism, and New Materialisms: Differences and Relations" (2013), p. 29; Miller, "Conclusion: Beyond the Human: Ontogenesis, Technology, and the Posthuman in Kubrick and Clarke's 2001" (2012), p. 164; Herbrechter, *Posthumanism: A Critical Analysis* (2013), pp. 2-3; Miah, "A Critical History of Posthumanism" (2008), p. 83; and Birnbacher, "Posthumanity, Transhumanism and Human Nature" (2008), p. 104.

[3] Such forms of posthumanism include the critical and cultural posthumanism pioneered by Haraway, Halberstam and Livingstone, Hayles, Badmington, and others. See, e.g., Haraway, "A Manifesto for Cyborgs: Science, Technology, and Socialist Feminism in the 1980s" (1985); Haraway, *Simians, Cyborgs, and Women: The Reinvention of Nature* (1991); *Posthuman Bodies*, edited by Halberstam & Livingstone (1995); Hayles, *How We Became Posthuman: Virtual Bodies in Cybernetics, Literature, and Informatics* (1999); Graham, *Representations of the Post/Human: Monsters, Aliens and Others in Popular Culture* (2002); Badmington, "Cultural Studies and the Posthumanities" (2006); and Herbrechter (2013).

[4] Such forms of posthumanism include philosophical posthumanism, bioconservatism, and transhumanism, which are analyzed in Miah (2008), pp. 73-74, 79-82, and Ferrando (2013), p. 29. Such approaches can be seen, for example, in Fukuyama, *Our Posthuman Future: Consequences of the*

sound and discerning forms of posthumanist thought is becoming increasingly important as society grapples with the ontological, ethical, legal, and cultural implications of emerging technologies that are generating new forms of posthumanized existence.

The establishing of conceptual links between organizational management and the idea of the 'posthuman' is nothing new. As early as 1978, management scholars Bourgeois, McAllister, and Mitchell had written that "Much of the organization theory literature from the posthuman relations era concentrates on defining which organizational structures, management styles, et cetera are most appropriate (effective) for different technologies and/or environmental contingencies."[5] Writing in 1996, Gephart drew on fictional depictions of cyborgs to envision an emerging 'Postmanagement Era' in which an organization's complex network of computerized systems – with its own synthetic values and logic – would become the true manager of an organization that no longer exists and acts for the sake of human beings. Although a human being might still appear to function as a 'manager' within such an organization, in reality she would be neither a manager nor a natural, biological human being; instead she would possess the form of a cyborg who has been permanently integrated into her employer's operational, financial, and technological systems and who has been weaponized for commercial ends – a being whose human agency has been dissolved until she becomes little more than a cold and lethally efficient "posthuman subject, ripping at flesh as part of her job."[6]

More recently, scholars have explored potential relationships between posthumanism and particular specialized fields within organizational theory and management. For example, Mara and Hawk consider the relationship of posthumanism to the technical communication that constitutes an important

Biotechnology Revolution (2002); Bostrom, "Why I Want to Be a Posthuman When I Grow Up" (2008); and other texts in *Medical Enhancement and Posthumanity*, edited by Gordijn & Chadwick (2008).

[5] Bourgeois et al., "The Effects of Different Organizational Environments upon Decisions about Organizational Structure" (1978), pp. 508-14. This allusion to the posthuman is not elaborated upon elsewhere in the text. The article describes an empirical study that was conducted to test hypotheses relating to the default behavior of managers when their organizations encounter "turbulent and threatening business environments" (p. 508).

[6] See Gephart, "Management, Social Issues, and the Postmodern Era" (1996), pp. 36-37, 41. Strictly speaking, Gephart's approach is more postmodernist than posthumanist. While there are areas of overlap between postmodernism and posthumanism, postmodernism generally posits a more nihilistic deconstruction of the notion of 'humanity,' while posthumanism seeks to transform and expand the historically anthropocentric concepts of personal agency and subjectivity to incorporate quasi-human, parahuman, and nonhuman entities. See Part One of this volume, "A Typology of Posthumanism: A Framework for Differentiating Analytic, Synthetic, Theoretical, and Practical Posthumanisms," and Herbrechter (2013).

form of information flow within contemporary organizations that are so dependent on technology. They note the evolving roles that organizations' human and nonhuman actors play in change management, organizational culture, human-computer interaction (HCI), and the integration of technology into the workplace within the context of a complex posthuman organizational ecology in which "it is no longer tenable to divide the world into human choice and technological or environmental determinism."[7] Barile, meanwhile, explores the impact that technologies for augmented reality play in creating 'posthuman consumers' by breaking down boundaries between the virtual and the actual and supplanting previous forms of HCI with "a new kind of interaction where the machines become softer and immaterial, emotions become contents, and places become media."[8]

Other scholars have sought to identify the ultimate drivers of the processes of posthumanization that are expected to increasingly impact organizations of all types. For example, Herbrechter notes the ongoing and intensifying 'technologization' of humanity, by which technoscientific forces that had previously constituted just one element of society attempt to gain economic and political power over all aspects of human culture.[9] Insofar as all organizations exist within human cultures, utilize technology, and are subject to economic and political forces, they become a participant in these dynamics of technologization and posthumanization. However, while the forces of technologization are undoubtedly real, they may not fully explain the rising prominence of posthuman dynamics and motifs within organizational life. Indeed, it has even been suggested that the popular notion of posthumanism may have been engineered as a sort of ruse generated by the power structures of postmodern neoliberal capitalism to pacify the masses with the hope or fear (or both) of a radically different future that looms just over the horizon.[10] According to that view, posthumanist imagery, themes, and philosophies are a mechanism employed by some organizations in order to facilitate the achievement of their strategic objectives.

While a diverse array of connections between posthumanism and organizational management has thus been hinted at for some time, it has not been

---

[7] Mara & Hawk, "Posthuman rhetorics and technical communication" (2009), pp. 1-3.

[8] Barile, "From the Posthuman Consumer to the *Ontobranding* Dimension: Geolocalization, Augmented Reality and Emotional Ontology as a Radical Redefinition of What Is Real" (2013), p. 101.

[9] See Herbrechter (2013), p. 19.

[10] See the discussion of such cynical interpretations of posthumanism in Herbrechter (2013), p. 80.

comprehensively or systematically explored. Much scholarship has been dedicated to understanding fields such as literature,[11] film,[12] computer games,[13] biomedical engineering,[14] and politics and economics[15] in light of posthumanist thought. However, efforts to apply posthumanist methodologies and insights to organizational management have remained relatively underdeveloped. This is striking, given the fact that many of the issues of interest to posthumanism have strong organizational repercussions.

In this text, we attempt to address this lacuna by presenting one approach to developing a comprehensive 'organizational posthumanism.' After formulating a definition for organizational posthumanism, we compare it to established forms of post-dualistic and post-anthropocentric posthumanist thought, arguing that it constitutes a type of 'hybrid posthumanism' that incorporates both analytic, synthetic, theoretical, and practical aspects. We then consider six organizational elements that will increasingly be impacted by the forces of posthumanization: namely, an organization's members, personnel structures, information systems, processes, physical and virtual spaces, and external environment. Finally, three main types of technologies that facilitate the development of organizational posthumanity are described; these are technologies for human augmentation and enhancement (including implantable computers, neuroprosthetic devices, virtual reality systems, genetic engineering, new forms of medicine, and life extension); technologies for synthetic agency (including social robotics, artificial intelligence, and artificial life); and technologies for building digital-physical ecosystems and networks (such as the Internet of Things). It is our hope that the questions raised and the framework formulated within this text can offer a useful starting point for those scholars and management practitioners who will address in an ever more explicit manner the increasingly important intersection of organizational life and posthumanist thought.

---

[11] See posthumanist analyses of literature in, e.g., Hayles (1999); *Posthumanist Shakespeares*, edited by Herbrechter & Callus (2012); and Thomsen, *The New Human in Literature: Posthuman Visions of Change in Body, Mind and Society after 1900* (2013).

[12] Examples can be found in the articles relating to cinema in *Posthuman Bodies* (1995); Short, *Cyborg Cinema and Contemporary Subjectivity* (2005); and Miller (2012).

[13] For such studies, see, e.g., Schmeink, "Dystopia, Alternate History and the Posthuman in Bioshock" (2009); Krzywinska & Brown, "Games, Gamers and Posthumanism" (2015); and Boulter, *Parables of the Posthuman: Digital Realities, Gaming, and the Player Experience* (2015).

[14] See, e.g., *Medical Enhancement and Posthumanity* (2008); Thacker, "Data made flesh: biotechnology and the discourse of the posthuman" (2003); and Lee, "Cochlear implantation, enhancements, transhumanism and posthumanism: some human questions" (2016).

[15] Examples of such analyses include Gray, *Cyborg Citizen: Politics in the Posthuman Age* (2002); Fukuyama (2002); and Cudworth & Hobden, "Complexity, ecologism, and posthuman politics" (2013).

II. Definition of Organizational Posthumanism

Having considered the nature of posthumanism and some links that have been suggested between posthumanism and the theory and management of organizations, we are in a position to explicitly formulate a systematic approach that applies posthumanist insights and methodologies to the study and management of organizations. This approach can be described as *organizational posthumanism.*

Lune defines an organization as "a group with some kind of name, purpose, and a defined membership" that possesses "a clear boundary between its inside and its outside" and which can take the form of either a formal organization with clearly defined roles and rules, an informal organization with no explicitly defined structures and processes, or a semi-formal organization that possesses nominal roles and guidelines that in practice are not always observed.[16] Meanwhile, Daft et al. define organizations as "(1) social entities that (2) are goal-directed, (3) are designed as deliberately structured and coordinated activity systems, and (4) are linked to the external environment."[17] Such organizations include businesses, nonprofit organizations, schools, religious groups, professional associations, political parties, governments, and military organizations. Other collections of human beings – such as cities, families, or the proponents of a particular philosophical perspective – share some of the characteristics of organizations but are not generally classified as such.

The very nature of organizations is changing as ongoing technological and social change reshapes the capacities and relationality of the human beings who belong to organizations and creates new kinds of entities (like social robots) that can engage in goal-directed social interaction with human beings and one another. Organizational posthumanism can aid us in making sense of – and, ideally, anticipating and controlling – such changes. By way of a formal definition, we would suggest that:

> Organizational posthumanism is an approach to analyzing, understanding, creating, and managing organizations that employs a post-anthropocentric and post-dualistic perspective; it recognizes that the emerging technologies which complement traditional biological human beings with new types of intelligent actors also transform the kinds of members, structures, dynamics, and roles that are available for organizations.

As we shall see, while organizational posthumanism shares elements in common with established disciplines such as philosophical posthumanism, critical posthumanism, and biopolitical posthumanism, it also possesses unique and contrasting elements that prevent it from being understood simply as a subfield of one of those disciplines. Rather, we would argue that as defined above,

---

[16] Lune, *Understanding Organizations* (2010), p. 2.

[17] Daft et al., *Organization Theory and Design* (2010), p. 10.

organizational posthumanism is better viewed as an independently conceptualized body of thought within posthumanism. When understood in the context of organizational and management theory, organizational posthumanism does not represent a new discipline, insofar as it still addresses historical topics of organizational structures, systems, and processes; however, it does constitute an entirely new perspective and set of methodologies – a new approach.

III. Classification of Organizational Posthumanism as a Type of Posthumanism

It is possible to categorize different forms of posthumanism into general types by employing a two-dimensional conceptual framework that classifies a form of posthumanism based on its understanding of posthumanity and the role or purpose for which the posthumanism was developed. With regard to its perspective on posthumanity, a form of posthumanism may be: 1) an *analytic posthumanism* that understands posthumanity as a sociotechnological reality that already exists in the contemporary world and which needs to be analyzed; or 2) a *synthetic posthumanism* that understands posthumanity as a collection of hypothetical future entities whose development can be either intentionally realized or prevented, depending on whether or not human society chooses to research and deploy certain transformative technologies. With regard to the purpose or role for which it was created, a form of posthumanism can be: 1) a *theoretical posthumanism* that seeks primarily to develop new knowledge and understanding; or 2) a *practical posthumanism* that seeks primarily to bring about some social, political, economic, or technological change in the world.[18] This framework yields five general types of posthumanism:

- **Analytic theoretical posthumanisms** seek to understand the posthumanized present and include fields like critical and cultural posthumanism. Such disciplines can collectively be understood as constituting a 'posthumanism of critique' that employs posthumanist methodologies to diagnose hidden anthropocentric biases and posthumanist aspirations contained within different fields of human activity.[19]
- **Synthetic theoretical posthumanisms** envision hypothetical forms of posthumanity and include such pursuits as philosophical posthumanism and many forms of science fiction. Such fields could be seen

[18] For a more detailed discussion of the distinctions between analytic, synthetic, theoretical, and practical posthumanisms, see Part One of this book, "A Typology of Posthumanism."

[19] For an example, see the critical posthumanism described in Herbrechter (2013).

as representing a 'posthumanism of imagination' that creatively conceptualizes future (or otherwise inexistent) posthumanities so that their implications can be explored.[20]

- **Analytic practical posthumanisms** seek to reshape the posthumanized present and include some forms of metahumanism and neohumanism. Such movements can be understood as constituting a 'posthumanism of conversion' that is aimed at changing hearts and minds and influencing the way in which human beings view and treat the world around themselves.[21]
- **Synthetic practical posthumanisms** seek to steer the processes that can generate a future posthumanity; they include such movements as transhumanism and bioconservatism. Such programs can be viewed as representing a 'posthumanism of control' that seeks to develop new technologies that give individuals control over their own posthumanization or to implement legal or economic controls to block the development of such technologies.[22]
- **Hybrid posthumanisms** that span all four spheres of the analytic, synthetic, practical, and theoretical include such phenomena as sociopolitical posthumanism and the metahumanism of Del Val and Sorgner. Such ventures can be understood as examples of a 'posthumanism of production' that develops a robust and rigorous theoretical framework that is then utilized to successfully generate concrete products or services within the contemporary world.[23]

By applying this framework, organizational posthumanism can be classified as a form of hybrid posthumanism that integrates strong analytic, synthetic, theoretical, and practical elements. We can consider each of these elements of organizational posthumanism in more detail.

---

[20] Regarding, e.g., posthumanist aspects of science fiction, see Short (2005); Goicoechea, "The Posthuman Ethos in Cyberpunk Science Fiction" (2008); Miller (2012); and Herbrechter (2013), pp. 115-17.

[21] Regarding different forms of metahumanism, see Ferrando (2013), p. 32. For the form of neohumanism developed by Sarkar, see Sarkar, "Neohumanism Is the Ultimate Shelter (Discourse 11)" (1982). A classification of different forms of metahumanism and neohumanism is found in Part One of this volume, "A Typology of Posthumanism."

[22] For examples, see Fukuyama (2002); Bostrom, "A History of Transhumanist Thought" (2005); and Bostrom (2008).

[23] For an instance of sociopolitical posthumanism as it relates to law, see Berman, "Posthuman Law: Information Policy and the Machinic World" (2002). For the form of metahumanism developed by Sorgner and Del Val, see Del Val & Sorgner, "A *Metahumanist* Manifesto" (2011), and Del Val et al., "Interview on the Metahumanist Manifesto with Jaime del Val and Stefan Lorenz Sorgner" (2011).

A. Theoretical Aspects

Organizational posthumanism is theoretical insofar as it involves efforts to understand the ways in which organizations' form and dynamics are being affected by (and are shaping) processes of posthumanization. Such work involves developing new conceptual frameworks that can explain and predict the unique ways in which organizations will become agents and objects of posthumanization and will exist as elements of a larger posthumanized ecosystem.

For example, scholars can explore the ways in which organizations' members, personnel structures, processes, information systems, physical and virtual spaces, and external environment will be altered by the integration of artificial general intelligences, sentient robotic swarms, sapient networks, neuroprosthetically augmented cyborgs, genetically engineered human beings, and other posthumanized entities into organizations whose membership was previously the exclusive domain of unmodified, 'natural' biological human beings. Such posthumanization may allow the creation of new organizational forms that were previously impossible while simultaneously rendering some traditional organizational forms ineffective or obsolete.

In its theoretical aspects, organizational posthumanism draws on and can inform fields such as organizational theory, systems theory, and cybernetics. It can work in parallel with sociopolitical posthumanism, which explores at a theoretical level the impact of posthumanization on legal, political, and economic systems and institutions. Similarly, organizational posthumanism can take up many existing lines of theoretical inquiry within fields such as philosophical, critical, and biopolitical posthumanism and science fiction and advance them in a way that is informed by a deeper concern for and insight into their implications at the organizational level.

For example, Miah notes posthumanism's longstanding interest in the blurring physical and cognitive boundaries between human beings and the tools that we use to accomplish work. Drawing on Mazlish, Miah notes that tools have historically served to extend human beings' capacities and freedom while simultaneously subjugating human beings to the organizational systems required for the tools' production and effective use.[24] Whereas tools can serve as an 'artificial skin' that mediates our relationship with our environment and offers us protection, they have also facilitated the creation of large, impersonal organizations in which human beings are reduced to functional bodies that provide some economic value. The creation of new tools such as neuroprosthetic devices is serving to make human beings "more machine-like, physically

---

[24] See Miah (2008), p. 82, and its discussion of Mazlish, *The Fourth Discontinuity: The Co-Evolution of Humans and Machines* (1993).

and cognitively," while the creation of increasingly autonomous tools such as artificial intelligences threatens to replace human beings altogether as components of some organizational systems.[25] Organizational posthumanism can develop new theoretical frameworks that shed light on such relationships between agent and instrument, between human 'employee' and nonhuman 'tool,' within the evolving context of posthumanized organizations.

B. Practical Aspects

Organizational posthumanism is also practical, insofar as its goal is not simply to understand at an abstract level the ways in which posthuman realities are affecting organizations but also to aid managers in proactively designing, creating, and maintaining organizations that can survive and thrive within novel competitive environments such as those emerging as a result of the posthumanization of our world. Just as sociopolitical posthumanism works to produce new legal, political, and economic systems that are adapted to emerging posthuman realities, so organizational posthumanism works to produce successfully posthumanized organizations – and, through them, to produce the goods, services, and other resources that such organizations release into the wider ecosystem. In its more practical aspects, organizational posthumanism draws on, shapes, and acts through disciplines like organizational design, organizational architecture, enterprise architecture, organization development, management cybernetics, and strategic management.

Research has already begun to explore the practical implications of technological posthumanization (though without necessarily naming the phenomenon as such) for areas such as strategic planning, business models, entrepreneurship, marketing, knowledge management, and customer relationship management (CRM);[26] change management, organizational culture, and organizational HCI;[27] potential roles for artificial intelligences in leading teams of human workers;[28] and the creation of neurocybernetically linked organizational systems.[29]

---

[25] Miah (2008), p. 82.

[26] See the thoughtful overview of the impacts of posthumanizing technologies on such areas in Berner, *Management in 20XX: What Will Be Important in the Future – A Holistic View* (2004).

[27] See Mara & Hawk (2009).

[28] See Gladden, "Leveraging the Cross-Cultural Capacities of Artificial Agents as Leaders of Human Virtual Teams" (2014); Gladden, "The Social Robot as 'Charismatic Leader': A Phenomenology of Human Submission to Nonhuman Power" (2014); and Gladden, "Managerial Robotics: A Model of Sociality and Autonomy for Robots Managing Human Beings and Machines" (2014).

[29] See Gladden, "Neural Implants as Gateways to Digital-Physical Ecosystems and Posthuman Socioeconomic Interaction" (2016).

C. Analytic Aspects

The fact that processes of posthumanization are expected to accelerate and expand in the future does not diminish the posthumanizing impacts that have already been felt and which every day are creating new opportunities and challenges for organizations. Organizational posthumanism is analytic, insofar as it strives to understand the changes to organizations that have already occurred as a result of such previous and ongoing processes of posthumanization. On the basis of such knowledge, managers and other organizational stakeholders can develop strategies and best practices to optimize the functioning of real-world organizations today.

For example, researchers in the field of organizational posthumanism might, for example, attempt to anticipate the implications of employing artificial general intelligences (AGIs) to fill roles as senior executives within otherwise human organizations.[30] Such efforts to imagine the eventual impacts of radically posthumanized far-future technological systems complement organizational posthumanism's efforts to analyze the impact that is already being felt on organizations by more rudimentary technologies for artificial intelligence, such as those that control industrial robots for assembly-line manufacturing,[31] automated systems for resource scheduling and planning,[32] web-based chatbots for basic interactions with customers, [33] and robotic sales associates for dispensing goods and services to customers.[34]

D. Synthetic Aspects

In addition to analyzing the kinds of posthumanized organizations that already exist today, organizational posthumanism seeks to envision the kinds of even more radically posthumanized organizations that may be able to exist in the future thanks to accelerating forces of technologization and other anticipated sociotechnological change.

In a sense, all long-term organizational decision-making involves a sort of 'futurology,' as stakeholders make decisions on the basis of their empirically

---

[30] See, e.g., Gladden, "The Social Robot as 'Charismatic Leader'" (2014).

[31] For an overview of such technologies, see, e.g., Perlberg, *Industrial Robotics* (2016).

[32] See, e.g., *Automated Scheduling and Planning: From Theory to Practice*, edited by Etaner-Uyar et al. (2013).

[33] Such technologies are described, e.g., in Perez-Marin & Pascual-Nieto, *Conversational Agents and Natural Language Interaction: Techniques and Effective Practices* (2011).

[34] See, e.g., the account from a consumer's perspective of interactions with such technologies in Nazario, "I went to Best Buy and encountered a robot named Chloe – and now I'm convinced she's the future of retail" (2015).

grounded projections, estimates, or intuitions about how an organization's external context is likely to evolve over time (e.g., as captured in a PESTLE analysis[35]) and how the impact of a decision is likely to reshape the organization's internal form and dynamics. Organizational posthumanism involves a specialized form of organizational futurology that attempts to conceptualize and predict the ways in which organizations in general (or one organization in particular) will be transformed by the dynamics of posthumanization or will be able to exploit those dynamics for their own strategic purposes.

Within organizational posthumanism, the analytic and theoretical effort to understand effective posthumanized organizations and the synthetic and practical effort to design and create them are thus joined as two sides of a single coin.

IV. Organizational Posthumanization as Reflected in Organizational Elements

One aspect of posthumanization is the emergence of a world in which natural human beings are joined by other kinds of entities such as cyborgs, social robots, AGIs, sapient networks, and artificial life-forms in serving as employees, collaborators, and consumers. This posthuman reality will increasingly be reflected in various aspects of organizational life. Particular implications of such posthumanization can be identified in the kinds of *members, structures, systems, processes, spaces,* and *external ecosystems* that organizations will possess.[36] Below we consider each of these elements.

A. Posthumanized Members

Traditionally, the members of organizations have been 'natural' biological human beings who have not been engineered or extensively enhanced with the aid of biomedical technologies. The membership of future organizations will comprise a much more diverse array of entities. It is expected that increasingly the members of organizations will, for example, also include:[37]

---

35 See Cadle et al., *Business Analysis Techniques: 72 Essential Tools for Success* (2010), pp. 3-6, for a description of various versions of this analytic tool.

36 Structures, processes, and systems constitute the three main elements within the 'congruence model' of organizational architecture as conceptualized by Nadler and Tushman. See Nadler & Tushman, *Competing by Design: The Power of Organizational Architecture* (1997), p. 47.

37 For an overview of the roles that such beings may play in future organizations, see Berner (2004). Discussions of specific types of posthumanized organizational members are found, e.g., in Bradshaw et al., "From Tools to Teammates: Joint Activity in Human-Agent-Robot Teams" (2009); Samani et al., "Towards Robotics Leadership: An Analysis of Leadership Characteristics and the Roles Robots Will Inherit in Future Human Society" (2012); Wiltshire et al., "Cybernetic Teams: Towards the Implementation of Team Heuristics in HRI" (2013); Gladden, "The Social Robot as 'Charismatic Leader'" (2014); Gladden, "The Diffuse Intelligent Other: An Ontology of

- Human beings possessing implantable computers (such as devices resembling subcutaneous smartphones)
- Human beings equipped with sensory, cognitive, or motor neuroprosthetics, including human beings who possess full cyborg bodies
- Genetically engineered human beings
- Human beings who are long-term users of virtual reality systems and whose interaction with other persons and their environment takes place largely within virtual worlds
- Social robots
- Artificial general intelligences
- Artificial life-forms
- Sapient networks
- Human and synthetic beings whose thoughts and volitions have been cybernetically linked to create 'hive minds'

Such members will be discussed in more detail later in this text, in our analysis of technological changes facilitating organizational posthumanization. From an organizational perspective, the capacities, vulnerabilities, needs, and forms of interaction demonstrated by such entities can differ radically from those of the natural human beings who have historically constituted an organization's membership. The use of posthuman entities (including artificial beings) to fill organizational roles as senior executives, product designers, or the providers of sensitive goods or services (such as health care or military activities) raises a range of complex ethical, legal, and information security questions.[38] Organizational posthumanism can investigate the theoretical constraints and possibilities for creating organizations that include such posthumanized members

Nonlocalizable Robots as Moral and Legal Actors" (2016); and Gladden, "Neural Implants as Gateways" (2016).

[38] For a discussion of questions that can arise when entrusting organizational roles and responsibilities to robots and AIs, see, e.g., Stahl, "Responsible Computers? A Case for Ascribing Quasi-Responsibility to Computers Independent of Personhood or Agency" (2006); Sparrow, "Killer Robots" (2007); Calverley, "Imagining a non-biological machine as a legal person" (2008); Grodzinsky et al., "Developing Artificial Agents Worthy of Trust: 'Would You Buy a Used Car from This Artificial Agent?'" (2011); Coeckelbergh, "Can We Trust Robots?" (2012); Datteri, "Predicting the Long-Term Effects of Human-Robot Interaction: A Reflection on Responsibility in Medical Robotics" (2013); Gladden, "The Social Robot as 'Charismatic Leader'" (2014); and Gladden, "The Diffuse Intelligent Other" (2016). Regarding questions that arise in the case of neurocybernetically enhanced human workers, see, e.g., McGee, "Bioelectronics and Implanted Devices" (2008); Koops & Leenes, "Cheating with Implants: Implications of the Hidden Information Advantage of Bionic Ears and Eyes" (2012); and Gladden, "Neural Implants as Gateways" (2016).

and can develop practical approaches for the management of organizations that incorporate them.

B. Posthumanized Structures

The types of internal and external structures that are available for use by organizations are expected be reshaped and expanded by emerging posthuman realities. When managing contemporary organizations, possible organizational forms identified by Horling and Lesser include hierarchies (which can be either simple, uniform, or multi-divisional), holarchies (or 'holonic organizations'), coalitions, teams, congregations, societies, federations (or 'federated systems'), matrix organizations, compound organizations, and sparsely connected graph structures (which may either possess statically defined elements or be an 'adhocracy').[39] Such structures have been developed over time to suit the particular characteristics of the members that constitute contemporary organizations – i.e., natural biological human beings. As organizations evolve to include members that possess radically different physical and cognitive capacities and novel ways of interacting with one another, the kinds of structures that are available to organize the work of these groups of members will change, and novel organizational structures are expected to become feasible and even necessary.[40]

For example, an organization composed of neuroprosthetically augmented human members may be able to link them through a decentralized network that enables the direct sharing of thoughts and sentiments between members' minds, allowing information to be disseminated in an instantaneous fashion and decisions to be made in a distributed and collective manner that is impossible for conventional human organizations.[41] The reporting and decision-

---

[39] Horling & Lesser, "A Survey of Multi-Agent Organizational Paradigms" (2004).

[40] For the sake of convenience, it is possible to refer to such developments as 'novel *personnel* structures' – however it must be kept in mind that the 'personnel' constituting such future organizations will not necessarily be human 'persons' but may include, e.g., such radically different types of entities as nanorobot swarms or sapient networks of computerized devices.

[41] Regarding the prospect of creating hive minds and neuroprosthetically facilitated collective intelligence, see, e.g., McIntosh, "The Transhuman Security Dilemma" (2010); Roden, *Posthuman Life: Philosophy at the Edge of the Human* (2014), p. 39; and Gladden, "Utopias and Dystopias as Cybernetic Information Systems: Envisioning the Posthuman Neuropolity" (2015). For a classification of different kinds of potential hive minds, see Chapter 2, "Hive Mind," in Kelly, *Out of Control: The New Biology of Machines, Social Systems and the Economic World* (1994); Kelly, "A Taxonomy of Minds" (2007); Kelly, "The Landscape of Possible Intelligences" (2008); Yonck, "Toward a standard metric of machine intelligence" (2012); and Yampolskiy, "The Universe of Minds" (2014). For critical perspectives on hive minds, see, e.g., Maguire & McGee, "Implantable brain chips? Time for debate" (1999); Bendle, "Teleportation, cyborgs and the posthuman ideology" (2002); and Heylighen, "The Global Brain as a New Utopia" (2002).

making structures of such an organization might reflect multidimensional cybernetic network topologies that were previously possible only for computerized systems (or some nonhuman animal species) but which could not be effectively employed within human organizations.[42] Organizational posthumanism can conceptualize such new possibilities and develop concrete recommendations regarding organizational structures that are especially well- or poorly suited for organizations comprising posthumanized members.

C. Posthumanized (Information) Systems

The word 'system' is used with different meanings in different organizational contexts. From the perspective of management cybernetics, an organization as a whole can be considered a 'viable system,' as can each of its constituent subsystems.[43] On the other hand, within the context of contemporary

---

[42] See, e.g., Gladden, "Utopias and Dystopias as Cybernetic Information Systems" (2015). Efforts by organizational posthumanists to envision and implement new kinds of posthumanized organizational structures should be distinguished from management approaches such as the Holacracy movement, which abolishes job titles and hierarchical structures for decision-making and authority and replaces them with largely self-organizing, self-guiding circles of employees. From the perspective of Holacracy, an organization can essentially be viewed as though it were a conventional electronic computer and each of the organization's human members were components of that computer. The *Holacracy Constitution* provides an organization with a complex set of decision-making rules and procedures that constitute the organization's 'operating system' and which – after this 'OS' has become sufficiently engrained in employees' interactions and decision-making patterns – allow new business processes to be implemented in the form of 'apps' which, in theory, can be downloaded and installed in the minds and behaviors of the organization's human employees in a manner similar to that of installing a new program on a desktop computer. See Robertson, *Holacracy: The New Management System for a Rapidly Changing World* (2015), pp. 9-14, and the *Holacracy Constitution v4.1* (2015).

Superficially, Holacracy shares some elements in common with posthumanism, insofar as it recognizes the fact that innovative new organizational structures that draw inspiration from sources other than traditional human institutions are increasingly becoming possible and even necessary. However, Holacracy diverges from the principles of organizational posthumanism by declining to acknowledge that the circle of intelligent actors within organizations is expanding to include entities other than natural biological human beings. Holacracy is essentially anthropocentric, insofar as it presumes that natural biological human beings are and will continue to be the lone relevant actors within organizations; it simply attempts to induce such human beings to behave as if they were electronic computer components rather than human persons. Such an approach may prove more effective in the future, if implantable computers, neurocybernetics, long-term immersive virtual environments, and other technologizing phenomena lead to the development of human workers that display sufficiently 'computronic' characteristics. (See Part Three of this volume, "The Posthuman Management Matrix: Understanding the Organizational Impact of Radical Biotechnological Convergence," for a discussion of such phenomena.) However, current attempts at implementing approaches such as Holacracy would appear to significantly underestimate the fundamental structural and behavioral differences that presently exist between human and synthetic agents.

[43] For cybernetic accounts of viable systems from a management perspective, see, e.g., Beer, *Brain*

organizational architecture, 'systems' are typically computerized information systems such as manufacturing systems that govern and constitute a physical assembly line, an internally hosted accounting database, a cloud-based HR management system, a public-facing website for handling retail transactions, or a social media platform for use in marketing and public relations.

Traditionally, the relationship of human employees to such systems has been relatively straightforward: human workers serve as the designers, programmers, data-entry specialists, and end users of the information systems, while the systems themselves are assigned the role of receiving, storing, and transmitting data securely and manipulating it in an efficient and accurate fashion, as instructed by human employees. However, the boundary between the electronic systems that store and process information and the human workers that use them are expected to increasingly blur as implantable computers, neuroprosthetic devices, and persistent virtual reality environments integrate human workers ever more intimately into organizational information systems at both the physical and cognitive levels.[44] Moreover, the growing sophistication of artificial intelligence platforms for use in data mining and other applications[45] is expected to increasingly create information systems that are self-organizing, self-analyzing, and even self-aware. Through the use of such systems, organizations may move beyond the era of Big Data and Smart Data and into an era of 'Sapient Data' in which information systems utilize human workers as tools rather than being utilized by them. Organizational posthumanism can offer critical perspectives regarding both the ontological and ethical aspects of such human-electronic systems as well as their practical implementation.

D. Posthumanized Processes

The essential processes found within an organization do not simply include those by which it directly generates the end products for which the organization is known – such as the actions used to physically assemble some device on an assembly line (for a consumer electronics company) or to generate

---

of the Firm (1981); Barile et al., "An Introduction to the Viable Systems Approach and Its Contribution to Marketing" (2012); and Gladden, "The Artificial Life-Form as Entrepreneur: Synthetic Organism-Enterprises and the Reconceptualization of Business" (2014).

[44] For an in-depth analysis of the ways in which such historical barriers between human workers and electronic information systems are being dissolved, see Part Three of this text, "The Posthuman Management Matrix."

[45] Regarding the prospects of developing autonomous AI systems for data mining, see, for example, Warkentin et al., "The Role of Intelligent Agents and Data Mining in Electronic Partnership Management" (2012); Bannat et al., "Artificial Cognition in Production Systems" (2011), pp. 152-55; and Wasay et al., "Queriosity: Automated Data Exploration" (2015).

sounds from musical instruments during a concert (for a symphony orchestra). An organization's fundamental processes also include all of those behaviors and dynamics through which resources (including human resources, financial resources, material resources, and information)[46] are acquired from the external environment, created internally, transmitted between different parts of the organization, combined or transformed, or released into the external environment – as well as all of the second-order processes by which those behaviors and dynamics are planned, led, organized, and controlled.[47] Such second-order processes include the use of the three key mechanisms of programming, feedback, and hierarchical supervision to coordinate the activities of an organization's members.[48] They also include compensation and incentive schemes that are used to reward and motivate desired behaviors on the part of an organization's members, as well as processes of career advancement which ensure that an organization's most talented and effective workers move into positions in which their abilities can be employed to their fullest potential.[49]

In the case of contemporary organizations that include only traditional biological human members, there exists a rich body of theory and best practices relating to the design and implementation of such processes. However, it is clear that the nature of these processes can change dramatically within a radically posthumanized organizational context. For example, some kinds of advanced robots and AIs may require no compensation at all – other than 'compensation' in the form of an electric power supply, physical maintenance and software upgrades, and other resources needed to ensure their continued operation. However, very sophisticated AGIs whose cognitive dynamics are based on those of human beings might request – and, as a practical matter, require – compensation in the form of intellectual stimulation, self-fulfillment, and generic financial resources (i.e., a paycheck) that an entity can spend as it sees fit to pursue its own personal goals or objectives in its spare time.[50] Similarly, neurocybernetically augmented human employees may be able to in-

---

[46] For the role of such resources in organizational dynamics, see, e.g., Pride et al., *Foundations of Business* (2014), p. 8., and Gladden, "The Artificial Life-Form as Entrepreneur" (2014).

[47] Planning, organizing, leading, and controlling are considered to be the four primary functions that must be performed by managers. See Daft, *Management* (2011).

[48] For a review of the scholarship on such mechanisms and their role in organizations, see Puranam et al., "Organization Design: The Epistemic Interdependence Perspective" (2012), p. 431.

[49] See Brickley et al., "Corporate Governance, Ethics, and Organizational Architecture" (2003), p. 43; Puranam et al. (2012); and Nadler & Tushman (1997), loc. 862, 1807.

[50] For an in-depth analysis of the prospects of developing AGIs with human-like cognitive capacities and psychological needs, see Friedenberg, *Artificial Psychology: The Quest for What It Means to Be Human* (2008).

stantly acquire new skills or capacities in ways that render traditional professional advancement schemes outdated and irrelevant, and such employees might demand new forms of compensation (such as lifetime technical support for neuroprosthetic devices that have been implanted to enable the fulfillment of their official organizational responsibilities[51]). Organizational posthumanism can develop theoretical accounts of such posthumanized processes as well as best practices to facilitate their management.

E. Posthumanized Spaces

The physical spaces in which an organization's members come together to plan and execute its activities have historically included venues such as factories, office buildings, warehouses, retail stores, farms, campuses, military bases, and other specialized locations. As organizations evolve and expand to include nonhuman members such as sapient networks or robotic swarms, the range of physical spaces in which such organizational members can (or need) to work will be similarly transformed. Moreover, building on the use of technologies such as telephony, email, instant messaging, and videoconferencing, even the traditional biologically human members of organizations will find themselves interacting in new posthumanized venues such as persistent virtual worlds. Within such new physical and virtual organizational spaces, one member of an organization may or may not always know whether the other intelligent members with which the member is interacting socially are natural biological human beings, neurocybernetically enhanced human beings, robots, AIs, or other kinds of entities.[52] Organizational posthumanism can engage with practitioners in the fields of architecture, facilities design, ergonomics, operations management, and logistics to create and operate posthumanized physical facilities for organizations functioning in such a deanthropocentrized context. With regard to the development and use of posthumanized virtual spaces, organizational posthumanism can provide a conceptual bridge by seeking out insights from fields as diverse as biocybernetics, HCI, psychology, anthropology, communications, philosophy of mind, computer game design, science fiction, and film and television studies to develop immersive multisensory worlds that serve as effective venues for organizational life.

F. Posthumanized External Environments and Ecosystems

An organization can be understood as a viable system that operates within a broader ecosystem (or 'suprasystem') that includes other competing or collaborating organizations as well as natural resources, potential consumers,

---

[51] See Gladden, "Neural Implants as Gateways" (2016).

[52] See Grodzinsky et al. (2011) and Gladden, "The Social Robot as 'Charismatic Leader'" (2014).

and other external environmental features.[53] These ecosystems are expected to take on an increasingly posthumanized nature. For example, new environmental elements might include other organizations that consist entirely of intelligent nonhuman members such as robotic swarms and societies of AIs. Similarly, a highly interconnected Internet of Things might be filled with informational resources that are no longer simply passive sets of data but which – through their integration with AI platforms – become intelligent, volitional, and potentially even sapient collections of data that act to pursue their own goals and interests.[54] The world's increasingly rich and complex digital-physical ecosystems might be populated by self-generating, self-propagating, highly adaptable memes in the form of evolvable computer worms or viruses that shape human popular culture as a whole and the thoughts and memories of individual human beings in particular, either through traditional forms of communication and social interaction or through the targeted reprogramming or technological manipulation of, for example, neurocybernetically augmented human beings.[55] The emergence of such new posthuman ecosystems is expected to significantly reshape the kinds of resources that organizations are able to obtain from their environments, the nature of collaboration and competition with external organizations, the types of consumers available to utilize the goods and services produced by an organization, and the organization's definition of long-term viability and success.

The roles that individual organizations play within societies may also be radically reshaped. For example, if future AIs and robotic systems are able to efficiently perform all of the functions of food production and preparation, health care, education, construction, transportation, energy production, retail sales, accounting, security, and other tasks that are needed for human beings and societies to thrive, there will no longer be a financial or operational need for organizations to employ human beings as workers in such roles. In that case, governments might take on the role of coordinating their human citizens' access to such superabundant resources, perhaps offering a 'universal

---

[53] Regarding viable systems and their environments, see, e.g., Beer (1981) and Gladden, "The Artificial Life-Form as Entrepreneur" (2014).

[54] For discussions of the theoretical and practical possibilities for and obstacles to the emergence of such systems, see, e.g., Gladden, "From Stand Alone Complexes to Memetic Warfare: Cultural Cybernetics and the Engineering of Posthuman Popular Culture" (2016), and Gladden, "The Artificial Life-Form as Entrepreneur" (2014).

[55] Regarding the growing possibilities that ideas and other forms of information might exist as actors that can propagate themselves through interaction with other nonhuman or human actors within complex posthumanized digital-physical ecosystems, see, e.g., Gladden, "From Stand Alone Complexes to Memetic Warfare" (2016), and Kowalewska, "Symbionts and Parasites – Digital Ecosystems" (2016).

basic income' redeemable in goods or services. The societal roles of governmental and commercial organizations would thus be dramatically transformed. On the other hand, widespread roboticization resulting in mass unemployment could potentially yield a loss of purpose for human beings, social unrest, violent revolution, and the oppression of the human species by automated systems; in this case, processes of posthumanization might result in 'dystopian' rather than 'utopian' organizational outcomes.[56] Organizational posthumanism can provide a theoretical bridge that links the consideration of posthumanization at an organizational level with that at a broader social or environmental level (as considered by fields such as economics, political science, sociology, evolutionary biology, or environmental science), while also developing concrete practices to aid organizations with optimizing their use of resources from and contribution of products to a posthumanized external environment.

V. Technological Changes Facilitating Organizational Posthumanization

While advanced technologies play an essential role in contemporary processes of posthumanization, they are not the only mechanisms through which such processes operate. As noted earlier, there exist many forms of 'posthumanism without technology.'[57] Such nontechnological critical or cultural posthumanism might focus, for example, on historical references to ghosts, angels, monsters, and semidivine heroes in theology and the arts and the ways in which they have long encouraged human beings to expand the boundaries of society to include a nonhuman 'other.'[58]

Posthumanized beings have always been part of organizations. Even if only tangentially, human organizations have always incorporated such quasi-human, parahuman, or nonhuman others. For example, the decision-making processes of Ancient Roman governmental and military organizations relied on augurs that were supposed by their practitioners to reveal the will of the gods.[59] According to the Catholic Church's traditional teaching on the

---

[56] For the debate on whether mass roboticization and the end of human employment as we know it is likely to generate utopian, dystopian, or less extreme social impacts, see, e.g., Sachs et al., "Robots: Curse or Blessing? A Basic Framework" (2015); Nourbakhsh, "The Coming Robot Dystopia" (2015); and Ford, *Rise of the Robots: Technology and the Threat of a Jobless Future* (2015). For longer-term interdisciplinary perspectives, see the texts in *Singularity Hypotheses*, edited by Eden et al. (2012).

[57] Herbrechter (2013), p. 157.

[58] Herbrechter (2013), pp. 2-3, 106. See also Graham (2002).

[59] See Hamilton, "What Is Roman Ornithomancy? A Compositional Analysis of an Ancient Roman Ritual" (2007), and Green, "Malevolent gods and Promethean birds: Contesting augury in

Communion of Saints, the organization of the Church incorporates both human members who are presently living on earth, members who have died but are still undergoing a purification, and members who have died and now contemplate God in His heavenly glory.[60] In a metaphorical sense, the 'ghost' of a company's beloved founder can continue to guide the company's actions even after his or her death, gazing watchfully from framed portraits on office walls and inspiring new generations of employees through aphorisms quoted reverently in the company's mission statement or employee handbook. And nonhuman others in the form of dogs, horses, and other animals have long been incorporated into human military organizations and businesses (e.g., family farms or circuses) in important roles as intelligent – if not sapient – agents.

Technologization is changing the nature of posthumanization. However, even critical posthumanists who argue that the processes of posthumanization have historically taken many forms unrelated to technological change will acknowledge that in today's world, the accelerating and intensifying technologization of humanity has become an essential – if not the most essential – driver of posthumanization.[61] Herbrechter notes that from the time of its prehistoric origins, humanity has always utilized technology. Indeed, it was only the creation of techniques and technologies for performing such tasks as making fire, hunting animals, and communicating information symbolically that humankind as such was able to develop; "Culture in a sense is therefore always 'technoculture', namely achieved and transmitted by technics."[62] However, the manner and extent of our integration with workplace technologies is now undergoing a qualitative transformation. Herbrechter suggests that the human operators of equipment are increasingly merging with their tools in order to manipulate them more effectively, thereby undergoing a process of cyborgization. But just as we are becoming more dependent on our technology, our technology is becoming less dependent on us – thanks to the growing sophistication of artificial intelligence and automated systems that can make decisions without any need for human input. Human agency is thus being attenuated by technology at the same time that the world of 'smart objects' is gaining its own agency.[63]

The new kinds of posthumanized beings produced through such technologization will become incorporated into human organizations in novel fashions. A ghost or saint or animal can indeed be 'incorporated' into the life and behaviors of an organization in meaningful ways – but not, for example, as an employee of the organization. The 'ghost' of a company's founder might offer

---

Augustus's Rome" (2009).

[60] See the *Catechism of the Catholic Church, Second Edition* (2016), pp. 249-250.

[61] See Herbrechter (2013), pp. 15, 6-7.

[62] Herbrechter (2013), p. 152.

[63] For a discussion of these simultaneous trends, see Herbrechter (2013), p. 150.

vague principles to guide decision-making but cannot determine which of three smartphone models to offer for sale in a particular country. A horse can transport a company's goods from place to place but cannot formulate the company's long-term business strategy. However, posthuman beings in the form of artificial intelligences, social robots, sentient (and even sapient) networks, and cyborgs *will* be able to do such things. Increasingly, such posthumanized entities will not simply operate at the fringes of an organization or in supporting roles that aid the decision-making of the organization's natural human members; such posthuman beings will instead increasingly fill critical roles as designers, producers, strategists, and decision-makers within organizations.[64]

While processes such as roboticization, cyborgization, and virtualization have not created the phenomenon of posthumanization, they are making its dynamics visible in new and more vivid ways.[65] Hayles suggests that some forms of 'uncritical' posthumanism (including strains of transhumanism and cybernetics) possess a naïvely technologized interpretation of these processes: such a perspective understands the human body as merely a prosthesis or computational substrate and the mind as a collection of informational patterns; it considers the biological organism of a human being, a social robot resembling a human being, and a computer simulation of a human being to be just three interchangeable manifestations of the same sort of viable system.[66] Critical posthumanists such as Hayles and Herbrechter reject such simplistic 'technoeuphoria' and argue that more rigorous critical posthumanist thought is necessary in order to understand, anticipate, and guide the processes of sociotechnological transformation that are challenging our concept of humanity and altering humanity's role in the world.[67] Organizational posthumanism is well-positioned to explore such questions of technological posthumanization in a way that marries the circumspectness of critical posthumanism with a strategic awareness of the fact that the ability to generate and embrace radical new forms of technological transformation is growing ever more important to the survival of organizations.

Three categories of posthumanizing technologies. For the purposes of this text, there are three broad categories of ongoing or anticipated technological developments that are contributing to posthumanization in especially

---

[64] An exploration of these possibilities can be found, e.g., in Samani et al. (2012) and Gladden, "The Social Robot as 'Charismatic Leader'" (2014).

[65] See Herbrechter (2013), p. 77.

[66] See Hayles (1999), pp. 2-3, and its discussion in Herbrechter (2013), p. 42.

[67] Herbrechter (2013), p. 200.

relevant ways: 1) technologies for human augmentation and enhancement, which include many forms of neuroprosthetics and genetic engineering; 2) technologies for synthetic agency, which include robotics, artificial intelligence, and artificial life; and 3) technologies for digital-physical ecosystems and networks that help create the environments within which and infrastructure through which human and artificial agents will interact.[68] We can consider these three types of technologies in turn.

A. Technologies for Human Augmentation and Enhancement

Technologies that are expected to alter the sensory, motor, and cognitive capacities of human beings include implantable computers, advanced neuroprosthetics, genetic engineering, and the use of immersive virtual reality systems.[69] The implementation of such technologies will result in a posthumanization of organizations' **members** (e.g., as an organization purposefully hires cyborgs to fill particular roles or the organization's current employees acquire cybernetic enhancements on their own initiative), **structures** (e.g., as implantable computers and communication devices allow workers to engage in new types of decision-making and reporting relationships), **systems** (e.g., by giving human workers new abilities to control, be controlled by, and otherwise interface with an organization's technological infrastructure), **processes** (e.g., by facilitating direct brain-to-brain communication and providing workers with in-body access to organizational databases), **spaces** (e.g., by allowing cyborg workers to operate in areas dangerous or inaccessible to natural human beings), and **external ecosystems** (e.g., by creating cyborg consumers that need new kinds of goods and services and external cyborg partners and consultants that can provide them). We can consider such posthumanizing technologies in more detail.

1. Implantable Computers

The universe of contemporary information and communications technology (ICT) includes a wide range of implantable devices such as passive RFID tags that are not in themselves computers but which can interact with com-

---

[68] For a discussion of the role of such technologies in posthumanization, see Herbrechter (2013), pp. 90-91, and its analysis of Graham (2002) and Graham, "Post/Human Conditions" (2004). Note that while we focus in this text on three kinds of posthumanizing technologization that have a particular impact on the form and dynamics of organizations, they are by no means the only kinds of technologization that will contribute to posthumanization. Technological developments in other fields such as agriculture, transportation, energy, space exploration, and the military will also likely contribute to the posthumanization of our world and the organizations within it.

[69] Such technologies are reviewed, e.g., in Bostrom (2008); Fukuyama (2002); Gray (2002); and Herbrechter (2013), pp. 90-91.

puters and serve as elements of computerized systems. However, an increasing number of implantable devices indeed constitute full-fledged computers that possess their own processor, memory, software, and input/output mechanisms and whose programming can be updated after they are implanted into the body of their human host. Among these are many implantable medical devices (IMDs) such as pacemakers, defibrillators, neuroprostheses including retinal and cochlear implants, deep brain stimulation (DBS) devices, body sensor networks (BSNs), and even some of the more sophisticated implantable RFID transponders.[70] A growing number of these implantable computers utilize sophisticated biocybernetic control loops that allow the physiological and cognitive activity of their host to be detected, processed, and interpreted for use in exercising real-time computer control.[71]

The implantable computers that have been developed to date typically serve a restorative or therapeutic medical purpose: they are used to treat a particular illness or restore to their user a sensory, motor, or cognitive ability that has been lost through illness or injury. Increasingly, though, implantable computers will be developed not to restore some regular human capacity that has been lost but to augment their users' physical or intellectual capacities in ways that exceed typical human abilities.[72] For example, implantable computers resembling miniaturized subcutaneous smartphones might provide their users with wireless communication capacities including access to cloud-based services.[73] The elective use of implantable computers for physical and cognitive augmentation will expand the market for such devices to broader segments of the population beyond those who currently rely on them to address medical conditions.[74]

2. Advanced Neuroprosthetics

Drawing on definitions offered by Lebedev and others, we can define a neuroprosthesis as a technological device that is integrated into the neural

---

[70] See Gasson et al., "Human ICT Implants: From Invasive to Pervasive" (2012); Gasson, "ICT Implants" (2008); and Gladden, *The Handbook of Information Security for Advanced Neuroprosthetics* (2015), pp. 19-20.

[71] See Fairclough, "Physiological Computing: Interfacing with the Human Nervous System" (2010), and Park et al., "The Future of Neural Interface Technology" (2009).

[72] Regarding the anticipated increasing use of implantable computers for purposes of human enhancement, see, e.g., Warwick & Gasson, "Implantable Computing" (2008); Berner (2004), p. 17; and Gladden, *The Handbook of Information Security for Advanced Neuroprosthetics* (2015), p. 28.

[73] For discussion of such a device, see Gladden, *The Handbook of Information Security for Advanced Neuroprosthetics* (2015), p. 93.

[74] See McGee (2008) and Gasson et al. (2012).

circuitry of a human being; such devices are often categorized as being sensory, motor, bidirectional sensorimotor, or cognitive.[75] While there is much overlap between implantable computers and neuroprosthetic devices, not all implantable computers interface directly with their host's neural circuitry and not all neuroprosthetic devices are implantable.[76]

The power and potential applications of neuroprosthetic devices are expected to grow significantly in the coming years. For example, it is anticipated that current types of retinal implants that demonstrate very limited functionality will be supplanted by future sensory neuroprosthetics such as artificial eyes[77] that give their human hosts the capacity to experience their environments in dramatic new ways, such as through the use of telescopic or night vision[78] or by presenting an augmented reality that overlays actual sense data with supplemental information from a neuroprosthetic device's computer.[79] A neuroprosthetic device could also allow all of the sense data experienced by a human mind to be recorded as a stream of digital data that can be played back on demand by other human beings, enabling them to vicariously experience the world as though they were temporarily occupying the body of the device's host. Similar technologies might allow a person to play back any of his or her own earlier sensory experiences with perfect fidelity or replace the sense data generated by his or her actual external environment with sense data depicting some fictional virtual world.[80]

Meanwhile, cognitive neuroprosthetic devices may offer their user the ability to create, delete, or otherwise edit memories stored within his or her brain's biological neural network; such abilities could be used, for example, to acquire new knowledge or skills or to erase existing fears.[81] Some scholars

---

[75] Such a classification is discussed in Lebedev, "Brain-Machine Interfaces: An Overview" (2014), and Gladden, *The Handbook of Information Security for Advanced Neuroprosthetics* (2015), pp. 21-22.

[76] For this distinction, see Gladden, *The Handbook of Information Security for Advanced Neuroprosthetics* (2015), p. 32.

[77] Regarding such possibilities, see Berner (2004), p. 17, and Koops & Leenes (2012).

[78] Such enhanced forms of vision are discussed, e.g., in Gasson et al. (2012) and Merkel et al., "Central Neural Prostheses" (2007).

[79] See Koops & Leenes (2012) and Gladden, *The Handbook of Information Security for Advanced Neuroprosthetics* (2015), pp. 32-33.

[80] Regarding such sensory playback and virtual reality systems, see Gladden, *The Handbook of Information Security for Advanced Neuroprosthetics* (2015), pp. 33, 156-57; Koops & Leenes (2012), pp. 115, 120, 126; Merkel et al. (2007); Robinett, "The Consequences of Fully Understanding the Brain" (2002); and McGee (2008), p. 217.

[81] Such possibilities build on experimental techniques and technologies that are currently being

envision the development of ingestible 'knowledge pills' whose contents (perhaps a swarm of networked nanorobots[82]) travel to the brain, where they manipulate neurons to create engrams containing particular memories.[83] Other researchers foresee the possibility of being able to simply download new skills or knowledge onto a memory chip implanted within the brain.[84] Cognitive neuroprosthetic devices might also be used to provide their human hosts with enhanced levels of intelligence[85] and creativity,[86] more desirable emotional dynamics and behavior,[87] enhanced conscious awareness (e.g., by reducing the need for sleep),[88] a strengthened or modified conscience,[89] and real-time assistance with decision-making to mitigate the impact of cognitive biases.[90]

---

tested in mice. See Han et al., "Selective Erasure of a Fear Memory" (2009); Ramirez et al., "Creating a False Memory in the Hippocampus" (2013); McGee (2008); Warwick, "The Cyborg Revolution" (2014), p. 267; and Gladden, *The Handbook of Information Security for Advanced Neuroprosthetics* (2015), p. 148.

[82] See Pearce, "The Biointelligence Explosion" (2012).

[83] For such possibilities, see Spohrer, "NBICS (Nano-Bio-Info-Cogno-Socio) Convergence to Improve Human Performance: Opportunities and Challenges" (2002).

[84] See McGee (2008) and Gladden, *The Handbook of Information Security for Advanced Neuroprosthetics* (2015), p. 33.

[85] Berner (2004), p. 17.

[86] Increases in creativity have been anecdotally reported to occur after the use of neuroprosthetics for deep brain stimulation. See Cosgrove, "Session 6: Neuroscience, brain, and behavior V: Deep brain stimulation" (2004); Gasson, "Human ICT Implants: From Restorative Application to Human Enhancement" (2012); Gladden, *The Handbook of Information Security for Advanced Neuroprosthetics* (2015), p. 149; Gladden, "Neural Implants as Gateways" (2016); and Gasson (2012), pp. 23-24.

[87] Regarding the intentional creation of emotional neuroprosthetics, see, e.g., Soussou & Berger, "Cognitive and Emotional Neuroprostheses" (2008). Effects on emotion have already been observed, for example, with devices used for deep brain stimulation. See Kraemer, "Me, Myself and My Brain Implant: Deep Brain Stimulation Raises Questions of Personal Authenticity and Alienation" (2011).

[88] Regarding efforts by the DARPA military research agency and others to develop neurotechnologies that can increase soldiers' alertness and reduce their need for sleep, see, e.g., Falconer, "Defense Research Agency Seeks to Create Supersoldiers" (2003); Moreno, "DARPA On Your Mind" (2004); Clancy, "At Military's Behest, Darpa Uses Neuroscience to Harness Brain Power" (2006); Wolf-Meyer, "Fantasies of extremes: Sports, war and the science of sleep" (2009); Kourany, "Human Enhancement: Making the Debate More Productive" (2013), pp. 992-93; and Gladden, *The Handbook of Information Security for Advanced Neuroprosthetics* (2015), p. 151.

[89] The conscience can be understood as a set of metavolitions, or desires about the kinds of volitions that a person wishes to possess. See Calverley (2008) and Gladden, *The Handbook of Information Security for Advanced Neuroprosthetics* (2015), pp. 151-52. To the extent that a neuroprosthetic device enhances processes of memory and emotion that allow for the development of the conscience, it may enhance one's ability to develop, discern, and follow one's conscience.

[90] Regarding the potential use of neuroprosthetic devices for such purposes, see Gladden, "Neural Implants as Gateways" (2016). For a description of common cognitive biases and their impact on

Similarly, a motor neuroprosthetic device might grant its user enhanced control over his or her existing biological body, expand the user's body to incorporate new devices (such as an exoskeleton or robotic vehicle) through body schema engineering, replace most of the user's existing biological body with electromechanical components to turn the individual into a cyborg,[91] allow the user to control external networked physical systems such as drones or 3D printers, or provide the host with a radically nonhuman body for use in sensing and manipulating a virtual environment.[92]

3. Virtual Reality

In principle, a virtual reality system may be capable of creating a fully immersive visual, auditory, olfactory, gustatory, and tactile environment that its human user would find impossible to qualitatively distinguish from the real world, if the system is capable of presenting either roughly 200 Gbps of raw sense data to the body's sensory organs (such as the retina, hair cells in the ear, and taste buds) through their external stimulation or roughly 250 Mbps of already-processed sense data in the form of direct electrochemical stimulation either of the nerves (such as the optic and cochlear nerves) that carry such data to the brain or of the relevant brain regions themselves.[93] Such fully immersive – and potentially continuous and long-term – virtual reality experiences could be facilitated through the use of advanced neuroprosthetic devices that provide a human brain with all of its sense data, perhaps aided by the use of genetic engineering to make the brain or sensory organs better suited to receive input from such devices.[94]

There is no logical necessity for these fully immersive virtual worlds to resemble our real world in all respects: within a virtual world, human beings might be given new kinds of sensory capacities[95] or even radically nonhuman bodies.[96] Moreover, the laws of physics and biology that hold sway within the real world need not apply in a virtual world; the designers of such worlds

organizational decision-making, see Kinicki & Williams, *Management: A Practical Introduction* (2010), pp. 217-19.

[91] See Lebedev (2014) and Berner (2004), p. 16.

[92] Gladden, "Cybershells, Shapeshifting, and Neuroprosthetics: Video Games as Tools for Posthuman 'Body Schema (Re)Engineering'" (2015).

[93] See Berner (2004), pp. 37-38, 45-47.

[94] On implantable systems for augmented or virtual reality, see Sandor et al., "Breaking the Barriers to True Augmented Reality" (2015), pp. 5-6. Regarding the theoretical possibilities and limits of such technologies, taking into account human physiological and psychological constraints, see Gladden, "Cybershells, Shapeshifting, and Neuroprosthetics" (2015).

[95] See Merkel et al. (2007).

[96] Such possibilities are explored in Gladden, "Cybershells, Shapeshifting, and Neuroprosthetics" (2015).

could formulate their own cultural, social, biological, physical, and even logical and ontological principles that govern or mediate the interactions of subjects and objects within a virtual world. For example, a world designer might decide that within a particular virtual world all human beings, all computers possessing artificial general intelligence, and some of the more intelligent forms of animals represented within it are able to instantaneously share their thoughts and emotions with one another through a form of 'telepathy,' thereby creating new kinds of communal creativity, thought, and agency.[97]

Such technologies could potentially have significant negative consequences; for example, particularly immersive and stimulating virtual environments may become addictive, with their users unable or unwilling to leave them.[98] Moreover, if a user possesses a permanently implanted virtual reality device that is able to alter or replaces its host's sensory perceptions, it may be impossible for the user to know which (if any) of the sense data that he or she is experiencing corresponds to some actual element of an external physical environment and which is 'virtual' or simply 'false'; such an individual may lose the ability (and perhaps desire) to distinguish between real and virtual experiences and worlds.[99]

4. Genetic Engineering, Medicine, and Life Extension

Notwithstanding the many serious questions about whether such applications are ontologically coherent and ethically acceptable, as a practical matter

---

[97] Such options available to the designers of virtual worlds in immersive and long-term multisensory VR environments are discussed in Gladden, "Cybershells, Shapeshifting, and Neuroprosthetics" (2015), and Gladden, "'Upgrading' the Human Entity: Cyberization as a Path to Posthuman Utopia or Digital Annihilation?" (2015).

[98] Regarding the ramifications of long-term immersion in virtual reality environments, see, e.g., Heim, *The Metaphysics of Virtual Reality* (1993); Koltko-Rivera, "The potential societal impact of virtual reality" (2005); and Bainbridge, *The Virtual Future* (2011). Regarding the danger of 'toxic immersion' in a virtual world, see Castronova, *Synthetic Worlds: The Business and Culture of Online Games* (2005). See also Berner (2004), p. 16, and Gladden, *The Handbook of Information Security for Advanced Neuroprosthetics* (2015), pp. 55-56.

[99] For the possibility that a device designed to receive raw data from an external environment could have that data replaced with other data transmitted from some external information system, see Koops & Leenes (2012). Regarding the possibility of neuroprosthetic devices being used to provide false data or information to their hosts or users, see McGee (2008), p. 221, and Gladden, *The Handbook of Information Security for Advanced Neuroprosthetics* (2015). For an analysis of the relationship between physical and virtual reality and ways in which entities can move between these worlds, see Kedzior, "How Digital Worlds Become Material: An Ethnographic and Netnographic Investigation in Second Life" (2014). For more general analyses of the phenomenon of virtual reality, see, e.g., *Communication in the Age of Virtual Reality*, edited by Biocca & Levy (1995); *Cybersociety 2.0: Revisiting Computer-Mediated Communication and Community*, edited by Jones (1998); and Lyon, "Beyond Cyberspace: Digital Dreams and Social Bodies" (2001).

scholars expect that new techniques for genetic engineering will eventually be used, for example, to produce a continually refreshed inventory of personalized replacement organs that can be implanted when their human host's previous organs 'wear out' – or even organs that regenerate themselves within their host's body.[100] It is also anticipated that gene therapy will be employed not simply to replace damaged body components with healthy replicas but to modify the form and functioning of an individual's body or to create new human beings who possess particularly desirable characteristics.[101]

Some scholars expect that the use of medical technologies for radical life extension will become more widespread even as the availability of such technologies remains restricted for legal, ethical, financial, or cultural reasons. Those individuals who possess access to such technologies may be allowed to extend their life indefinitely (in whatever form such a life might take) and may be permitted and expected to choose the time of their own death.[102]

Genetic engineering may also be used to create new forms of sensory, motor, or computing devices within the human body. For example, a neuroprosthetic device need not be electronic in nature: ongoing developments in fields such as genetic engineering, synthetic biology, bionanotechnology, and biomolecular computing are expected to make possible the creation of neuroprosthetic devices that are partially or wholly composed of biological material (perhaps based on the DNA of the device's host) or other non-electronic components.[103] Other advances in medical technology may involve the use of more traditional electronics and robotics. For example, a swarm of nanorobots that has been injected or ingested may travel to a specific location within the body to perform surgery, clean clogged arteries, or modify or stimulate neurons to

[100] See Berner (2004), p. 61, and Ferrando (2013), p. 27.

[101] For a range of perspectives on such possibilities, see, e.g., Berner (2004), p. 17; Panno, *Gene Therapy: Treating Disease by Repairing Genes* (2005); Mehlman, *Transhumanist Dreams and Dystopian Nightmares: The Promise and Peril of Genetic Engineering* (2012); Bostrom, "Human Genetic Enhancements: A Transhumanist Perspective" (2012); Lilley, *Transhumanism and Society: The Social Debate over Human Enhancement* (2013); and De Melo-Martín, "Genetically Modified Organisms (GMOs): Human Beings" (2015).

[102] For a discussion of various approaches to human life extension, see Koene, "Embracing Competitive Balance: The Case for Substrate-Independent Minds and Whole Brain Emulation" (2012). See also Berner (2004), pp. 16-17, and Ferrando (2013), p. 27.

[103] Such technologies are discussed, e.g., in Ummat et al., "Bionanorobotics: A Field Inspired by Nature" (2005); Andrianantoandro et al., "Synthetic biology: new engineering rules for an emerging discipline" (2006); Cheng & Lu, "Synthetic biology: an emerging engineering discipline" (2012); Lamm & Unger, *Biological Computation* (2011); and Berner (2004), pp. 15, 18, 31, 61-62. For a hybrid biological-electronic interface device that includes a network of cultured neurons, see Rutten et al., "Neural Networks on Chemically Patterned Electrode Arrays: Towards a Cultured Probe" (2007). Hybrid biological-electronic interface devices are also discussed by Stieglitz in "Restoration of Neurological Functions by Neuroprosthetic Technologies: Future Prospects and Trends towards Micro-, Nano-, and Biohybrid Systems" (2007).

create new information within neural networks.[104] Ingestible robotic pills might be used to evaluate an individual's internal biological processes and to administer precise dosages of drugs according to complex criteria.[105]

More futuristic and contentious is the concept of 'mind uploading' as a means of extending the life (or if not the life, then in some sense the 'agency') of a particular human being by somehow copying or transferring the structures and processes of his or her mind from their original biological substrate to a new electronic form – for example, by gradually replacing all of a brain's original biological neurons with electronic artificial neurons. Many scholars argue that while it may, for example, be possible to copy the data that comprise the contents of a mind's memories to some external system, it is impossible to transfer or extend the conscious awareness of the mind itself in such a fashion. Nevertheless, some transhumanist proponents of mind uploading argue that such a process would not truly destroy the consciousness or essence of its human host – and that even if it did, they would be willing to transform their own bodies in this fashion, insofar as it might provide a bridge that would allow them to duplicate their memories and patterns of mental activity in a robotic or computerized body that could survive indefinitely.[106]

B. Technologies for Synthetic Agency: Robotics, AI, and Artificial Life

Ongoing rapid developments are expected in those fields such as robotics, artificial intelligence, and artificial life that involve the creation of entities that possess artificial agency and which are able to receive data from their environment, process information, select a course of action, and act to influence their world. For example, research within the field of artificial intelligence is expected to yield artificial agents that possess human-like levels of intelligence, creativity, learning capacity, sociality, and cultural knowledge and which will eventually claim to possess consciousness and their own spirituality.[107] Such artificial agents might be capable of serving as charismatic leaders

---

[104] Medical and other applications of such technologies are discussed in Spohrer (2002); Berner (2004), pp. 18, 76; Pearce (2012); and Ferrando (2013), p. 27.

[105] Berner (2004), p. 76.

[106] For different perspectives on techniques such as the use of artificial neurons to gradually replace the natural biological neurons within a living human brain as a means of effecting 'mind uploading,' see Moravec, *Mind Children: The Future of Robot and Human Intelligence* (1990); Hanson, "If uploads come first: The crack of a future dawn" (1994); Proudfoot, "Software Immortals: Science or Faith?" (2012); Koene (2012); Pearce (2012); and Ferrando (2013), p. 27.

[107] Regarding the prospect of robots and AIs that possess truly human-like cognitive capacities, see Friedenberg (2008) and Berner (2004), pp. 16-17, 38. For discussion of robots that interact socially with human beings, see Breazeal, "Toward sociable robots" (2003); Kanda and Ishiguro, *Human-Robot Interaction in Social Robotics* (2013); *Social Robots and the Future of Social Relations*, edited by Seibt et al. (2014); *Social Robots from a Human Perspective*, edited by Vincent et al.

of human beings by utilizing their powers of persuasion, inspiration, and interpersonal attractiveness,[108] and they may be able to draw on their social capacities and cultural knowledge to serve, for example, as the managers of vast global virtual teams of human workers.[109]

Significant changes are also expected regarding the physical substrates upon which robots and AI platforms are based, as it becomes possible to design systems utilizing components that are increasingly miniaturized, spatially dispersed, and biological; no longer will an artificially intelligent software-based system be chained to the electronic physical substrate found in traditional computers.[110] Entirely new kinds of robots and AI systems may become possible thanks to emerging technologies for physical neural networks,[111] photonic computing, quantum computing, the use of DNA for digital data storage and computing, and other kinds of biocomputing.[112] Thanks to advances in nanorobotics, robots will come to outnumber human beings and become truly ubiquitous: through the use of piezoelectric components, nanoscale switches and sensors can be created that require no electrical power source, allowing clouds of nanorobots to float on the air and fill the space around us with an invisible mesh of sensors, actuators, and information-processors.[113] Such swarms of customized nanorobots might be sent into dangerous environments to aid with disaster relief or to conduct military operations,[114] and moving beyond

---

(2015); and *Social Robots: Boundaries, Potential, Challenges*, edited by Nørskov (2016). Regarding elements that must be present in order for a computerized device to develop its own spirituality, see, e.g., Geraci, "Spiritual robots: Religion and our scientific view of the natural world" (2006); Nahin, "Religious Robots" (2014); and Section 6.2.3.2 on "Religion for Robots" in Yampolskiy, *Artificial Superintelligence: A Futuristic Approach* (2015).

[108] See Gladden, "The Social Robot as 'Charismatic Leader'" (2014).

[109] Regarding potential managerial roles for robots and AIs, see Samani & Cheok, "From human-robot relationship to robot-based leadership" (2011); Samani et al. (2012); and Gladden, "Leveraging the Cross-Cultural Capacities of Artificial Agents" (2014). Regarding the possibility of 'supersocial' AIs that can simultaneously maintain social relations with massive numbers of human colleagues or subordinates, see, e.g., Gladden, "Managerial Robotics" (2014).

[110] Regarding the evolving physical form of robots, see, e.g., Gladden, "The Diffuse Intelligent Other" (2016), and Berner (2004), p. 16.

[111] Regarding AIs that utilize physical neural networks rather than running as an executable software program on a conventional computer employing a Von Neumann architecture, see, e.g., Snider, "Cortical Computing with Memristive Nanodevices" (2008); Versace & Chandler, "The Brain of a New Machine" (2010); and *Advances in Neuromorphic Memristor Science and Applications*, edited by Kozma et al. (2012).

[112] For discussion of DNA-based and biological computing, see, e.g., Berner (2004), pp. 15, 18, 31, 61-62; Ummat et al. (2005); Andrianantoandro et al. (2006); Lamm & Unger (2011); Church et al., "Next-generation digital information storage in DNA" (2012); and Cheng & Lu (2012).

[113] Berner (2004), pp. 16, 18, 38, 40-41.

[114] See Coeckelbergh, "From Killer Machines to Doctrines and Swarms, or Why Ethics of Military

today's relatively simple 3D printing systems, portable (perhaps even handheld) manufacturing facilities could be created that employ specialized swarms of nanorobots to produce highly sophisticated physical goods.[115]

Ongoing developments in the fields of synthetic biology, bionanotechnology, biologically inspired robotics, soft robotics, evolutionary robotics, and artificial life are expected to result in robotic systems whose structures and dynamics resemble those of living organisms and ecosystems or are even composed of biological material. For example, researchers envision the development of robotic systems controlled not by a traditional CPU-based computer but by a synthetic brain;[116] autonomous robots that can learn, adapt, reproduce themselves, and evolve through competition for resources within a digital-physical ecosystem;[117] autonomous computer networks that function as a living entity[118] that possesses its own immune system and whose remaining networked components are able to automatically take over the work of a member computer that has been disconnected or destroyed;[119] and software programs that can repair damage to themselves or even reprogram themselves to accomplish a new purpose, as well as computer chips or entire robots that can intentionally repair or automatically heal damage to themselves.[120] Emerging technologies are expected to eventually allow the development of 'biological operating systems' for groups of cells and entire organisms as well as the design of entirely new species[121] that could be understood alternatively as either artificial biological organisms or biological robots.

Together, technologies that create advanced synthetic agents such as social robots, artificial general intelligences, and artificial life-forms are expected to drive an ongoing posthumanization of organizations' **members** (e.g., by allowing such nonhuman entities to serve as organizational members alongside or instead of human beings), **structures** (e.g., by allowing optimized decision-

---

Robotics Is Not (Necessarily) About Robots" (2011), and Berner (2004), pp. 16-17.

[115] Berner (2004), p. 17.

[116] See Warwick (2014) and Berner (2004), p. 17.

[117] See Gladden, "The Artificial Life-Form as Entrepreneur" (2014), and Berner (2004), pp. 16, 18.

[118] Regarding collectively conscious computer networks, see Callaghan, "Micro-Futures" (2014). For a future Internet that is technically 'self-aware' (if not subjectively conscious), see Galis et al., "Management Architecture and Systems for Future Internet Networks" (2009), pp. 112-13. A sentient Internet is also discussed in Porterfield, "Be Aware of Your Inner Zombie" (2010), p. 19. For a future Internet whose degree of self-awareness resembles that of a living entity, see Hazen, "What is life?" (2006). See also Gladden, "The Artificial Life-Form as Entrepreneur" (2014).

[119] See Berner (2004), pp. 17, 31.

[120] Berner (2004), pp. 17-18. Regarding self-maintenance and self-healing as one capacity that robotic systems must possess in order to be fully autonomous, see Gladden, "The Diffuse Intelligent Other" (2016).

[121] Berner (2004), pp. 16, 61. See also the discussion in Friedenberg (2008), pp. 201-03, of essential elements that must be present in order for an artificial entity to be 'alive,' which are based on the criteria for biological life presented in Curtis, *Biology* (1983).

making and reporting structures designed through genetic algorithms that are free from human cognitive biases and limitations), **systems** (e.g., by allowing the development of organizational systems that are operated by synthetic beings with high speed and accuracy, without the need for human workers to enter data or access information through the slow and error-prone processes of reading printed text), **processes** (e.g., by allowing an organization's synthetic members to analyze data and make decisions faster, more accurately, or more imaginatively than is possible for human beings), **spaces** (e.g., by eliminating the need for physical facilities whose atmosphere, temperature, radiation levels, and other characteristics can sustain human life), and **external ecosystems** (e.g., by creating external resource-providers and consumers that are synthetic beings whose needs and capacities differ widely from those of human beings).

C. Technologies for Digital-Physical Ecosystems and Networks: Connectivity, Relationships, and Knowledge

Many technological changes are either underway or expected that do not relate exclusively to human or artificial agents but which instead shape the larger networks and ecosystems within which all intelligent agents interact. Through the incorporation into the Internet of all public knowledge that has been generated by the human species, the expansion of the Internet of Things to encompass a growing variety and number of networked devices (including ubiquitous sensors conducting real-time surveillance),[122] and the use of RFID or other technologies to assign a unique identifier to any physical object, cyberspace can in effect become a virtual representation of the entire world.[123] Successor networks to the current-day Internet may serve as a mesh that creates a digital-physical ecosystem tying together all kinds of intelligent agents that are able to access the network through biological, electronic, or other means, including unmodified 'natural' human beings, genetically engineered human beings, human beings with extensive cybernetic augmentations, human minds that dwell permanently within virtual realities, social robots, artificially intelligent software, nanorobot swarms, and sapient networks.[124] Within such vast, complex digital ecosystems, most communication will no

[122] This evolution in the Internet of Things is discussed in Evans, "The Internet of Everything: How More Relevant and Valuable Connections Will Change the World" (2012).

[123] See Berner (2004), pp. 18, 35, and Gladden, "Utopias and Dystopias as Cybernetic Information Systems" (2015).

[124] Cybernetic networks that can link such entities are discussed in Gladden, "Utopias and Dystopias as Cybernetic Information Systems" (2015).

longer involve human beings but will take place between networked devices,[125] as real-time data mining is performed by automated systems to continually unearth new theoretical, historical, and predictive knowledge.[126] Some researchers expect that so close will be the symbiotic[127] integration of computerized networks with their natural environment that it may be possible to 'reboot' entire ecosystems as needed, in order to save or improve the lives of their inhabitants.[128]

In particular, neuroprosthetic devices may serve as gateways that unite the human and electronic inhabitants of a digital-physical ecosystem, allowing their human hosts to participate in new kinds of technologically mediated social relations and structures that were previously impossible – perhaps including new forms of merged agency[129] or cybernetic networks that display utopian (or dystopian) characteristics that are not possible for non-neuroprosthetically-enabled societies.[130] Neuroprosthetic devices may also link hosts or users in ways that form communication and information systems[131] that can generate greater collective knowledge, skills, and wisdom than are possessed by any individual member of the system.[132] Because this ubiquitous digital-physical mesh of networked neuroprosthetic devices, sensors, actuators, data pools, and servers will allow human and synthetic minds to exchange thoughts with one another in a manner that seems direct, instantaneous, and

---

[125] See Berner (2004), p. 18, and Evans (2012).

[126] See Berner (2004), p. 32. Existing semi-automated data-mining processes are described, e.g., in Giudici, *Applied Data Mining: Statistical Methods for Business and Industry* (2003), and Provost & Fawcett, *Data Science for Business* (2013), p. 7. Regarding the prospects of developing more fully autonomous AI systems for data mining, see, for example, Warkentin et al. (2012); Bannat et al. (2011), pp. 152-55; and Wasay et al. (2015).

[127] For a philosophical exploration (drawing on Actor-Network Theory) of ways in which nonhuman and human actors coexisting within digital-physical ecosystems might enter into 'symbioses' that are not simply metaphorical but are instead true symbiotic relationships, see Kowalewska (2016).

[128] This possibility is raised in Berner (2004), p. 16.

[129] See McGee (2008), p. 216, and Koops & Leenes (2012), pp. 125, 132.

[130] Different forms that such societies might take are discussed in Gladden, "Utopias and Dystopias as Cybernetic Information Systems" (2015).

[131] The intentional or *ad hoc* creation of such systems is discussed, e.g., in McGee (2008), p. 214; Koops & Leenes (2012), pp. 128-29; Gasson (2012), p. 24; and Gladden, "'Upgrading' the Human Entity" (2015).

[132] The dynamics through which this can occur are discussed, e.g., in Wiener, *Cybernetics: Or Control and Communication in the Animal and the Machine* (1961), loc. 3070ff., 3149ff.; Gladden, "Utopias and Dystopias as Cybernetic Information Systems" (2015); and Gladden, *The Handbook of Information Security for Advanced Neuroprosthetics* (2015), pp. 160-61.

unmediated and to control physical systems and objects and virtual environments, it will create what is, for practical purposes, a 'quasi-magical' world in which beings demonstrate functional telepathy and telekinesis.[133]

Such technological change will not only result in a posthumanization of the larger **external ecosystems** within which organizations exist; it will also spur an ongoing posthumanization of organizations' **members** (e.g., by increasing or decreasing members' sensory input, span of motor control, and social interaction with other intelligent nodes within the environment), **structures** (e.g., by allowing decision-making and reporting relations to be overlaid on top of naturally existing cybernetic relationships created between members within the environment), **systems** (e.g., by providing free or fee-based public information systems that can be utilized by an organization), **processes** (e.g., by allowing an organization to develop its own customized processes or exploit SaaS-based approaches that utilize the environment's publically accessible cloud infrastructure), and **spaces** (e.g., by creating ready-made physical and virtual spaces that an organization can move into and adapt for its own ends).

VI. Conclusion

The relationship of posthumanist thought to organizational studies and management is a topic that is increasingly worth exploring, thanks largely to the ongoing acceleration and intensification of technological change that is fashioning a new organizational context which can appropriately be described as 'posthuman.' Within this text, we have attempted to advance the development of this new sphere of academic inquiry and management practice by presenting one approach to formulating a systematic organizational posthumanism.

We began by noting that established forms of posthumanism could be divided into analytic types that view posthumanity as an existing sociotechnological reality that is best understood from a post-dualist and post-anthropocentric perspective and synthetic types that view posthumanity as a kind of future entity whose creation can either be intentionally brought about or avoided. Similarly, established forms of posthumanism can be understood as either theoretical or practical in nature, depending on whether their goal is to expand human knowledge or generate some concrete impact in the world. We have argued that organizational posthumanism combines analytic, synthetic, theoretical, and practical elements as a type of hybrid posthumanism. It is analytic and theoretical insofar as it attempts to identify and understand the

[133] See Berner (2004), pp. 16-17, 38; Gladden, "Cybershells, Shapeshifting, and Neuroprosthetics" (2015); and the potential indistinguishability of advanced technology and magic, as famously discussed in Clarke, "Hazards of Prophecy: The Failure of Imagination" (1973), p. 36.

ways in which contemporary organizations' structures and dynamics are being affected by emerging sociotechnological realities, and it is synthetic and practical insofar as its goal is to fashion a new 'posthuman entity' not in the form of a genetically or neuroprosthetically augmented human being but in the form of organizations that can survive and thrive within a rapidly evolving posthumanized ecosystem. Building on concepts from the field of organizational architecture, six particular aspects of organizations were identified that are likely to be impacted by ongoing posthumanization: namely, an organization's members, structures, information systems, processes, physical and virtual spaces, and external environment. Finally, we explored the manner in which technologies for human augmentation and enhancement, synthetic agency, and the construction of digital-physical ecosystems and networks are expected to increasingly drive the development of organizational posthumanity. It is our hope that this investigation of the ways in which a current and emerging posthumanity is transforming the shape, dynamics, and roles of organizations will both raise new questions and offer a path to developing creative insights that can inform the work of those who seek to understand the nature of organizations and those who are charged with managing them now and in the future.

References

Advances in Neuromorphic Memristor Science and Applications, edited by Robert Kozma, Robinson E. Pino, and Giovanni E. Pazienza. Dordrecht: Springer Science+Business Media, 2012.

Andrianantoandro, Ernesto, Subhayu Basu, David K. Karig, and Ron Weiss. "Synthetic biology: new engineering rules for an emerging discipline." *Molecular Systems Biology* 2, no. 1 (2006).

Automated Scheduling and Planning: From Theory to Practice, edited by A. Şima Etaner-Uyar, Ender Özcan, and Neil Urquhart. Springer Berlin Heidelberg, 2013.

Badmington, Neil. "Cultural Studies and the Posthumanities." In *New Cultural Studies: Adventures in Theory,* edited by Gary Hall and Claire Birchall, pp. 260-72. Edinburgh: Edinburgh University Press, 2006.

Bainbridge, William Sims. *The Virtual Future.* London: Springer, 2011.

Bannat, Alexander, Thibault Bautze, Michael Beetz, Juergen Blume, Klaus Diepold, Christoph Ertelt, Florian Geiger, et al. "Artificial Cognition in Production Systems." *IEEE Transactions on Automation Science and Engineering* 8, no. 1 (2011): 148-74.

Barile, Nello. "From the Posthuman Consumer to the *Ontobranding* Dimension: Geolocalization, Augmented Reality and Emotional Ontology as a Radical Redefinition of What Is Real." *intervalla: platform for intellectual exchange,* vol. 1 (2013). http://www.fus.edu/intervalla/images/pdf/9_barile.pdf. Accessed May 18, 2016.

Barile, S., J. Pels, F. Polese, and M. Saviano. "An Introduction to the Viable Systems Approach and Its Contribution to Marketing," *Journal of Business Market Management* 5(2) (2012): 54-78 (2012).

Beer, Stafford. *Brain of the Firm.* 2nd ed. New York: John Wiley, 1981.

Bendle, Mervyn F. "Teleportation, cyborgs and the posthuman ideology." *Social Semiotics* 12, no. 1 (2002): 45-62.

Berman, Sandra. "Posthuman Law: Information Policy and the Machinic World." *First Monday* 7, no. 12 (2002). http://www.ojphi.org/ojs/index.php/fm/article/view/1011/932. Accessed March 15, 2016.

Berner, Georg. *Management in 20XX: What Will Be Important in the Future – A Holistic View.* Erlangen: Publicis Corporate Publishing, 2004.

Birnbacher, Dieter. "Posthumanity, Transhumanism and Human Nature." In *Medical Enhancement and Posthumanity,* edited by Bert Gordijn and Ruth Chadwick, pp. 95-106. The International Library of Ethics, Law and Technology 2. Springer Netherlands, 2008.

Bostrom, Nick. "A History of Transhumanist Thought." *Journal of Evolution and Technology* vol. 14, no. 1 (2005). http://jetpress.org/volume14/bostrom.html.

Bostrom, Nick. "Human Genetic Enhancements: A Transhumanist Perspective." In *Arguing About Bioethics,* edited by Stephen Holland, pp. 105-15. New York: Routledge, 2012.

Bostrom, Nick. "Why I Want to Be a Posthuman When I Grow Up." In *Medical Enhancement and Posthumanity,* edited by Bert Gordijn and Ruth Chadwick, pp. 107-37. The International Library of Ethics, Law and Technology 2. Springer Netherlands, 2008.

Boulter, Jonathan. *Parables of the Posthuman: Digital Realities, Gaming, and the Player Experience.* Detroit: Wayne State University Press, 2015.

Bourgeois, III, L.J., Daniel W. McAllister, and Terence R. Mitchell. "The Effects of Different Organizational Environments upon Decisions about Organizational Structure." *Academy of Management Journal,* vol. 21, no. 3 (1978), pp. 508-14.

Bradshaw, Jeffrey M., Paul Feltovich, Matthew Johnson, Maggie Breedy, Larry Bunch, Tom Eskridge, Hyuckchul Jung, James Lott, Andrzej Uszok, and Jurriaan van Diggelen. "From Tools to Teammates: Joint Activity in Human-Agent-Robot Teams." In *Human Centered Design,* edited by Masaaki Kurosu, pp. 935-44. Lecture Notes in Computer Science 5619. Springer Berlin Heidelberg, 2009.

Breazeal, Cynthia. "Toward sociable robots." *Robotics and Autonomous Systems* 42 (2003): 167-75.

Brickley, James A., Clifford W. Smith, and Jerold L. Zimmerman. "Corporate Governance, Ethics, and Organizational Architecture." *Journal of Applied Corporate Finance* 15, no. 3 (2003): 34-45.

Cadle, James, Debra Paul, and Paul Turner. *Business Analysis Techniques: 72 Essential Tools for Success.* Swindon: British Informatics Society Limited, 2010.

Callaghan, Vic. "Micro-Futures." Presentation at Creative-Science 2014, Shanghai, China, July 1, 2014.

Calverley, D.J. "Imagining a non-biological machine as a legal person." *AI & SOCIETY* 22, no. 4 (2008): 523-37.

Castronova, Edward. *Synthetic Worlds: The Business and Culture of Online Games.* Chicago: The University of Chicago Press, 2005.

Catechism of the Catholic Church, Second Edition. Washington, DC: United States Conference of Catholic Bishops, 2016.

Cheng, Allen A., and Timothy K. Lu. "Synthetic biology: an emerging engineering discipline." *Annual Review of Biomedical Engineering* 14 (2012): 155-78.

Church, George M., Yuan Gao, and Sriram Kosuri. "Next-generation digital information storage in DNA." *Science* 337, no. 6102 (2012): 1628.

Clancy, Frank. "At Military's Behest, Darpa Uses Neuroscience to Harness Brain Power." *Neurology Today* 6, no. 2 (2006): 4-8.

Clarke, Arthur C. "Hazards of Prophecy: The Failure of Imagination." In *Profiles of the Future: An Inquiry into the Limits of the Possible,* revised edition. New York: Harper & Row, 1973.

Coeckelbergh, Mark. "Can We Trust Robots?" *Ethics and Information Technology* 14 (1) (2012): 53-60. doi:10.1007/s10676-011-9279-1.

Coeckelbergh, Mark. "From Killer Machines to Doctrines and Swarms, or Why Ethics of Military Robotics Is Not (Necessarily) About Robots." *Philosophy & Technology* 24, no. 3 (2011): 269-78.

Communication in the Age of Virtual Reality, edited by Frank Biocca and Mark R. Levy. Hillsdale, NJ: Lawrence Erlbaum Associates, Publishers, 1995.

Cosgrove, G.R. "Session 6: Neuroscience, brain, and behavior V: Deep brain stimulation." Meeting of the President's Council on Bioethics. Washington, DC, June 24-25, 2004. https://bioethicsarchive.georgetown.edu/pcbe/transcripts/june04/session6.html. Accessed June 12, 2015.

Cudworth, Erika, and Stephen Hobden. "Complexity, ecologism, and posthuman politics." *Review of International Studies* 39, no. 03 (2013): 643-64.

Curtis, H. *Biology,* 4th edition. New York: Worth, 1983.

Cybersociety 2.0: Revisiting Computer-Mediated Communication and Community, edited by Steven G. Jones. Thousand Oaks: Sage Publications, 1998.

Daft, Richard L. *Management.* Mason, OH: South-Western / Cengage Learning, 2011.

Daft, Richard L., Jonathan Murphy, and Hugh Willmott. *Organization Theory and Design.* Andover, Hampshire: Cengage Learning EMEA, 2010.

Datteri, E. "Predicting the Long-Term Effects of Human-Robot Interaction: A Reflection on Responsibility in Medical Robotics." *Science and Engineering Ethics* vol. 19, no. 1 (2013): 139-60.

De Melo-Martín, Inmaculada. "Genetically Modified Organisms (GMOs): Human Beings." In *Encyclopedia of Global Bioethics*, edited by Henk ten Have. Springer Science+Business Media Dordrecht. Version of March 13, 2015. doi: 10.1007/978-3-319-05544-2_210-1. Accessed January 21, 2016.

Del Val, Jaime, and Stefan Lorenz Sorgner. "A *Metahumanist* Manifesto." *The Agonist* IV no. II (Fall 2011). http://www.nietzschecircle.com/AGONIST/2011_08/METAHUMAN_MANIFESTO.html. Accessed March 2, 2016.

Del Val, Jaime, Stefan Lorenz Sorgner, and Yunus Tuncel. "Interview on the Metahumanist Manifesto with Jaime del Val and Stefan Lorenz Sorgner." *The Agonist* IV no. II (Fall 2011). http://www.nietzschecircle.com/AGONIST/2011_08/Interview_Sorgner_Stefan-Jaime.pdf. Accessed March 2, 2016.

Evans, Dave. "The Internet of Everything: How More Relevant and Valuable Connections Will Change the World." Cisco Internet Solutions Business Group: Point of View, 2012. https://www.cisco.com/web/about/ac79/docs/innov/IoE.pdf. Accessed December 16, 2015.

Fairclough, S.H. "Physiological Computing: Interfacing with the Human Nervous System." In *Sensing Emotions*, edited by J. Westerink, M. Krans, and M. Ouwerkerk, pp. 1-20. Philips Research Book Series 12. Springer Netherlands, 2010.

Falconer, Bruce. "Defense Research Agency Seeks to Create Supersoldiers," *Government Executive*, November 10, 2003. http://www.govexec.com/defense/2003/11/defense-research-agency-seeks-to-create-supersoldiers/15386/. Accessed May 22, 2016.

Ferrando, Francesca. "Posthumanism, Transhumanism, Antihumanism, Metahumanism, and New Materialisms: Differences and Relations." *Existenz: An International Journal in Philosophy, Religion, Politics, and the Arts* 8, no. 2 (Fall 2013): 26-32.

Ford, Martin. *Rise of the Robots: Technology and the Threat of a Jobless Future*. New York: Basic Books, 2015.

Friedenberg, Jay. *Artificial Psychology: The Quest for What It Means to Be Human*, Philadelphia: Psychology Press, 2008.

Fukuyama, Francis. *Our Posthuman Future: Consequences of the Biotechnology Revolution*. New York: Farrar, Straus, and Giroux, 2002.

Galis, Alex, Spyros G. Denazis, Alessandro Bassi, Pierpaolo Giacomin, Andreas Berl, Andreas Fischer, Hermann de Meer, J. Srassner, S. Davy, D. Macedo, G. Pujolle, J. R. Loyola, J. Serrat, L. Lefevre, and A. Cheniour. "Management Architecture and Systems for Future Internet Networks." In *Towards the Future Internet: A European Research Perspective*, edited by Georgios Tselentis, John Domingue, Alex Galis, Anastasius Gavras, David Hausheer, Srdjan Krco, Volkmar Lotz, and Theodore Zahariadis, pp. 112-22. IOS Press, 2009.

Gasson, M.N. "Human ICT Implants: From Restorative Application to Human Enhancement." In *Human ICT Implants: Technical, Legal and Ethical Considerations*, edited by Mark N. Gasson, Eleni Kosta, and Diana M. Bowman, pp. 11-28. Information Technology and Law Series 23. T. M. C. Asser Press, 2012.

Gasson, M.N. "ICT implants." In *The Future of Identity in the Information Society*, edited by S. Fischer-Hübner, P. Duquenoy, A. Zuccato, and L. Martucci, pp. 287-95. Springer US, 2008.

Gasson, M.N., Kosta, E., and Bowman, D.M. "Human ICT Implants: From Invasive to Pervasive." In *Human ICT Implants: Technical, Legal and Ethical Considerations*, edited by Mark N. Gasson, Eleni Kosta, and Diana M. Bowman, pp. 1-8. Information Technology and Law Series 23. T. M. C. Asser Press, 2012.

Gephart, Jr., Robert P. "Management, Social Issues, and the Postmodern Era." In *Postmodern Management and Organization Theory*, edited by David M. Boje, Robert P. Gephart, Jr., and Tojo Joseph Thatchenkery, pp. 21-44. Thousand Oaks, CA: Sage Publications, 1996.

Geraci, Robert M. "Spiritual robots: Religion and our scientific view of the natural world." *Theology and Science* 4, issue 3 (2006). doi: 10.1080/14746700600952993.

Giudici, P. *Applied Data Mining: Statistical Methods for Business and Industry*. Wiley, 2003.

Gladden, Matthew E. "The Artificial Life-Form as Entrepreneur: Synthetic Organism-Enterprises and the Reconceptualization of Business." In *Proceedings of the Fourteenth International Conference on the Synthesis and Simulation of Living Systems*, edited by Hiroki Sayama, John Rieffel, Sebastian Risi, René Doursat and Hod Lipson, pp. 417-18. The MIT Press, 2014.

Gladden, Matthew E. "Cybershells, Shapeshifting, and Neuroprosthetics: Video Games as Tools for Posthuman 'Body Schema (Re)Engineering'." Keynote presentation at the Ogólnopolska Konferencja Naukowa Dyskursy Gier Wideo, Facta Ficta / AGH, Kraków, June 6, 2015.

Gladden, Matthew E. "The Diffuse Intelligent Other: An Ontology of Nonlocalizable Robots as Moral and Legal Actors." In *Social Robots: Boundaries, Potential, Challenges*, edited by Marco Nørskov, pp. 177-98. Farnham: Ashgate, 2016.

Gladden, Matthew E. "From Stand Alone Complexes to Memetic Warfare: Cultural Cybernetics and the Engineering of Posthuman Popular Culture." Presentation at the 50 Shades of Popular Culture International Conference. Facta Ficta / Uniwersytet Jagielloński, Kraków, February 19, 2016.

Gladden, Matthew E. *The Handbook of Information Security for Advanced Neuroprosthetics*, Indianapolis: Synthypnion Academic, 2015.

Gladden, Matthew E. "Leveraging the Cross-Cultural Capacities of Artificial Agents as Leaders of Human Virtual Teams." *Proceedings of the 10th European Conference on Management Leadership and Governance*, edited by Visnja Grozdanić, pp. 428-35. Reading: Academic Conferences and Publishing International Limited, 2014.

Gladden, Matthew E. "Managerial Robotics: A Model of Sociality and Autonomy for Robots Managing Human Beings and Machines." *International Journal of Contemporary Management* 13, no. 3 (2014), pp. 67-76.

Gladden, Matthew E. "Neural Implants as Gateways to Digital-Physical Ecosystems and Posthuman Socioeconomic Interaction." In *Digital Ecosystems: Society in the*

Digital Age, edited by Łukasz Jonak, Natalia Juchniewicz, and Renata Włoch, pp. 85-98. Warsaw: Digital Economy Lab, University of Warsaw, 2016.

Gladden, Matthew E. "The Social Robot as 'Charismatic Leader': A Phenomenology of Human Submission to Nonhuman Power." In *Sociable Robots and the Future of Social Relations: Proceedings of Robo-Philosophy 2014*, edited by Johanna Seibt, Raul Hakli, and Marco Nørskov, pp. 329-39. Frontiers in Artificial Intelligence and Applications 273. IOS Press, 2014.

Gladden, Matthew E. "'Upgrading' the Human Entity: Cyberization as a Path to Posthuman Utopia or Digital Annihilation?" Lecture in the Arkana Fantastyki lecture cycle, Centrum Informacji Naukowej i Biblioteka Akademicka (CINiBA), Katowice, May 27, 2015.

Gladden, Matthew E. "Utopias and Dystopias as Cybernetic Information Systems: Envisioning the Posthuman Neuropolity." *Creatio Fantastica* nr 3 (50) (2015).

Goicoechea, María. "The Posthuman Ethos in Cyberpunk Science Fiction." *CLCWeb: Comparative Literature and Culture* 10, no. 4 (2008): 9. http://docs.lib.purdue.edu/cgi/viewcontent.cgi?article=1398&context=clcweb. Accessed May 18, 2016.

Graham, Elaine. "Post/Human Conditions." *Theology & Sexuality* 10:2 (2004), pp. 10-32.

Graham, Elaine. *Representations of the Post/Human: Monsters, Aliens and Others in Popular Culture.* Manchester: Manchester University Press, 2002.

Gray, Chris Hables. *Cyborg Citizen: Politics in the Posthuman Age*, London: Routledge, 2002.

Green, Steven J. "Malevolent gods and Promethean birds: Contesting augury in Augustus's Rome." In *Transactions of the American Philological Association*, vol. 139, no. 1, pp. 147-167. The Johns Hopkins University Press, 2009.

Grodzinsky, F.S., K.W. Miller, and M.J. Wolf. "Developing Artificial Agents Worthy of Trust: 'Would You Buy a Used Car from This Artificial Agent?'" *Ethics and Information Technology* 13, no. 1 (2011): 17-27.

Hamilton, S. "What Is Roman Ornithomancy? A Compositional Analysis of an Ancient Roman Ritual." *Proceedings of the Multidisciplinary Graduate Research Conference.* Lethbridge: University of Lethbridge Graduate Students Association, 2007.

Han, J.-H., S.A. Kushner, A.P. Yiu, H.-W. Hsiang, T. Buch, A. Waisman, B. Bontempi, R.L. Neve, P.W. Frankland, and S.A. Josselyn. "Selective Erasure of a Fear Memory." *Science* 323, no. 5920 (2009): 1492-96.

Hanson, R. "If uploads come first: The crack of a future dawn." *Extropy* 6, no. 2 (1994): 10-15.

Haraway, Donna. "A Manifesto for Cyborgs: Science, Technology, and Socialist Feminism in the 1980s." *Socialist Review* vol. 15, no. 2 (1985), pp. 65-107.

Haraway, Donna. *Simians, Cyborgs, and Women: The Reinvention of Nature.* New York: Routledge, 1991.

Hayles, N. Katherine. *How We Became Posthuman: Virtual Bodies in Cybernetics, Literature, and Informatics*, Chicago: University of Chicago Press, 1999.

Hazen, Robert. "What is life?", *New Scientist* 192, no. 2578 (2006): 46-51.

Heim, Michael. *The Metaphysics of Virtual Reality.* New York: Oxford University Press, 1993.

Herbrechter, Stefan. *Posthumanism: A Critical Analysis.* London: Bloomsbury, 2013. [Kindle edition.]

Heylighen, Francis. "The Global Brain as a New Utopia." In *Renaissance der Utopie. Zukunftsfiguren des 21. Jahrhunderts*, edited by R. Maresch and F. Rötzer. Frankfurt: Suhrkamp, 2002.

Holacracy Constitution v4.1. Spring City, PA: HolacracyOne, 2015. http://www.holacracy.org/wp-content/uploads/2015/07/Holacracy-Constitution-v4.1.pdf. Accessed May 21, 2016.

Horling, Bryan, and Victor Lesser. "A Survey of Multi-Agent Organizational Paradigms." *The Knowledge Engineering Review* 19, no. 04 (2004): 281-316.

Kanda, Takayuki, and Hiroshi Ishiguro. *Human-Robot Interaction in Social Robotics.* Boca Raton: CRC Press, 2013.

Kedzior, Richard. *How Digital Worlds Become Material: An Ethnographic and Netnographic Investigation in Second Life.* Economics and Society: Publications of the Hanken School of Economics Nr. 281. Helsinki: Hanken School of Economics, 2014.

Kelly, Kevin. "The Landscape of Possible Intelligences." *The Technium*, September 10, 2008. http://kk.org/thetechnium/the-landscape-o/. Accessed January 25, 2016.

Kelly, Kevin. *Out of Control: The New Biology of Machines, Social Systems and the Economic World.* Basic Books, 1994.

Kelly, Kevin. "A Taxonomy of Minds." *The Technium*, February 15, 2007. http://kk.org/thetechnium/a-taxonomy-of-m/. Accessed January 25, 2016.

Kinicki, Angelo, and Brian Williams. *Management: A Practical Introduction*, 5th edition. New York: McGraw Hill, 2010.

Koene, Randal A. "Embracing Competitive Balance: The Case for Substrate-Independent Minds and Whole Brain Emulation." In *Singularity Hypotheses*, edited by Amnon H. Eden, James H. Moor, Johnny H. Søraker, and Eric Steinhart, pp. 241-67. The Frontiers Collection. Springer Berlin Heidelberg, 2012.

Koltko-Rivera, Mark E. "The potential societal impact of virtual reality." *Advances in virtual environments technology: Musings on design, evaluation, and applications* 9 (2005).

Koops, B.-J., and R. Leenes. "Cheating with Implants: Implications of the Hidden Information Advantage of Bionic Ears and Eyes." In *Human ICT Implants: Technical, Legal and Ethical Considerations*, edited by Mark N. Gasson, Eleni Kosta, and Diana M. Bowman, pp. 113-34. Information Technology and Law Series 23. T. M. C. Asser Press, 2012.

Kourany, J.A. "Human Enhancement: Making the Debate More Productive." *Erkenntnis* 79, no. 5 (2013): 981-98.

Kowalewska, Agata. "Symbionts and Parasites – Digital Ecosystems." In *Digital Ecosystems: Society in the Digital Age*, edited by Łukasz Jonak, Natalia Juchniewicz, and Renata Włoch, pp. 73-84. Warsaw: Digital Economy Lab, University of Warsaw, 2016.

Kraemer, Felicitas. "Me, Myself and My Brain Implant: Deep Brain Stimulation Raises Questions of Personal Authenticity and Alienation." *Neuroethics* 6, no. 3 (May 12, 2011): 483-97. doi:10.1007/s12152-011-9115-7.

Krzywinska, Tanya, and Douglas Brown. "Games, Gamers and Posthumanism." In *The Palgrave Handbook of Posthumanism in Film and Television*, pp. 192-201. Palgrave Macmillan UK, 2015.

Lamm, Ehud, and Ron Unger. *Biological Computation*, Boca Raton: CRC Press, 2011.

Lebedev, M. "Brain-Machine Interfaces: An Overview." *Translational Neuroscience* 5, no. 1 (March 28, 2014): 99-110.

Lee, Joseph. "Cochlear implantation, enhancements, transhumanism and posthumanism: some human questions." *Science and engineering ethics* 22, no. 1 (2016): 67-92.

Lilley, Stephen. *Transhumanism and Society: The Social Debate over Human Enhancement.* Springer Science & Business Media, 2013.

Lune, Howard. *Understanding Organizations*, Cambridge: Polity Press, 2010.

Lyon, David. "Beyond Cyberspace: Digital Dreams and Social Bodies." In *Education and Society*, third edition, edited by Joseph Zajda, pp. 221-38. Albert Park: James Nicholas Publishers, 2001.

Maguire, Gerald Q., and Ellen M. McGee. "Implantable brain chips? Time for debate." *Hastings Center Report* 29, no. 1 (1999): 7-13.

Mara, Andrew, and Byron Hawk. "Posthuman rhetorics and technical communication." *Technical Communication Quarterly* 19, no. 1 (2009): 1-10.

Mazlish, Bruce. *The Fourth Discontinuity: The Co-Evolution of Humans and Machines.* New Haven: Yale University Press, 1993.

McGee, E.M. "Bioelectronics and Implanted Devices." In *Medical Enhancement and Posthumanity*, edited by Bert Gordijn and Ruth Chadwick, pp. 207-24. The International Library of Ethics, Law and Technology 2. Springer Netherlands, 2008.

McIntosh, Daniel. "The Transhuman Security Dilemma." *Journal of Evolution and Technology* 21, no. 2 (2010): 32-48.

Medical Enhancement and Posthumanity, edited by Bert Gordijn and Ruth Chadwick. The International Library of Ethics, Law and Technology 2. Springer Netherlands, 2008.

Mehlman, Maxwell J. *Transhumanist Dreams and Dystopian Nightmares: The Promise and Peril of Genetic Engineering.* Baltimore: The Johns Hopkins University Press, 2012.

Merkel, R., G. Boer, J. Fegert, T. Galert, D. Hartmann, B. Nuttin, and S. Rosahl. "Central Neural Prostheses." In *Intervening in the Brain: Changing Psyche and Society*, 117-60. Ethics of Science and Technology Assessment 29. Springer Berlin Heidelberg, 2007.

Miah, Andy. "A Critical History of Posthumanism." In *Medical Enhancement and Posthumanity*, edited by Bert Gordijn and Ruth Chadwick, pp. 71-94. The International Library of Ethics, Law and Technology 2. Springer Netherlands, 2008.

Miller, Jr., Gerald Alva. "Conclusion: Beyond the Human: Ontogenesis, Technology, and the Posthuman in Kubrick and Clarke's 2001." In *Exploring the Limits of the Human through Science Fiction*, pp. 163-90. American Literature Readings in the 21st Century. Palgrave Macmillan US, 2012.

Moravec, Hans. *Mind Children: The Future of Robot and Human Intelligence*. Cambridge: Harvard University Press, 1990.

Moreno, Jonathan. "DARPA On Your Mind." *Cerebrum* vol. 6 issue 4 (2004): 92-100.

Nadler, David, and Michael Tushman. *Competing by Design: The Power of Organizational Architecture*. Oxford University Press, 1997. [Kindle edition.]

Nahin, Paul J. "Religious Robots." In *Holy Sci-Fi!*, pp. 69-94. Springer New York, 2014.

Nazario, Marina. "I went to Best Buy and encountered a robot named Chloe – and now I'm convinced she's the future of retail." *Business Insider*, October 23, 2015. http://www.businessinsider.com/what-its-like-to-use-best-buys-robot-2015-10. Accessed May 20, 2016.

Nourbakhsh, Illah Reza. "The Coming Robot Dystopia." *Foreign Affairs* 94, no. 4 (2015): 23.

Panno, Joseph. *Gene Therapy: Treating Disease by Repairing Genes*. New York: Facts on File, 2005.

Park, M.C., M.A. Goldman, T.W. Belknap, and G.M. Friehs. "The Future of Neural Interface Technology." In *Textbook of Stereotactic and Functional Neurosurgery*, edited by A.M. Lozano, P.L. Gildenberg, and R.R. Tasker, 3185-3200. Heidelberg/Berlin: Springer, 2009.

Pearce, David. "The Biointelligence Explosion." In *Singularity Hypotheses*, edited by A.H. Eden, J.H. Moor, J.H. Søraker, and E. Steinhart, pp. 199-238. The Frontiers Collection. Berlin/Heidelberg: Springer, 2012.

Perez-Marin, D., and I. Pascual-Nieto. *Conversational Agents and Natural Language Interaction: Techniques and Effective Practices*. Hershey, PA: IGI Global, 2011.

Perlberg, James. *Industrial Robotics*. Boston: Cengage Learning, 2016.

Porterfield, Andrew. "Be Aware of Your Inner Zombie." *ENGINEERING & SCIENCE* (Fall 2010): 14-19.

Posthuman Bodies, edited by Judith Halberstam and Ira Livingstone. Bloomington, IN: Indiana University Press, 1995.

Posthumanist Shakespeares, edited by Stefan Herbrechter and Ivan Callus. Palgrave Shakespeare Studies. Palgrave Macmillan UK, 2012.

Pride, W., R. Hughes, and J. Kapoor. *Foundations of Business*, 4e. Stamford, CT: Cengage Learning, 2014.

Proudfoot, Diane. "Software Immortals: Science or Faith?" In *Singularity Hypotheses*, edited by Amnon H. Eden, James H. Moor, Johnny H. Søraker, and Eric Steinhart, pp. 367-92. The Frontiers Collection. Springer Berlin Heidelberg, 2012.

Provost, Foster, and Tom Fawcett. *Data Science for Business*. Sebastopol, CA: O'Reilly Media, Inc., 2013.

Puranam, Phanish, Marlo Raveendran, and Thorbjørn Knudsen. "Organization Design: The Epistemic Interdependence Perspective." *Academy of Management Review* 37, no. 3 (July 1, 2012): 419-40.

Ramirez, S., X. Liu, P.-A. Lin, J. Suh, M. Pignatelli, R.L. Redondo, T.J. Ryan, and S. Tonegawa. "Creating a False Memory in the Hippocampus." *Science* 341, no. 6144 (2013): 387-91.

Robertson, Brian J. *Holacracy: The New Management System for a Rapidly Changing World.* New York: Henry Holt, 2015.

Robinett, W. "The consequences of fully understanding the brain." In *Converging Technologies for Improving Human Performance: Nanotechnology, Biotechnology, Information Technology and Cognitive Science*, edited by M.C. Roco and W.S. Bainbridge, pp. 166-70. National Science Foundation, 2002.

Roden, David. *Posthuman Life: Philosophy at the Edge of the Human*. Abingdon: Routledge, 2014.

Rutten, W. L. C., T. G. Ruardij, E. Marani, and B. H. Roelofsen. "Neural Networks on Chemically Patterned Electrode Arrays: Towards a Cultured Probe." In *Operative Neuromodulation*, edited by Damianos E. Sakas and Brian A. Simpson, pp. 547-54. Acta Neurochirurgica Supplements 97/2. Springer Vienna, 2007.

Sachs, Jeffrey D., Seth G. Benzell, and Guillermo LaGarda. "Robots: Curse or Blessing? A Basic Framework." NBER Working Papers Series, Working Paper 21091. Cambridge, MA: National Bureau of Economic Research, 2015.

Samani, Hooman Aghaebrahimi, and Adrian David Cheok. "From human-robot relationship to robot-based leadership." In *2011 4th International Conference on Human System Interactions (HSI)*, pp. 178-81. IEEE, 2011.

Samani, Hooman Aghaebrahimi, Jeffrey Tzu Kwan Valino Koh, Elham Saadatian, and Doros Polydorou. "Towards Robotics Leadership: An Analysis of Leadership Characteristics and the Roles Robots Will Inherit in Future Human Society." In *Intelligent Information and Database Systems*, edited by Jeng-Shyang Pan, Shyi-Ming Chen, and Ngoc Thanh Nguyen, pp. 158-65. Lecture Notes in Computer Science 7197. Springer Berlin Heidelberg, 2012.

Sandor, Christian, Martin Fuchs, Alvaro Cassinelli, Hao Li, Richard Newcombe, Goshiro Yamamoto, and Steven Feiner. "Breaking the Barriers to True Augmented Reality." arXiv preprint, *arXiv:1512.05471 [cs.HC]*, December 17, 2015. http://arxiv.org/abs/1512.05471. Accessed January 25, 2016.

Sarkar, Prabhat Rainjan. "Neohumanism Is the Ultimate Shelter (Discourse 11)." In *The Liberation of Intellect: Neohumanism*. Kolkata: Ananda Marga Publications, 1982.

Schmeink, Lars. "Dystopia, Alternate History and the Posthuman in Bioshock." *Current Objectives of Postgraduate American Studies* 10 (2009). http://copas.uni-regensburg.de/article/viewArticle/113/137. Accessed May 19, 2016.

Short, Sue. *Cyborg Cinema and Contemporary Subjectivity*. New York: Palgrave Macmillan, 2005.

Singularity Hypotheses, edited by A.H. Eden, J.H. Moor, J.H. Søraker, and E. Steinhart. The Frontiers Collection. Berlin/Heidelberg: Springer, 2012.

Snider, Greg S. "Cortical Computing with Memristive Nanodevices." *SciDAC Review* 10 (2008): 58-65.

Social Robots and the Future of Social Relations, edited by Johanna Seibt, Raul Hakli, and Marco Nørskov. Amsterdam: IOS Press, 2014.

Social Robots from a Human Perspective, edited by Jane Vincent, Sakari Taipale, Bartolomeo Sapio, Giuseppe Lugano, and Leopoldina Fortunati. Springer International Publishing, 2015.

Social Robots: Boundaries, Potential, Challenges, edited by Marco Nørskov. Farnham: Ashgate Publishing, 2016.

Soussou, Walid V., and Theodore W. Berger. "Cognitive and Emotional Neuroprostheses." In *Brain-Computer Interfaces*, pp. 109-23. Springer Netherlands, 2008.

Sparrow, R. "Killer Robots." *Journal of Applied Philosophy* vol. 24, no. 1 (2007): 62-77.

Spohrer, Jim. "NBICS (Nano-Bio-Info-Cogno-Socio) Convergence to Improve Human Performance: Opportunities and Challenges." In *Converging Technologies for Improving Human Performance: Nanotechnology, Biotechnology, Information Technology and Cognitive Science*, edited by M.C. Roco and W.S. Bainbridge, pp. 101-17. Arlington, Virginia: National Science Foundation, 2002.

Stahl, B. C. "Responsible Computers? A Case for Ascribing Quasi-Responsibility to Computers Independent of Personhood or Agency." *Ethics and Information Technology* 8, no. 4 (2006): 205-13.

Stieglitz, Thomas. "Restoration of Neurological Functions by Neuroprosthetic Technologies: Future Prospects and Trends towards Micro-, Nano-, and Biohybrid Systems." In *Operative Neuromodulation*, edited by Damianos E. Sakas, Brian A. Simpson, and Elliot S. Krames, 435-42. Acta Neurochirurgica Supplements 97/1. Springer Vienna, 2007.

Thacker, Eugene. "Data made flesh: biotechnology and the discourse of the posthuman." *Cultural Critique* 53, no. 1 (2003): 72-97.

Thomsen, Mads Rosendahl. *The New Human in Literature: Posthuman Visions of Change in Body, Mind and Society after 1900*. London: Bloomsbury, 2013.

Ummat, Ajay, Atul Dubey, and Constantinos Mavroidis. "Bionanorobotics: A Field Inspired by Nature." In *Biomimetics: Biologically Inspired Technologies*, edited by Yoseph Bar-Cohen, pp. 201-26. Boca Raton: CRC Press, 2005.

Versace, Massimiliano, and Ben Chandler. "The Brain of a New Machine." *IEEE spectrum* 47, no. 12 (2010): 30-37.

Warkentin, Merrill, Vijayan Sugumaran, and Robert Sainsbury. "The Role of Intelligent Agents and Data Mining in Electronic Partnership Management." *Expert Systems with Applications* 39, no. 18 (2012): 13277-88.

Warwick, K. "The Cyborg Revolution." *Nanoethics* 8 (2014): 263-73.

Warwick, K., and M. Gasson. "Implantable Computing." In *Digital Human Modeling*, edited by Y. Cai, pp. 1-16. Lecture Notes in Computer Science 4650. Berlin/Heidelberg: Springer, 2008.

Wasay, A., M. Athanassoulis, and S. Idreos. "Queriosity: Automated Data Exploration." In *2015 IEEE International Congress on Big Data (BigData Congress)*, pp. 716-19, 2015. doi:10.1109/BigDataCongress.2015.116.

Wiener, Norbert. *Cybernetics: Or Control and Communication in the Animal and the Machine*, second edition. Cambridge, MA: The MIT Press, 1961. [Quid Pro ebook edition for Kindle, 2015.]

Wiltshire, Travis J., Dustin C. Smith, and Joseph R. Keebler. "Cybernetic Teams: Towards the Implementation of Team Heuristics in HRI." In *Virtual Augmented and Mixed Reality. Designing and Developing Augmented and Virtual Environments*, edited by Randall Shumaker, pp. 321-30. Lecture Notes in Computer Science 8021. Springer Berlin Heidelberg, 2013.

Wolf-Meyer, Matthew. "Fantasies of extremes: Sports, war and the science of sleep." *BioSocieties* 4, no. 2 (2009): 257-71.

Yampolskiy, Roman V. "The Universe of Minds." arXiv preprint, *arXiv:1410.0369 [cs.AI]*, October 1, 2014. http://arxiv.org/abs/1410.0369. Accessed January 25, 2016.

Yampolskiy, Roman V. *Artificial Superintelligence: A Futuristic Approach*. Boca Raton: CRC Press, 2015.

Yonck, Richard. "Toward a standard metric of machine intelligence." *World Future Review* 4, no. 2 (2012): 61-70.

Chapter Three

The Posthuman Management Matrix: Understanding the Organizational Impact of Radical Biotechnological Convergence[1]

Abstract. In this text we present the Posthuman Management Matrix, a model for understanding the ways in which organizations of the future will be affected by the blurring – or even dissolution – of boundaries between human beings and computers. In this model, an organization's employees and consumers can include two different kinds of agents (human and artificial) who may possess either of two sets of characteristics (anthropic or computer-like); the model thus defines four types of possible entities. For millennia, the only type of relevance for management theory and practice was that of human agents who possess anthropic characteristics – i.e., natural human beings. During the 20th Century, the arrival of computers and industrial robots made relevant a second type: that of artificial agents possessing computer-like characteristics.

Management theory and practice have traditionally overlooked the remaining two types of possible entities – human agents possessing computer-like physical and cognitive characteristics (which can be referred to as 'cyborgs') and artificial agents possessing anthropic physical and cognitive characteristics (which for lack of a more appropriate term might be called 'bioroids') – because such agents did not yet exist to serve as employees or consumers for organizations. However, in this text we argue that ongoing developments in neuroprosthetics, genetic engineering, virtual reality, robotics, and artificial intelligence are indeed giving rise to such types

---

[1] This chapter was originally published in Gladden, Matthew E., *Sapient Circuits and Digitalized Flesh: The Organization as Locus of Technological Posthumanization*, pp. 133-201, Indianapolis: Defragmenter Media, 2016.

of agents and that new spheres of management theory and practice will be needed to allow organizations to understand the operational, legal, and ethical issues that arise as their pools of potential workers and customers evolve to include human beings whose bodies and minds incorporate ever more computerized elements and artificial entities that increasingly resemble biological beings.

By analyzing the full spectrum of human, computerized, and hybrid entities that will constitute future organizations, the Posthuman Management Matrix highlights ways in which established disciplines such as cybernetics, systems theory, organizational design, and enterprise architecture can work alongside new disciplines like psychological engineering, AI resource management, metapsychology, and exoeconomics to help organizations anticipate and adapt to posthumanizing technological and social change.

I. Introduction

Facilitated by ongoing technological developments in fields like neuroprosthetics, genetic engineering, social robotics, nanorobotics, and artificial intelligence, a growing convergence between sapient biological entities like human beings and electronic computerized systems is underway. Looking beyond the current reality in which human beings interact with technological instruments that mediate so many of our daily activities, researchers anticipate a future in which human persons themselves *become* technological instruments. Human beings who display carefully engineered architectures,[2] electromechanical physical components,[3] software-guided cognitive processes,[4] and digitally mediated interactions[5] will increasingly resemble computers – and they

[2] See, e.g., Canton, "Designing the future: NBIC technologies and human performance enhancement" (2004); De Melo-Martín, "Genetically Modified Organisms (GMOs): Human Beings" (2015); Nouvel, "A Scale and a Paradigmatic Framework for Human Enhancement" (2015); and Bostrom, "Human Genetic Enhancements: A Transhumanist Perspective" (2012). Regarding 'brain engineering,' see Gross, "Traditional vs. modern neuroenhancement: notes from a medico-ethical and societal perspective" (2011).

[3] Regarding expected future growth in the use of implantable electronic neuroprosthetic devices for purposes of human enhancement, see, e.g., McGee, "Bioelectronics and Implanted Devices" (2008), and Gasson, "Human ICT Implants: From Restorative Application to Human Enhancement" (2012).

[4] For the potential use of an electronic 'brain pacemaker' to regulate cognitive activity, see Naufel, "Nanotechnology, the Brain, and Personal Identity" (2013). Regarding possible manipulation of the human brain's activity through the use of computerized neuroprosthetic devices, see Viirre et al., "Promises and perils of cognitive performance tools: A dialogue" (2008), and Heinrichs, "The promises and perils of non-invasive brain stimulation" (2012).

[5] See, e.g., *Communication in the Age of Virtual Reality*, edited by Biocca & Levy (1995); *Cybersociety 2.0: Revisiting Computer-Mediated Communication and Community*, edited by Jones (1998); and Lyon, "Beyond Cyberspace: Digital Dreams and Social Bodies" (2001).

will share digital-physical ecosystems with computerized systems whose biological or biomimetic components,[6] evolutionary processes,[7] unpredictable neural networks,[8] and physically mediated social relations[9] cause them to ever more closely resemble human beings.

Such technological and social changes will be so transformative in their effects that they can be understood as creating a world best described as *posthuman*.[10] Within such a post-anthropocentric and post-dualistic environment,[11] it will no longer be natural biological human beings alone who seek

---

[6] See, e.g., Ummat et al., "Bionanorobotics: A Field Inspired by Nature" (2005); Andrianantoandro et al., "Synthetic biology: new engineering rules for an emerging discipline" (2006); Cheng & Lu, "Synthetic biology: an emerging engineering discipline" (2012); Lamm & Unger, *Biological Computation* (2011); Church et al., "Next-generation digital information storage in DNA" (2012); and Berner, *Management in 20XX: What Will Be Important in the Future – A Holistic View* (2004), pp. 15, 18, 31, 61-62.

[7] For a discussion of evolutionary robotics and evolvable robotic hardware, see Friedenberg, *Artificial Psychology: The Quest for What It Means to Be Human* (2008), pp. 206-10.

[8] Regarding factors that make it difficult to analyze or predict the behavior of artificially intelligent systems – especially of distributed artificial intelligences (DAIs) displaying emergent behavior – see Friedenberg (2008), pp. 31-32. For a discussion of the behavior of physical artificial neural networks, see, e.g., Snider, "Cortical Computing with Memristive Nanodevices" (2008); Versace & Chandler, "The Brain of a New Machine" (2010); and *Advances in Neuromorphic Memristor Science and Applications*, edited by Kozma et al. (2012).

[9] For robots that interact socially with human beings, see, e.g., Breazeal, "Toward sociable robots" (2003); Kanda & Ishiguro, *Human-Robot Interaction in Social Robotics* (2013); *Social Robots and the Future of Social Relations*, edited by Seibt et al. (2014); *Social Robots from a Human Perspective*, edited by Vincent et al. (2015); and *Social Robots: Boundaries, Potential, Challenges*, edited by Marco Nørskov (2016). For robots that interact socially with one another, see, e.g., Arkin & Hobbs, "Dimensions of communication and social organization in multi-agent robotic systems" (1993); Barca & Sekercioglu, "Swarm robotics reviewed" (2013); and Brambilla et al., "Swarm robotics: a review from the swarm engineering perspective" (2013).

[10] The processes of posthumanization that expand the boundaries of society to include entities other than natural biological human beings as traditionally understood include the age-old forces of *nontechnological posthumanization* (as reflected in works of critical and cultural posthumanism and fantasy literature) and the newly emerging and intensifying forces of *technological posthumanization*, which is the focus of this text and is explored in works of biopolitical posthumanism, philosophical posthumanism, and science fiction. Regarding nontechnological posthumanization, see, e.g., Graham, *Representations of the Post/Human: Monsters, Aliens and Others in Popular Culture* (2002); Badmington, "Cultural Studies and the Posthumanities" (2006); and Herbrechter, *Posthumanism: A Critical Analysis* (2013). Regarding technological posthumanization, see, e.g., Fukuyama, *Our Posthuman Future: Consequences of the Biotechnology Revolution* (2002); Bostrom, "Why I Want to Be a Posthuman When I Grow Up" (2008); and other texts in *Medical Enhancement and Posthumanity*, edited by Gordijn & Chadwick (2008). For an overview of the forms of posthumanism that take these phenomena as their objects of study and practice, see Ferrando, "Posthumanism, Transhumanism, Antihumanism, Metahumanism, and New Materialisms: Differences and Relations" (2013), and our classification scheme in Part One of this text, "A Typology of Posthumanism: A Framework for Differentiating Analytic, Synthetic, Theoretical, and Practical Posthumanisms."

[11] See Ferrando (2013).

out and create meaning through their exercise of imagination, reason, volition, and conscience; instead the world will likely include a bewildering array of sources of intelligent agency that create meaning through their networks and relations.[12] The implications for organizational management of this dawning 'Posthuman Age' are expected to be vast, and yet they have not yet been comprehensively explored from a theoretical perspective.

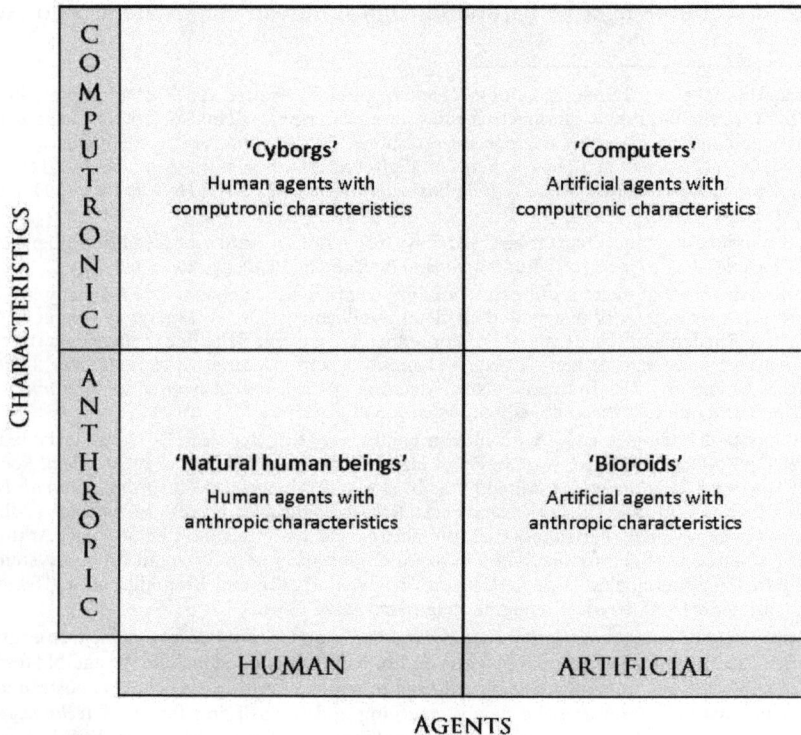

CHARACTERISTICS		
COMPUTRONIC	**'Cyborgs'** Human agents with computronic characteristics	**'Computers'** Artificial agents with computronic characteristics
ANTHROPIC	**'Natural human beings'** Human agents with anthropic characteristics	**'Bioroids'** Artificial agents with anthropic characteristics
	HUMAN	ARTIFICIAL
	AGENTS	

Fig. 1: The Posthuman Management Matrix delineates four types of entities, each of which may be of greater or lesser relevance for the practice of organizational management at a particular point in human history.

In an effort to advance such study, in this text we develop the Posthuman Management Matrix, a two-dimensional model designed to aid management scholars and practitioners in analyzing and anticipating the impacts of posthumanizing technological and social change on organizations. We begin by

[12] See Ferrando (2013).

showing that the agents that are relevant to organizational management can be divided into two varieties (human and artificial agents) and that the traits possessed by a particular agent fall into one of two kinds (which we refer to as 'anthropic' and 'computronic' characteristics[13]). The Matrix thus delineates four general types of possible entities that can potentially serve as workers or consumers for businesses and other organizations. These types of entities are: human agents possessing anthropic characteristics (whom we can refer to simply as 'natural human beings'); artificial agents possessing computronic characteristics (or in other words, conventional 'computers'); human agents possessing computronic characteristics (whom we can refer to as 'cyborgs'); and artificial agents possessing anthropic characteristics (which, for lack of a better term, can be referred to as 'bioroids'[14]). An overview of the four quadrants of the Posthuman Management Matrix and the types of entities that they represent is contained in Figure 1.

The Matrix is then utilized to analyze management theory and practice as they have existed prior to this emerging age of radical technological posthumanization. Beginning from the dawn of human history, the only type of entity relevant to management theory and practice was long that of human agents who possess anthropic characteristics – or in other words, natural human beings who have not been modified through the use of technologies such as neuroprosthetic augmentation or genetic engineering. Only with the arrival of electronic information-processing systems and simple industrial robots in the 20th Century did a second type of entity become broadly relevant for organizational management: that of the artificial agent that possesses computronic characteristics, or the 'computer.'[15] Integrating such computerized systems into an organization of human workers is not an easy task, and management disciplines such as enterprise architecture, IT management, and infor-

[13] In this text we use the portmanteau 'computronic' to refer to physical structures, behaviors, or other phenomena or characteristics which in recent decades have commonly been associated with *computers* and *electronic* devices. This builds on earlier uses of the word found, e.g., in Turner, "The right to privacy in a computronic age" (1970), and Rankin, "Business Secrets Across International Borders: One Aspect of the Transborder Data Flow Debate" (1985).

[14] For use of the term 'bioroid' in an engineering context, see Novaković et al., "Artificial Intelligence and Biorobotics: Is an Artificial Human Being our Destiny?" (2009). Regarding the use of the term in speculative fiction, see, e.g., Pulver, *GURPS Robots* (1995), pp. 74-81, where 'bioroid' is a portmanteau derived explicitly from 'biological android.'

[15] For early examples of workplace robotics explored from the perspective of management theory and practice, see, e.g., Thompson, "The Man-Robot Interface in Automated Assembly" (1976), and Goodman & Argote, "New Technology and Organizational Effectiveness" (1984).

mation security have emerged that provide conceptual frameworks and practical tools for successfully coordinating the actions of human and artificial agents to create effective organizations.[16]

The largest portion of this text is dedicated to employing the Matrix as a means of investigating the remaining two types of entities – 'cyborgs' and 'bioroids' – that have heretofore received relatively little serious attention within the field of management but which are set to become ever more prevalent as workers, managers, consumers, and other organizational stakeholders, thanks to the accelerating and intensifying processes of technological posthumanization. We suggest that it will not be possible to adequately understand and manage the many complex operational, legal, and ethical issues that arise from adopting such posthuman agents as employees or customers simply by relying on existing fields such as HR management, IT management, or enterprise architecture. The radically expanded universe of posthuman agents that will participate in the life of organizations will require the development of new spheres of theory and practice that can address the unique forms, behaviors, strengths, and weaknesses of such agents, along with the ways in which they will combine to create rich and complex cybernetic networks and digital-physical ecosystems. Our exploration of these questions concludes by contemplating the sorts of transdisciplinary management approaches that might be able to successfully account for such organizational systems in which natural human beings, genetically engineered persons, individuals possessing extensive neuroprosthetic augmentation, human beings who spend all of their time dwelling in virtual worlds, social robots, artificially intelligent software, nanorobot swarms, and sentient or sapient networks work together in physical and virtual environments to achieve organizational goals.[17]

Through this formulation, application, and discussion of the Posthuman Management Matrix, we hope to highlight the challenges that await management scholars and practitioners in an increasingly posthumanized world and to suggest one possible conceptual framework that can aid us in making sense of and responding to these challenges.

---

[16] For a review of enterprise architecture frameworks, see Magoulas et al., "Alignment in Enterprise Architecture: A Comparative Analysis of Four Architectural Approaches" (2012), and Rohloff, "Framework and Reference for Architecture Design" (2008); for a practical overview of organizational design, see Burton et al., *Organizational Design: A Step-by-Step Approach* (2015); for an overview of information security, see Rao & Nayak, *The InfoSec Handbook* (2014).

[17] See Gladden, *The Handbook of Information Security for Advanced Neuroprosthetics* (2015), pp. 95-96.

II. Formulating the Posthuman Management Matrix

We would suggest that it is useful to analyze the impact of posthumanizing social and technological change on organizational management through a two-dimensional conceptual framework that creates a coherent tool for identifying, understanding, and anticipating organizational transformations that will occur as a result of the convergences described in this text. We can refer to this proposed framework as the 'Posthuman Management Matrix.' Our hope is that such a model can serve as both a theoretical framework for management scholars as well as a practical tool for management practitioners. The Posthuman Management Matrix comprises two dimensions: the horizontal dimension is that of an 'agent' and the vertical dimension is that of an agent's 'characteristics.' We can consider each of these dimensions in turn.

A. The Matrix's Horizontal Dimension: The Kind of Agent

There are many types of entities and phenomena that must be managed by organizations; however, many of them do not possess or manifest their own agency. Such non-agents include financial assets, land, raw materials, intellectual property, contracts, policies and procedures, and other elements of organizational life that are not capable of gathering data from their environment, processing information, and selecting a course of action.[18]

On the other hand, there are many kinds of agents[19] that may actively participate in an organization's activities; these include typical adult human beings, some kinds of domesticated animals (which, for example, can be employed in particular roles within the fields of agriculture, law enforcement, and entertainment), many types of autonomous and semiautonomous robots, and artificially intelligent software programs that run on particular computing platforms. Note that in order to qualify as an agent, an entity does not need to need to possess the same kind of sapience as a typical adult human being; relatively simple automated systems (such as an assembly-line robot or the software managing an automated customer-service telephone line) can be described as agents, even if they do not possess full human-like artificial general intelligence. Conversely, not all human beings can be considered agents from the managerial perspective, even if they are considered to be legal persons and moral patients; for example, an adult human being who is in a coma and whose mind is not able to receive sensory input, process information, and select and

[18] Within the context of enterprise architecture, for example, both agents and non-agents can be understood generically as 'entities' that play particular 'roles' in various 'activities' within an organization; see Caetano et al., "A Role-Based Enterprise Architecture Framework" (2009).

[19] For an overview of biological, robotic, and software-based agents and their key characteristics of autonomy, social ability, reactivity, and proactivity, see Tweedale & Jain, "Agent Oriented Programming" (2011).

act upon particular courses of action would not be considered an 'agent' in the organizational sense employed here.

Much ongoing research and debate is taking place regarding questions of whether and to what extent collective entities can be considered agents. It is a matter of contention whether a social organization such as a country or a swarm of insects can possess its own 'agency' distinct from the agency of all the individuals that constitute it.[20] In some cases, the law recognizes certain types of social entities (e.g., states or corporations) as possessing a sort of agency independent of that of their human constituents, although different conclusions may be formulated when viewing such entities from an ontological or moral rather than a legal perspective. Similarly, some automated artificial agents have been designed in such a way that they are in fact multi-agent systems composed of a number of smaller subsystems and components that are themselves agents. In such cases, the agency possessed by a multi-agent system as a whole is typically of a different sort from that possessed by its individual components. More complex is the case of large computer-facilitated networks (e.g., the Internet) that can, in a certain sense, be said to select and act upon particular courses of action and whose 'decisions' are shaped by the activities of individual human and artificial agents that have access to the network and who participate in its sensorimotor and information-processing actions.[21]

Traditionally, facilities such as office buildings or warehouses would not in themselves have qualified as 'agents,' even though they were home to the activities of large numbers of agents and contained an extensive technological infrastructure of mechanical, electrical, and other components that were regularly manipulated by those agents as part of their work. However, the rise of the Internet of Things and smart buildings means that in some cases an office

---

[20] Regarding questions about the nature and degree of agency and decision-making responsibility that can be possessed by robotic swarms or networks, see, e.g., Coeckelbergh, "From Killer Machines to Doctrines and Swarms, or Why Ethics of Military Robotics Is Not (Necessarily) About Robots" (2011), pp. 274-75, and Gladden, "The Diffuse Intelligent Other: An Ontology of Nonlocalizable Robots as Moral and Legal Actors" (2016).

[21] Regarding collectively conscious networks and a "post-internet sentient network," see Callaghan, "Micro-Futures" (2014). Regarding a future Internet that is 'self-aware' in a technical and technological sense, even if it is not subjectively conscious, see Galis et al., "Management Architecture and Systems for Future Internet Networks" (2009), pp. 112-13. A sentient Internet is also discussed in Porterfield, "Be Aware of Your Inner Zombie" (2010), p. 19. For a future Internet that is self-aware as a sort of potentially living entity, see Hazen, "What is life?" (2006). Regarding the growing prevalence of robotic systems that comprise networks and swarms – rather than autonomous unitary robots – and the distributed or unclear nature of decision-making and responsibility in such systems, see Coeckelbergh (2011), pp. 272-75, and Gladden, "The Diffuse Intelligent Other" (2016).

building or production facility that includes sufficient sensory and motor components controlled by a computerized system can potentially be understood as a single coherent 'agent.' A similar phenomenon is now occurring with vehicles, which may be considered agents if they possess self-driving capabilities or other forms of AI.[22]

For purposes of the Posthuman Management Matrix, we can divide the broad spectrum of agents that are relevant to contemporary organizational management into two main categories: human beings (described below as 'human agents') and robots or other artificially intelligent computing systems (described below as 'artificial agents').[23]

1. Human Agents

Human agents are intelligent and sapient actors whose agency is grounded in and exercised through the actions of a biological human brain. Throughout history, such human agents have been the primary (and often only) agents constituting human organizations. Human beings possess a distinct set of biological, psychological, social, and cultural properties that have been extensively studied by disciplines including biology, psychology, anthropology, sociology, economics, history, philosophy, theology, political science, and organizational management.

2. Artificial Agents

Artificial agents represent a relatively new kind of intelligent actor that has emerged during recent decades and which has the potential to carry out particular tasks or roles within a human organization. Although the universe of artificial agents comprises a diverse array of entities with a broad variety of forms and functions, artificial agents are similar in that: 1) they all possess some means of receiving data from their environment, a means of processing information, and a means of acting on their environment; and 2) the physical substrate within which their agency subsists is not a natural biological human brain.

An artificial agent often takes the form of a piece of software being executed by some physical computational substrate such as a desktop computer, mobile device, server, robot, or network of distributed devices.[24] However,

---

[22] Regarding the ethical implications of creating autonomous driverless vehicles that can exercise their own agency, see Goodall, "Ethical decision making during automated vehicle crashes" (2014).

[23] The simplified schema presented by the Posthuman Management Matrix thus omits, for example, the explicit consideration of domesticated animals as potential workplace agents.

[24] Each particular instantiation of such a sensorimotor-cognitive system can be understood as a unique artificial agent; thus technically, the same piece of AI software run on two different computers (or even on the same computer on two different occasions) can be understood as two

other examples exist that do not involve the execution of a conventional software program; these include artificial neural networks that are not run as a software program on a conventional CPU-based computer but which comprise a network of physical artificial neurons.[25]

B. The Matrix's Vertical Dimension: An Agent's Characteristics

From the perspective of organizational management, there are two broad sets of characteristics that a contemporary agent might display: 'anthropic characteristics' are those that are traditionally possessed by human beings, and 'computronic characteristics' are those traditionally possessed by artificial agents such as robots or artificially intelligent software. We can consider these two suites of characteristics in greater detail.

1. Anthropic Characteristics

Anthropic characteristics constitute that array of traits which throughout history has been possessed by and associated with human beings. These characteristics are reflected in: 1) an entity's physical form; 2) its capacity for and use of intelligence; and 3) its social interaction with other intelligent agents. Below we use these three perspectives to identify and describe some of the key anthropic characteristics.

a. Physical Form

The physical form of an agent possessing anthropic characteristics demonstrates a number of notable traits. Such an agent is:

Composed of biological components. The body of a human being is naturally composed of biological material and not mechanical or electronic components. The qualities of such biological material place limits on the kinds of work that human employees can perform. For example, it is impossible for human beings to work in areas of extreme heat, cold, or radiation without extensive protection, nor is it possible for a human employee to work for hundreds of consecutive hours without taking breaks for sleep or meals or to use the restroom.

---

different artificial agents. (See Wiener, *Cybernetics: Or Control and Communication in the Animal and the Machine* (1961), loc. 2402ff., for the idea that a human brain with all of its short- and long-term memories are "not the complete analogue of the computing machine but rather the analogue of a single run on such a machine" – something which, by definition, cannot be duplicated in another substrate.) However, the term 'artificial agent' is also used in a looser sense to refer to a hardware-software platform comprising a particular piece of hardware and the AI software that it executes rather than to each separate execution of that software.

[25] See, e.g., Friedenberg (2008), pp. 17-36, for a discussion of different physical models that do not necessarily require a conventional Von Neumann computer architecture.

Alive. In order to function as an agent within an organization, a human being (and the biological subsystems that constitute its body) must be alive. As a living organism, a human being possesses a metabolism that requires a continual supply of resources (e.g., oxygen, water, and food) from the external environment as well as the ability to emit waste products into the environment in order for the individual to survive.[26]

Non-engineered. The basic physical form of a particular human being is determined largely by genotypic factors that are a result of randomized inheritance of genetic material from the individual's biological parents; the individual's particular physical characteristics are not intentionally selected or fabricated by a genetic engineer.[27]

Non-upgradeable. There are many congenital medical conditions that can be treated through conventional surgical procedures, medication, the use of traditional prosthetics, or other therapies. The application of such technologies could be understood as a form of 'augmentation' or 'enhancement' of one's body as it was naturally formed; however, such technologies are more commonly understood as 'restorative' approaches, insofar as they do not grant an individual physical elements or capacities that surpass those possessed by a typical human being.[28] Historically, human beings have not been subject to the sort of radical physical 'upgradeability' that might involve, for example, the implantation of additional memory capacity into the brain, an alteration of the rate of electrochemical communication between neurons to increase the brain's 'processing speed,' the addition of new sensory capacities (e.g., infrared vision), or the addition of new or different limbs or actuators (e.g., wheels instead of legs).[29] This differs from the case of contemporary computers, which often can easily be upgraded through the addition or replacement of physical components.

Confined to a limited lifespan. Although the lifespan of a particular human being can be shortened or extended to some degree as a result of environmental, behavioral, or other factors, the human organism is generally un-

[26] In considering a definition for artificial life, Friedenberg (2008), pp. 201-03, draws on the criteria for biological life presented in Curtis, *Biology* (1983): namely, a living being manifests organization, metabolism, growth, homeostasis, adaptation, response to stimuli, and reproduction.

[27] Although, for example, factors such as diet, exercise and training, environmental conditions, and medicines and medical procedures can extensively modify the form of a human body, the extent to which an existing biological human body can be restructured before ceasing to function is nonetheless relatively limited.

[28] See Gasson (2012).

[29] See Gladden, "Cybershells, Shapeshifting, and Neuroprosthetics: Video Games as Tools for Posthuman 'Body Schema (Re)Engineering'" (2015).

derstood to possess a finite biological lifespan that cannot be extended indefinitely through natural biological means.[30] A human being that has exceeded its maximum lifespan is no longer alive (i.e., it will have expired) and it cannot be repaired and revived by technological means to make it available once again for future organizational use.

Manifesting a developmental cycle. The physical structure and capacities of a human being do not remain unchanged from the moment of an individual's conception to the moment of his or her death; instead, a human being's physical form and abilities undergo continuous change as the individual develops through a cycle of infancy, adolescence, adulthood, and senescence.[31] From the perspective of organizational management, human beings are only capable of serving as employees, partners, or consumers during particular phases of this developmental cycle, and the unique strengths and weaknesses displayed by human workers vary as they move through the developmental cycle.

Possessing a unitary local body. A particular human being occupies or comprises a particular physical biological body. Because this body is unitary – consisting of a single spatially compact unit – a human being is able to inhabit only one space at a given time; a human being cannot simultaneously be physically present in multiple cities, for example.[32]

Possessing a permanent substrate. Although to some limited extent it is possible to modify or replace physical components of a human body, it is not possible for a human being to exchange his or her entire body for another.[33] The body with which a human being was born will – notwithstanding the natural changes that occur as part of its lifelong developmental cycle or any minor intentional modifications – serve as a single permanent substrate within which all of the individual's information processing and cognition will occur and in which all of the individual's sensory and motor activity will take place until the end of his or her life.

---

[30] For a discussion and comparison of biologically and nonbiologically based efforts at human life extension, see Koene, "Embracing Competitive Balance: The Case for Substrate-Independent Minds and Whole Brain Emulation" (2012).

[31] See Thornton, *Understanding Human Development: Biological, Social and Psychological Processes from Conception to Adult Life* (2008), and the *Handbook of Psychology, Volume 6: Developmental Psychology*, edited by Lerner et al. (2003).

[32] For a discussion of different types of bodies and their relation to an entity's degree of locality, see Gladden, "The Diffuse Intelligent Other" (2016).

[33] For complications relating to proposed body-replacement techniques such as mind uploading, see Proudfoot, "Software Immortals: Science or Faith?" (2012); for particular problems that would result from the attempt to adopt a nonhuman body, see Gladden, "Cybershells, Shapeshifting, and Neuroprosthetics" (2015).

Unique and identifiable. A human being's body creates (or at least, plays a necessary role in creating) a single identity for the individual that persists over time, throughout the person's life. The fact that each human body is unique and is identifiable to other human beings (e.g., such a body is not invisible, microscopic, or 'flickering' in and out of existence from moment to moment) means that it is possible to associate human actions with a particular human being who performed them.[34]

b. Intelligence

The information-processing mechanisms and behaviors of an agent possessing anthropic characteristics demonstrate a number of significant traits. Such an agent is:

Sapient and self-aware. A typical human adult possesses a subjective conscious experience that is not simply sensations of physical reality but a conceptual 'awareness of' and 'awareness that.' These characteristics are not found, for example, in infants or in adult human beings suffering from certain medical conditions. In a sense, a typical adult human being can be said to possess sapient self-awareness as a capacity even when the individual is unconscious (e.g., during sleep), although in that moment the capacity is latent and is not being actively utilized or experienced.[35]

Autonomous. Broadly speaking, adult human beings are considered to possess a high degree of autonomy.[36] Through the regular action of its mind and body, a human being is able to secure energy sources and information from its external environment, set goals, make decisions, perform actions, and even (to a limited extent) repair damage that might occur to itself during the course of its activities, all without direct external guidance or control by other human agents. Human beings which, for example, are still infants, are suffering from physical or cognitive impairments (such as being in a coma), or are operating in a hostile or unfamiliar environment may not be able to function with the same degree of autonomy.

---

[34] For an overview of philosophical questions relating to personal identity, see Olson, "Personal Identity" (2015).

[35] For a discussion of such issues, see, e.g., Siewert, "Consciousness and Intentionality" (2011); Fabbro et al., "Evolutionary aspects of self-and world consciousness in vertebrates" (2015); and Boly et al., "Consciousness in humans and non-human animals: recent advances and future directions" (2013).

[36] For a definition of autonomy applicable to agents generally, see Bekey, *Autonomous Robots: From Biological Inspiration to Implementation and Control* (2005), p. 1. Regarding ways of classifying different levels of autonomy, see Gladden, "Managerial Robotics: A Model of Sociality and Autonomy for Robots Managing Human Beings and Machines" (2014).

Metavolitional. Volitionality relates to an entity's ability to self-reflexively shape the intentions that guide its actions.[37] An entity is nonvolitional when it possesses no internal goals or 'desires' for achieving particular outcomes nor any expectations or 'beliefs' about how performing certain actions would lead to particular outcomes. An entity is volitional if it combines goals with expectations: in other words, it can possess an intention,[38] which is a mental state that comprises both a desire and a belief about how some act that the entity is about to perform can contribute to fulfilling that desire.[39] Meanwhile, typical adult human beings can be described as metavolitional: they possess what scholars have referred to as a 'second-order volition,' or an intention *about* an intention.[40] In human beings, this metavolitionality manifests itself in the form of conscience: as a result of possessing a conscience, human agents are able to determine that they do not wish to possess some of the intentions that they are currently experiencing, and they can resolve to change those intentions.

Educated. The cognitive processes and knowledge of a human being are shaped through an initial process of concentrated learning and formal and informal education that lasts for several years and through an ongoing process of learning that lasts throughout the individual's lifetime.[41] Human beings can learn empirically through the firsthand experience of interacting with their environment or by being taught factual information or theoretical knowledge. A human being cannot instantaneously 'download' or 'import' a large body of information into his or her memory in the way that a data file can be copied to a computer's hard drive.

Processing information through a neural network. Some information processing takes part in other parts of the body (e.g., the transduction of proximal stimuli into electrochemical signals by neurons in the sensory organs); however, the majority of a human being's information processing is performed by the neural network comprising interneurons in the individual's brain.[42] The brain constitutes an immensely large and intricate neural network, and despite ongoing advances in the field of neuroscience, profound

---

[37] For a discussion of the volitionality of agents, see Calverley, "Imagining a non-biological machine as a legal person" (2008), pp. 529-535, and Gladden, "The Diffuse Intelligent Other" (2016).

[38] The term 'intentionality' is often employed in a philosophical sense to describe an entity's ability to possess mental states that are directed toward (or 'about') some object; that is a broader phenomenon than the possession of a particular 'intention' as defined here.

[39] Calverley (2008), p. 529.

[40] Calverley (2008), pp. 533-35.

[41] See Thornton (2008), and *Handbook of Psychology, Volume 6* (2003).

[42] For example, see Gladden, *The Handbook of Information Security for Advanced Neuroprosthetics* (2015), pp. 148-49.

mysteries remain regarding the structure and behavior of this neural network's components and of the network as a whole.[43] The mechanisms by which this neural network processes the data provided by sensory input and stored memories to generate motor output and new memories are highly nonlinear and complex; they are not directly comparable to the process of a CPU-based computer running an executable software program.

Emotional. The possession and manifestation of emotions is not an extraneous supplement (or obstacle) to the rational decision-making of human beings but is instead an integral component of it. Some researchers suggest that the possession of emotions is necessary in order for an embodied entity to demonstrate general intelligence at a human-like level.[44]

Cognitively biased. Human beings are subject to a common set of cognitive biases that distort individuals' perceptions of reality and cause them to arrive at decisions that are objectively illogical and suboptimal.[45] While in earlier eras such biases may have created an evolutionary advantage that aided the survival of those beings that possessed them (e.g., by providing them with heuristics that allowed them to quickly identify and avoid potential sources of danger), these biases cause contemporary human workers to err when evaluating factual claims or attempting to anticipate future events or manage risk. To some extent, such biases can be counteracted through conscious awareness, training, and effort.

Possessing a flawed memory. The human mind does not store a perfect audiovisual record of all the sensory input, thoughts, and imaginings that it experiences during a human being's lifetime. The brain's capacities for both the retention and recall of information are limited. Not only are memories stored in a manner which from the beginning is compressed, impressionistic, and imperfect, but memories also degrade over time.[46] Historically, the only way to transfer memories stored within one human mind to another human mind has been for the memories to be described and expressed through some social mechanism such as oral speech or written text.

Demonstrating unpredictable behavior. All human beings demonstrate basic similarities in their behavior, and individual human beings possess

---

[43] For example, significant outstanding questions remain about the potentially holonomic nature of memory storage within the brain and the role of inter- and intraneuronal structures in memory creation and storage; see, e.g., Longuet-Higgins, "Holographic Model of Temporal Recall" (1968); Pribram, "Prolegomenon for a Holonomic Brain Theory" (1990); and Pribram & Meade, "Conscious Awareness: Processing in the Synaptodendritic Web – The Correlation of Neuron Density with Brain Size" (1999).

[44] See Friedenberg (2008), pp. 179-200.

[45] For an overview of human cognitive biases in relation to organizational management, see Kinicki & Williams, *Management: A Practical Introduction* (2010), pp. 217-19.

[46] See Dudai, "The Neurobiology of Consolidations, Or, How Stable Is the Engram?" (2004).

unique personalities, habits, and psychological and medical conditions that allow their reactions to particular stimuli or future behavior to be predicted with some degree of likelihood; however, it is not possible to predict with full precision, accuracy, and certainty the future actions of a particular human being.

Not capable of being hacked electronically. Because human beings possess biological rather than electronic components and their minds conduct information processing through the use of an internal physical neural network rather than a conventional executable software program stored in binary digital form, it is not possible for external adversaries or agents to hack into a human being's body and information-processing system in order to control sensory, motor, or cognitive activities or to access, steal, or manipulate the individual's thoughts or memories using the same electronic hacking techniques that are applied to the hardware or software of electronic computers and computer-based systems.[47]

c. Social Interaction

An agent possessing anthropic characteristics demonstrates a number of noteworthy traits relating to social interaction. Such an agent is:

Social. Human beings display social behaviors, engage in isolated and short-term social interactions, and participate in long-term social relations that evolve over time and are shaped by society's expectations for the social roles to be filled by a particular individual.[48] Although the social content and nature of complex communicative human actions such as speaking and writing are obvious, even such basic activities such as standing, walking, and breathing have social aspects, insofar as they can convey intentions, emotions, and attitudes toward other human beings.

Cultural. Human beings create and exist within unique cultures that include particular forms of art, literature, music, architecture, history, sports and recreation, technology, ethics, philosophy, and theology. Such cultures also develop and enforce norms regarding the ways in which organizations such as businesses should or should not operate.[49]

---

[47] The human mind is subject to other kinds of 'hacking' such as social engineering; see Rao & Nayak (2014).

[48] Regarding the distinction between social behaviors, interactions, and relations, see Vinciarelli et al., "Bridging the Gap between Social Animal and Unsocial Machine: A survey of Social Signal Processing" (2012), and Gladden, "Managerial Robotics" (2014).

[49] Regarding the critical role that organizational culture plays, e.g., in the management of enterprise architecture, see Aier, "The Role of Organizational Culture for Grounding, Management, Guidance and Effectiveness of Enterprise Architecture Principles" (2014), and Hoogervorst, "Enterprise Architecture: Enabling Integration, Agility and Change" (2004).

Spiritual. Human beings broadly manifest a search for and recognition of transcendent reality and ultimate purpose of a form that is described by organized religions and other spiritual and philosophical systems as well as nurtured by the idiosyncratic beliefs and sentiments of individual human beings. Recently researchers have sought to identify biological mechanisms that enable or facilitate the development and expression of such spirituality.[50]

Political. In order to regulate their shared social existence and create conditions that allow for productivity, prosperity, peace, and the common good, human beings have developed political systems for collective defense, decision-making, and communal action. Political activity typically involves a kind and degree of reasoning, debate, strategic thinking, risk assessment, prioritization of values, and long-term planning that is not found, for example, within the societies of nonhuman animals.[51]

An economic actor. In contemporary societies, an individual human being is typically not able to personally produce all of the goods and services needed for his or her survival and satisfaction, and he or she does not have the desire or ability to personally consume all of the goods or services that he or she produces. In order to transform the goods and services that a human being produces into the goods and services that he or she desires to have, human beings engage in economic exchange with one another. Within contemporary societies, businesses and other organizations play critical roles in facilitating such economic interaction.[52]

A legal person. An adult human being is typically recognized by the law as being a legal person who bears responsibility for his or her decisions and actions. In some cases, relevant distinctions exist between legal persons, moral subjects, and moral patients. For example, an adult human being who is conscious and not suffering from psychological or biological impairments would typically be considered both a legal person who is legally responsible for his or her actions as well as a moral subject who bears moral responsibility for those actions. An infant or an adult human being who is in a coma might be considered a legal person who possesses certain legal rights, even though a legal guardian may be appointed to make decisions on the person's behalf;

[50] For example, see Emmons, "Is spirituality an intelligence? Motivation, cognition, and the psychology of ultimate concern" (2000).

[51] Thus Aristotle's assertion that "man is by nature a political animal" (Aristotle, *Politics*, Book 1, Section 1253a). Regarding different perspectives on the organization of animal societies and the possible evolutionary origins of politics in human societies, see, e.g., *Man Is by Nature a Political Animal: Evolution, Biology, and Politics*, edited by Hatemi & McDermott (2011); Alford & Hibbing, "The origin of politics: An evolutionary theory of political behavior" (2004); Clark, *The Political Animal: Biology, Ethics and Politics* (1999); and *Primate Politics*, edited by Schubert & Masters (1991).

[52] For example, see Samuelson & Marks, *Managerial Economics* (2012), Chapter 11.

such a person is not (at the moment) a moral agent who undertakes actions for which he or she bears moral responsibility but is still a 'moral patient' whom other human beings have an obligation to care for and to not actively harm.[53]

2. Computronic Characteristics

Computronic characteristics constitute the collection of traits that have traditionally been possessed by the kinds of computers utilized by organizations, including mainframes, servers, desktop computers, laptop computers, and mobile devices, as well as more specialized devices such as supercomputers, satellites, assembly-line robots, automated guided vehicles, and other computerized systems based on a conventional Von Neumann architecture. These characteristics are reflected in: 1) an entity's physical form; 2) its capacity for and use of intelligence; and 3) its social interaction with other intelligent agents. Below we use these three perspectives to identify and describe some of the key computronic characteristics. It may be noted that in most cases they are very different from – and frequently the opposite of – the anthropic characteristics traditionally associated with human beings.

a. Physical Form

The physical form of an agent possessing computronic characteristics demonstrates a number of notable traits. Such an agent is:

Composed of electronic components. A conventional computer is typically composed of mass-produced electronic components that are durable and readily repairable and whose behavior can easily be analyzed and predicted.[54] Such components are often able to operate in conditions of extreme heat, cold, pressure, or radiation in which biological matter would not be able to survive and function. Such components can be built to a large or microscopic scale, depending on the intended purpose of a particular computer. The ability to manufacture electronic components to precise specifications with little variation means that millions of copies of a single artificial agent can be produced that are functionally identical.

---

[53] Regarding distinctions between legal persons, moral subjects, and moral patients – especially in the context of comparing human and artificial agents – see, e.g., Wallach & Allen, *Moral machines: Teaching robots right from wrong* (2008); Gunkel, *The Machine Question: Critical Perspectives on AI, Robots, and Ethics* (2012); Sandberg, "Ethics of brain emulations" (2014); and Rowlands, *Can Animals Be Moral?* (2012).

[54] For an in-depth review of the historical use of electronic components in computers as well as an overview of emerging possibilities for (non-electronic) biological, optical, and quantum computing, see Null & Lobur, *The Essentials of Computer Organization and Architecture* (2006). Regarding the degree to which the failure of electronic components can be predicted, see Băjenescu & Bâzu, *Reliability of Electronic Components: A Practical Guide to Electronic Systems Manufacturing* (1999).

Not alive. A conventional computer is not alive: it is not created through processes of biological reproduction, and its form and basic functionality are not shaped by a DNA- or RNA-based genotype; nor does the computer itself grow and reproduce.[55] A computer must typically receive energy from the external environment in the form of an electrical power supply that has been specifically prepared by its human operators and which meets exact specifications;[56] the computer does not possess a metabolism that allows it to assimilate raw materials that it obtains from the environment and convert them into energy and structural components, repair damage and grow, and emit waste products into the environment (apart from byproducts such as heat – which is a significant concern in microprocessor and computer design – and stray electromagnetic radiation such as radio waves).[57]

Intentionally designed. Historically, the structure and basic capacities of a computer are not the result of the inheritance of randomized genetic code from biological parents or from other processes of biological reproduction. Instead, all elements and aspects of a traditional computer's physical form and basic functionality are intentionally planned and constructed by human scientists, engineers, manufacturers, and programmers in order to enable the computer to successfully perform particular tasks.[58]

Upgradeable and expandable. The physical structure and capacities of computers are easily expandable through the addition of internal components or external peripheral devices. Such upgrades allow a computer to receive, for example, new sensory mechanisms, new forms of actuators for manipulating the external environment, an increase in processing speed, an increase in random-access memory, or an increase in the size of a computer's available space for the nonvolatile long-term storage of data.[59]

Not limited to a maximum lifespan. A typical computer does not possess a maximum lifespan beyond which it cannot be made to operate. As a practical matter, individual computers may eventually become obsolete because their functional capacities are inadequate to perform tasks that the computer's owner or operator needs it to perform or because cheaper, faster, and more

---

[55] Curtis (1983) cited seven requisites for a biological entity to be considered alive (organization, metabolism, growth, homeostasis, adaptation, response to stimuli, and reproduction), which Friedenberg (2008), pp. 201-03, also considers to be relevant when attempting to determine whether an artificial entity is alive.

[56] Exceptions would include, e.g., solar-powered computing devices.

[57] Such emissions by computers also create information security concerns; see, e.g., Gladden, *The Handbook of Information Security for Advanced Neuroprosthetics* (2015), p. 116.

[58] See, e.g., Dumas, *Computer Architecture: Fundamentals and Principles of Computer Design* (2006).

[59] See, e.g., Mueller, *Upgrading and Repairing PCs, 20th Edition* (2012).

powerful types of computers have become available to carry out those tasks. Similarly, the failure of an individual component within a computer may render it temporarily nonfunctional. However, the ability to repair, replace, upgrade, or expand a computer's physical components means that a computer's operability can generally be maintained indefinitely, if its owner or operator wishes to do so.[60]

Possessing a stable and restorable form. A computer's physical form is highly stable: although a computer's components can be physically upgraded or altered by the device's owner or operator, a computer does not physically upgrade or alter itself without its owner or operator's knowledge or permission.[61] A computer does not undergo the sort of developmental cycle of conception, growth, maturity, and senescence demonstrated by biological organisms. In general, the physical alterations made to a computer are reversible: a chip that has been installed to increase the computer's RAM can be removed; a peripheral device that has been added can be disconnected. This allows a computer to be restored to a previous physical and functional state.

Potentially multilocal. It is possible for a computer to – like a human being – possess a body that comprises a single unitary, spatially compact physical unit: computerized devices such as a typical desktop computer, smartphone, assembly-line robot, or server may possess a physical form that is clearly distinct from the device's surrounding environment and which is located in only a single place at any given time. However, other computers can – unlike a human being – possess a body comprising disjoint, spatially dispersed elements that exist physically in multiple locations at the same time. The creation of such computerized entities comprising many spatially disjoint and dispersed 'bodies' has been especially facilitated in recent decades by the development of the diverse networking technologies that undergird the Internet and, now, the nascent Internet of Things.[62] The destruction, disabling, or disconnection of one of these bodies that contributes to the form of such an entity may not cause the destruction of or a significant degradation of functionality for the computerized entity as a whole.

---

[60] For an overview of issues relating to computer reliability, availability, and lifespan, see Siewiorek & Swarz, *Reliable Computer Systems: Design and Evaluation* (1992), and Băjenescu & Bâzu (1999).

[61] An exception would be the case of computer worms or viruses that can cause a computer to disable or damage some of its internal components or peripheral devices without the owner or operator's knowledge. See, for example, Kerr et al., "The Stuxnet Computer Worm: Harbinger of an Emerging Warfare Capability" (2010).

[62] Regarding the Internet of Things, see Evans, "The Internet of Everything: How More Relevant and Valuable Connections Will Change the World" (2012). For one aspect of the increasingly networked nature of robotics and AI, see Coeckelbergh (2011). Regarding multilocal computers, see Gladden, "The Diffuse Intelligent Other" (2016).

Possessing an exchangeable substrate. Because they are stored in an electronic digital form that can easily be read and written, the data that constitute a particular computer's operating system, applications, configuration settings, activity logs, and other information that has been received, generated, or stored by the device can easily be copied to different storage components or to a different computer altogether. This means that the computational substrate or 'body' of a given computerized system can be replaced with a new body without causing any functional changes in the system's memory or behavior. In the case of computerized systems that are typically accessed remotely (e.g., a cloud-based storage device accessed through the Internet), a system's hardware could potentially be replaced by copying the device's data to a new device without remote users or operators ever realizing that the system's physical computational substrate had been swapped.[63]

Possessing an unclear basis for identity. It is unclear wherein the unique identity of a conventional computer or computerized entity subsists, or even if such an identity exists.[64] A computer's identity does not appear to be tied to any critical physical component, as such components can be replaced or altered without destroying the computer. Similarly, a computer's identity does not appear to be tied to a particular set of digital data that comprises the computer's operating system, applications, and user data, as that data can be copied with perfect fidelity to other devices, creating computers that are functionally clones of one another.

b. Intelligence

The information-processing mechanisms and behaviors of an agent possessing computronic characteristics demonstrate a number of significant traits. Such an agent is:

Non-sapient. A conventional computer does not possess sapient self-awareness or a subjective conscious experience of reality.[65]

Semiautonomous or nonautonomous. For computerized devices such as robots, autonomy can be understood as the state of being "capable of operating

---

[63] The ability to replace or reconfigure remote networked hardware without impacting web-based end users is widely exploited to offer cloud-based services employing the model of infrastructure as a service (IaaS), platform as a service (PaaS), or software as a service (SaaS); for more details, see the *Handbook of Cloud Computing*, edited by Furht & Escalante (2010).

[64] For a discussion of philosophical issues relating to personal identity, see Olson (2015); see also Friedenberg (2008), p. 250.

[65] Regarding different perspectives on the characteristics that a computer or other artificial system would need to have in order for it to possess sapient self-awareness and a subjective conscious experience of reality, see Friedenberg (2008), pp. 163-78.

in the real-world environment without any form of external control for extended periods of time."[66] Such autonomy does not simply involve the ability to perform cognitive tasks like setting goals and making decisions; it also requires an entity to successfully perform physical activities such as securing energy sources and carrying out self-repair without human intervention. Applying this definition, we can say that current computerized devices are typically either nonautonomous (e.g., telepresence robots that are fully controlled by their human operators) or semiautonomous (e.g., robots that require 'continuous assistance' or 'shared control' in order to fulfill their intended purpose).[67] Although some contemporary computerized systems can be understood as 'autonomous' with regard to fulfilling their intended purpose – in that they can receive sensory input, process information, make decisions, and perform actions without direct human control – they are not autonomous in the full sense of the word, insofar as they are generally not capable of, for example, securing energy sources within the environment or repairing physical damage to themselves.[68]

Volitional. Many conventional computerized devices are nonvolitional, meaning that they possess no internal goals or 'desires' for achieving particular outcomes nor any expectations or 'beliefs' about how performing certain actions would lead to such outcomes. However, many contemporary computerized devices – including a wide variety of robots used in commercial contexts – are volitional. As noted earlier, an entity is volitional if it combines goals with expectations; in other words, it can possess an intention, which is a mental state that comprises both a desire and a belief about how some act that the agent is about to perform can contribute to fulfilling that desire.[69] For example, a therapeutic social robot might possess the goal of evoking a positive emotional response in its human user, and its programming and stored information tells it that by following particular strategies for social interaction it is likely to evoke such a response.[70]

Programmed. A conventional computer does not 'learn' through experience; it does not undergo a long-term formative process of education in order to acquire new knowledge or information. Instead, a computer has software programs and data files copied onto its storage media, thereby instantaneously gaining new capacities and the possession of new information.[71] Alternatively, a computer may be directly programmed or configured by a human operator.

---

[66] Bekey (2005), p. 1.

[67] See Murphy, *Introduction to AI Robotics* (2000).

[68] Gladden, "The Diffuse Intelligent Other" (2016).

[69] Calverley (2008), p. 529.

[70] Gladden, "The Diffuse Intelligent Other" (2016).

[71] For a discussion of the ways in which the electronic components of traditional computers

Processing information by means of a CPU. A conventional contemporary computer (e.g., a desktop computer or smartphone) is based on a Von Neumann architecture comprising memory, I/O devices, and one or more central processing units connected by a communication bus.[72] Although one can be made to replicate the functioning of the other, the linear method by which such a CPU-based system processes information is fundamentally different from the parallel processing method utilized by a physical neural network such as that constituted by the human brain.[73]

Lacking emotion. A traditional computer does not possess emotions that are grounded in the current state of the computer's body, are consciously experienced by the computer, and influence the contents of its decisions and behavior.[74] Although a piece of software may run more slowly or have some features disabled when executed on particular computers, the nature of the software's decision-making is not influenced by factors of mood, emotion, or personality that are determined by a computer's hardware. A software program will typically either run or not run on a given computer; if it runs at all, it will run in a manner that is determined by the internal logic and instructions contained within the software code and not swayed or distorted by that computer's particular physical state.

Free from cognitive biases. A conventional computer is not inherently subject to human-like cognitive biases, as its decisions and actions are determined by the logic and instructions contained within its operating system and application code and not by the use of evolved heuristic mechanisms that are a core element of human psychology.[75]

---

carry out the work of and are controlled by executable programs – as well as an overview of the ways in which alternative architectures such as that of the neural network can allow computers to learn through experience – see Null & Lobur (2006). A more detailed presentation of the ways in which neural networks can be structured and learn is found in Haykin, *Neural Networks and Learning Machines* (2009). For a review of forms of computer behavior whose activity can be hard to predict (e.g., the actions of some forms of evolutionary algorithms or neural networks) as well as other forms of biological or biologically inspired computing, see Lamm & Unger (2011).

[72] See Friedenberg (2008), pp. 27-29.

[73] See Friedenberg (2008), pp. 30-32.

[74] For the distinction between the relatively straightforward phenomenon of computers possessing 'emotion' simply as a function versus the more doubtful possibility that computers could undergo 'emotion' as a conscious experience, see Friedenberg (2008), pp. 191-200.

[75] It is possible, however, for a computer to indirectly demonstrate human-like cognitive biases if the human programmers who designed a computer's software were not attentive to such considerations and inadvertently programmed the software to behave in a manner that manifests such biases. For a discussion of such issues, see, e.g., Friedman & Nissenbaum, "Bias in Computer Systems" (1997).

Possessing nonvolatile digital memory. Many conventional computers are able to store data in a stable electronic digital form that is practically lossless, does not degrade rapidly over time, can be copied to other devices or media and backed up with full fidelity, and does not require a continuous power supply in order to preserve the data.[76]

Demonstrating predictable and analyzable behavior. Computerized devices can be affected by a wide range of component failures and bugs resulting from hardware or software defects or incompatibilities. However, because a typical computer is controlled by discrete linear executable code that can be easily accessed – and because there exist diagnostic software, software debugging techniques, established troubleshooting practices, and methods for simulating a computer's real-world behaviors in development and testing environments – it is generally easier to analyze and reliably predict the behavior of a computer than that of, for example, a human being.[77]

Capable of being hacked electronically. Computerized systems are vulnerable to a wide variety of electronic hacking techniques and other attacks that can compromise the confidentiality, integrity, and availability of information that is received, generated, stored, or transmitted by a system or can result in unauthorized parties gaining complete control over the system.[78]

c. Social Interaction

An agent possessing computronic characteristics demonstrates a number of noteworthy traits relating to social interaction. Such an agent is:

Nonsocial or semisocial. Conventional computers may display social behaviors and engage in short-term, isolated social interactions with human beings or other computers, but they do not participate in long-term social relations that deepen and evolve over time as a result of their experience of such engagement and which are shaped by society's expectations for social roles to be filled by the participants in such relations.[79]

---

[76] Regarding the creation, storage, and transfer of digital data files by computers and other electronic devices, see, e.g., Austerberry, *Digital Asset Management* (2013), and Coughlin, *Digital Storage in Consumer Electronics: The Essential Guide* (2008).

[77] Even the behavior of sophisticated 'artificially intelligent' computerized systems can be easy to predict and debug, if it is controlled by a conventional executable program rather than, e.g., the actions of a physical artificial neural network. For a discussion of different models for generating artificial intelligence through hardware and software platforms, see Friedenberg (2008), pp. 27-36.

[78] For an overview of such possibilities (as well as related preventative practices and responses), see Rao & Nayak (2014).

[79] Although there already exist telepresence robots (e.g., Ishiguro's Geminoids) that manifest highly sophisticated, human-like levels of sociality, such sociality is technically possessed not by

Lacking culture. Although a large number of computers can be linked to form networks that may constitute a form of computerized society, such aggregations of conventional computers do not create their own cultures.[80]

Lacking spirituality. Conventional computers do not search for a connection with some transcendental truth or reality in order to provide meaning or purpose to their existence; they do not engage in contemplation, meditation, or prayer.[81]

Apolitical. Conventional computers do not directly participate as members of human or artificial political systems. Some computerized systems (e.g., some swarm robots as components in multi-agent systems) participate in social interactions, and even social relations and group governance structures, but they do not generally create political systems of the sort common among human populations.[82]

An economic participant. Conventional computers typically do not function independently within the real-world human economy as autonomous economic actors, although they participate in the economy in many other ways. Computers do not own or exchange their own financial or other assets, nor do they purchase goods or services for their own consumption, although computers may serve as agents that initiate and execute transactions on behalf of human beings or organizations.[83]

---

the robot itself but by the hybrid human-robotic system that it forms with its human operator. Regarding such issues, see Vinciarelli et al. (2012) and Gladden, "Managerial Robotics" (2014).

[80] Regarding prerequisites for artificial entities or systems to produce their own culture (or collaborate with human beings in the production of a shared human-artificial culture), see, e.g., Payr & Trappl, "Agents across Cultures" (2003).

[81] Regarding elements that would need to be present in order for a computerized device to develop its own spirituality (rather than to simply have some spiritual value attributed to it by human beings), see, e.g., Geraci, "Spiritual robots: Religion and our scientific view of the natural world" (2006); Nahin, "Religious Robots" (2014); Section 6.2.3.2 on "Religion for Robots" in Yampolskiy, *Artificial Superintelligence: A Futuristic Approach* (2015); and Kurzweil, *The Age of Spiritual Machines: When Computers Exceed Human Intelligence* (2000).

[82] Regarding ways in which advanced multi-agent systems (such as those found in swarm robotics) might potentially implement patterns of social interaction and organization that resemble or are explicitly based on human political behaviors and structures, see, e.g., McBurney & Parsons, "Engineering democracy in open agent systems" (2003); Ferber et al., "From agents to organizations: an organizational view of multi-agent systems" (2004); and Sorbello et al., "Metaphor of Politics: A Mechanism of Coalition Formation" (2004).

[83] For example, regarding the increasing sophistication of automated trading systems that are capable of teaching themselves and improving their investment strategies over time, without direct instruction from human beings, and the growing use of 'robo-advisors' to manage financial assets on behalf of human owners, see Scopino, "Do Automated Trading Systems Dream of Manipulating the Price of Futures Contracts? Policing Markets for Improper Trading Practices by Algorithmic Robots" (2015), and Sharf, "Can Robo-Advisors Survive A Bear Market?" (2015).

Property, not a legal person. A conventional computer is a piece of property that is typically owned by a specific human being or organization; a computer is not itself a legal person that possesses a recognized set of rights and responsibilities.[84]

III. Using the Matrix to Analyze the Traditional Practice of Organizational Management

Our two-dimensional Posthuman Management Matrix contains quadrants that describes four types of entities that could potentially be participants in or objects of the activities of organizations such as businesses and which – if they exist – would need to be accounted for by management theory and practice. As illustrated in Figure 1, these four potential types of entities are:

- **Human agents possessing anthropic characteristics**, which we can refer to as '**natural human beings**,' insofar as they have not been significantly enhanced or modified through the use of technologies such as neuroprosthetics or genetic engineering.
- **Artificial agents possessing computronic characteristics**, which we can refer to simply as '**computers**.' Such entities include conventional desktop and laptop computers, mainframes, web servers, and smartphones and other mobile devices whose software allows them to exercise a limited degree of agency.
- **Human agents possessing computronic characteristics**, which we can refer to as '**cyborgs**.' In the sense in which the term is employed in this text, a cyborg is a human being whose body includes some 'artificial components,'[85] however these components do not necessarily need to be electromechanical in nature (as in the case of contemporary neuroprosthetic devices); the artificial elements could be structures or systems composed of biological material that are not typically found in natural human beings and which are the result of genetic engineering.
- **Artificial agents possessing human characteristics**, which we can refer to as '**bioroids**.' Terms such as 'android' or 'humanoid robot'

---

[84] Stahl suggests that a kind of limited 'quasi-responsibility' can be attributed to conventional computers and computerized systems. In this model, it is a computer's human designers, programmers, or operators who are typically responsible for the computer's actions; declaring a particular computer to be 'quasi-responsible' for some action that it has performed serves as a sort of moral and legal placeholder, until the computer's human designers, programmers, and operators can be identified and ultimate responsibility for the computer's actions assigned to the appropriate human parties. See Stahl, "Responsible Computers? A Case for Ascribing Quasi-Responsibility to Computers Independent of Personhood or Agency" (2006).

[85] See Novaković et al. (2009).

could potentially be employed to describe such entities, however these terms are often used to imply that a robot has a human-like physical form, without necessarily possessing human-like psychology, cognitive capacities, or biological components. Similarly, the term 'biorobot' could be employed, but it is often used to refer to robots that mimic animals like insects or fish whose physical form and cognitive capacities have little in common with those of human beings. We choose to employ the term 'bioroid' (whose origins lie primarily in the field of science fiction rather than engineering)[86] insofar as it evokes the image of an artificially engineered agent that possesses human-like cognitive capacities and psychology, biological or biologically inspired components, and a physical form that allows it to engage in human-like social behaviors and interactions but which is not necessarily humanoid.

Prior to the development of computers as a practical organizational technology in the 20th Century, it was historically only the lower left quadrant of the Posthuman Management Matrix that was of relevance to organizational managers. Indeed, not only were natural human beings as a practical matter the only available employees and customers, but they were also generally considered to be the only *potential* employees and customers with which the scholarly discipline of management would ever need to concern itself. The possibility that organizations might someday employ and serve entities that were not human agents possessing anthropic characteristics was not studied as a theoretical possibility; the theory and practice of management were concerned only with understanding and managing the activities of natural human beings. Within that context, fields such as economics, organizational psychology, and human resource management played key roles.

Eventually, with the development of increasingly sophisticated computers over the course of the 20th Century and up through the present day, management scholars and practitioners began to realize the need to expand the theoretical and practical scope of management to include new subdisciplines that could guide the creation, implementation, and management of artificial agents

---

86 For uses of the term 'bioroid' in science fiction literature and roleplaying games, see, e.g., Pulver (1995), pp. 74-81, where 'bioroid' is used explicitly as a portmanteau derived from 'biological android'; Surbrook, *Kazei-5* (1998), pp. 64, 113; Pulver, *Transhuman Space* (2002), p. 12, where 'bioroid' refers to "living beings functionally similar to humans, but assembled using tissue engineering and 'biogenesis' nanotechnology, and educated using accelerated learning techniques"; *Appleseed*, directed by Aramaki (2010); Martinez, "Bodies of future memories: the Japanese body in science fiction anime" (2015); Litzsinger, *Android: Netrunner* (2012); and Duncan, "Mandatory Upgrades: The Evolving Mechanics and Theme of Android: Netrunner" (2014). For a reference to the fictional use of the term 'bioroid' in an engineering context, see Novaković et al. (2009).

such as manufacturing robots or server farms controlled by load-balancing software.[87] Because such artificial agents possessed structures, behaviors, and organizational roles that were quite different from those of human agents, existing disciplines such as psychology and HR management did not provide adequate or relevant tools for the oversight of such systems; instead, new fields such as computer science, electronics engineering, robotics, and IT management began to aid organizational managers in designing, implementing, and maintaining such systems that comprise artificial agents possessing computronic characteristics. As a result of such developments, a second quadrant of the Posthuman Management Matrix became not only relevant but critical for the successful management of contemporary organizations.

Despite this experience in which a previously disregarded quadrant of the Posthuman Management Matrix quickly assumed major theoretical and practical importance for organizations, the remaining two quadrants of the Matrix have remained largely neglected within the field of organizational management – as though there existed an implicit presumption that these areas define sets that would continue to remain empty or that these quadrants would only become relevant for organizational management at a date so far in the future that it would be a misallocation of time and resources for management scholars and practitioners to concern themselves with such possibilities now.

Figure 2 thus depicts the field of management as it largely exists today: a field in which centuries-old management traditions relating to natural human beings have recently been supplemented by new theory and practice that address the rise of conventional computers – but in which the possibility and organizational significance of cyborgs and bioroids remain, from a management perspective, largely unexplored.[88]

We can now consider in more detail these four types of entities described by the Posthuman Management Matrix as they have been understood by the field of organizational management from its historical origins up to the present day.

---

[87] The development of such disciplines and practices was spurred in part by the experience of organizations that made large investments in IT systems in the 1980s, only to discover that simply purchasing exotic new IT equipment would not, in itself, generate desired gains in productivity unless such equipment were thoughtfully aligned with and integrated into an organization's larger business plan, strategies, and processes. See Magoulas et al. (2012), p. 89, and Hoogervorst (2004), p. 16.

[88] For some time, the design, implementation, and implications of human agents possessing computronic characteristics and artificial agents possessing anthropic characteristics have been the subject of intense research and contemplation across a broad range of fields, from computer science and robotics to philosophy of mind and philosophy of technology, ethics, and science fiction; here we are only noting that – notwithstanding the work of a small number of future-oriented management scholars – the field of management has not yet taken up such topics as subjects worthy of (or even demanding) serious consideration.

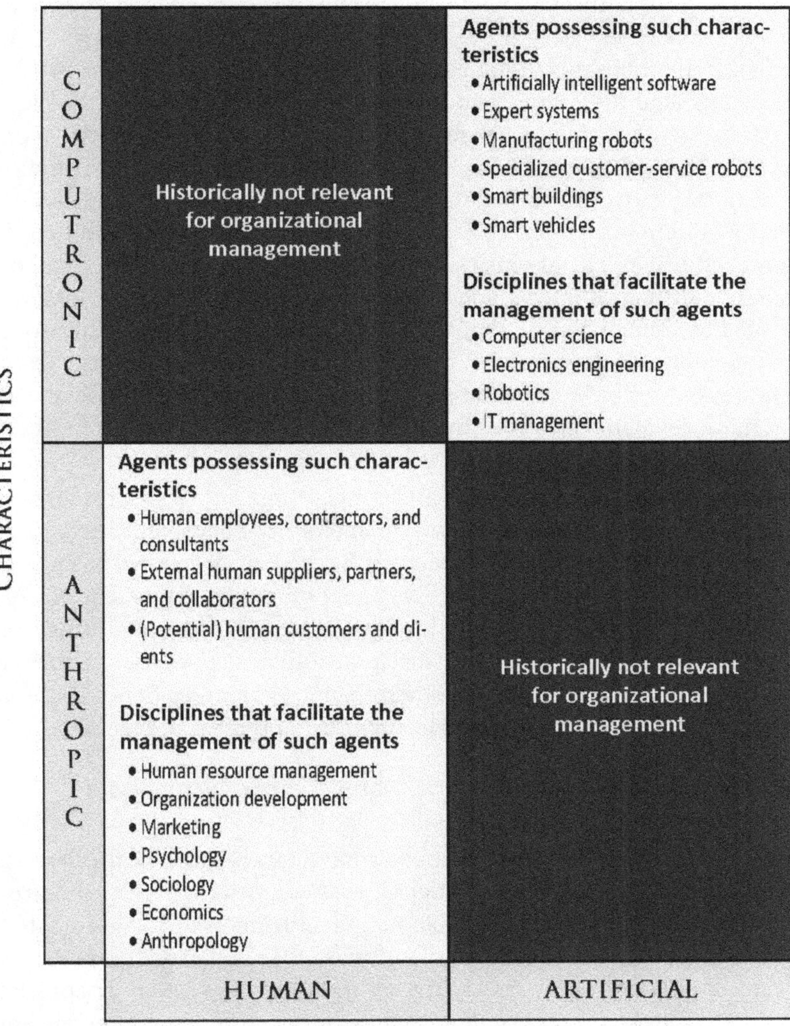

Fig. 2: The Posthuman Management Matrix displaying the two types of entities that have been relevant in recent decades for the theory and practice of organizational management, along with two types of entities that historically have not been considered relevant.

A. Human Agents with Anthropic Characteristics ('Natural Human Beings')

The actions of natural human beings – and the knowledge of how to anticipate and guide their activities – have formed the critical foundation upon which all human organizations have historically been built. Even before the dawn of artificial intelligence and the creation of the first artificial agents, nonhuman agents such as domesticated farm animals have played a supporting role in the activities of some human organizations. However, the overwhelming majority of roles within such organizations – including all of those leadership and management roles requiring strategic thinking and long-term planning, ethical and legal sensitivity, negotiation skills, risk management approaches, and the use of oral and written communication – have historically been filled by human beings, who have always been (and been understood as) human agents who possess anthropic characteristics. Human organizations such as businesses have relied on such human beings as their CEOs and executives, midlevel managers, frontline employees, consultants, partners and suppliers, competitors, and actual or potential customers and clients.

In order to plan, organize, lead, and control[89] the activities of such natural human beings that are found both within and outside of organizations, a number of academic disciplines and practices have been developed over the last century and more that can facilitate and support the management of organizations. Such disciplines include HR management, marketing, and organization development, along with other disciplines such as psychology, sociology, economics, anthropology, cultural studies, and ergonomics that have broader aims and applications but which can help inform organizational management.

B. Artificial Agents with Computronic Characteristics ('Computers')

Over the last half-century, computers have taken on critical roles within the lives of many organizations. Such agents comprise assembly-line robots used for painting or welding, flexible manufacturing systems, automated security systems, and a broad range of software that possesses some degree of artificial intelligence and runs as part of an operating system or application on servers, desktop computers, mobile devices, and other computerized equip-

[89] Planning, organizing, leading, and controlling are recognized as the four key functions that must be performed by managers. See Daft, *Management* (2011).

ment. Such artificial agents may schedule tasks and optimize the use of physical and electronic resources;[90] transport materials within production facilities;[91] assemble components to produce finished products;[92] interact directly with customers on automated customer-service phone lines, through online chat interfaces, and at physical kiosks to initiate and perform transactions and offer information and support;[93] monitor systems and facilities to detect physical or electronic intrusion attempts;[94] initiate and execute financial transactions within online markets;[95] and carry out data mining in order to evaluate an applicant's credit risk, identify suspected fraud, and decide what personalized offers and advertisements to display to a website's visitors.[96] In order to manage the activities of artificial agents possessing computronic characteristics, one can draw on insights from a number of disciplines and practices that have been developed over the last few decades, including computer science, electronics engineering, robotics, and IT management.

While human beings still play key roles as leaders, strategists, and managers within organizations, in many cases they are no longer capable of carrying out their work without the engagement and support of the artificial agents that permeate an organization's structures, processes, and systems in so many

---

[90] For an overview of methods that can be employed for such purposes, see Pinedo, *Scheduling: Theory, Algorithms, and Systems* (2012). For more specific discussions of the use of artificial agents (and especially multi-agent systems) for such ends, see, e.g., Ponsteen & Kusters, "Classification of Human and Automated Resource Allocation Approaches in Multi-Project Management" (2015); Merdan et al., "Workflow scheduling using multi-agent systems in a dynamically changing environment" (2013); and Xu et al., "A Distributed Multi-Agent Framework for Shared Resources Scheduling" (2012).

[91] See, e.g., Ullrich, *Automated Guided Vehicle Systems: A Primer with Practical Applications* (2015), and *The Future of Automated Freight Transport: Concepts, Design and Implementation*, edited by Priemus & Nijkamp (2005).

[92] See, e.g., *Agent-Based Manufacturing: Advances in the Holonic Approach*, edited by Deen (2003); *Intelligent Production Machines and Systems*, edited by Pham et al. (2006); and *Industrial Applications of Holonic and Multi-Agent Systems*, edited by Mařík et al. (2015).

[93] See, e.g., Ford, *Rise of the Robots: Technology and the Threat of a Jobless Future* (2015), and McIndoe, "Health Kiosk Technologies" (2010).

[94] Regarding the automation of intrusion detection and prevention systems, see Rao & Nayak (2014), pp. 226, 235, 238.

[95] See Philips, "How the Robots Lost: High-Frequency Trading's Rise and Fall" (2012); Scopino (2015); and Sharf (2015).

[96] Giudici, *Applied Data Mining: Statistical Methods for Business and Industry* (2003); Provost & Fawcett, *Data Science for Business* (2013), p. 7; and Warkentin et al., "The Role of Intelligent Agents and Data Mining in Electronic Partnership Management" (2012), p. 13282.

ways.[97] For many organizations, the sudden disabling or loss of such artificial agents would be devastating, as the organizations have become dependent on artificial agent technologies to perform critical tasks that cannot be performed by human beings with the same degree of speed, efficiency, or power.

C. Human Agents with Computronic Characteristics ('Cyborgs')

Historically, all human beings have been human agents that possess anthropic characteristics. From the perspective of organizational management, the set of human agents possessing computronic characteristics has been seen as empty; such beings are not yet understood to widely exist, and it is presumed that there is no special need to take them into account as potential employees, partners, or clients when considering a business's short-term objectives and operations. Although emerging posthumanizing technologies are beginning to create cases of human agents who indeed possess limited computronic characteristics, the number, nature, and scope of such cases of the 'cyborgization' of human agents is still relatively small, and from the managerial perspective most organizations have been able to simply ignore such cases, as though the category of the cyborg were not yet applicable or relevant to their organizational mission and objectives.[98] Because human agents possessing extensive computronic characteristics do not yet exist as a large population of beings who can serve as employees, partners, or customers for organizations, it is not surprising that organizations do not yet possess specialized practices or academic disciplines that they can rely on to aid them in the management of such entities.

D. Artificial Agents with Anthropic Characteristics ('Bioroids')

The artificial agents that have been broadly deployed and which are relevant for organizational management are generally artificial agents possessing computronic characteristics. While scientists and engineers are making great

---

[97] Within the 'congruence model' of organizational architecture developed by Nadler and Tushman, structures, processes, and systems constitute the three main elements of an organization that must be considered. See Nadler & Tushman, *Competing by Design: The Power of Organizational Architecture* (1997), p. 47, and the discussion of these elements within a posthumanized organizational context in Part Two of this volume, on "Organizational Posthumanism."

[98] Fleischmann argues, for example, that within human society there is an inexorable trend that will eventually result in full cyborg-cyborg interaction in the form of social relations among beings who are human-electronic hybrids – human beings whose biological organism possesses extensive and intimate internal interfaces with neuroprosthetic devices. Current phenomena like the widespread interaction of human beings who are dependent on (and interact through) mobile devices such as smartphones are one step along that trajectory. See Fleischmann, "Sociotechnical Interaction and Cyborg–Cyborg Interaction: Transforming the Scale and Convergence of HCI" (2009).

strides toward developing artificial agents that possess anthropic characteristics, at present such systems are experimental and exist largely in laboratory settings.[99] As a practical matter, within most organizations the category of bioroids is still treated as though it were an empty set; organizations have generally not seen the need to consider such entities when planning their objectives and operations. As with the cyborgs described above, because bioroids have historically not existed as potential employees, partners, or customers for organizations, it is unsurprising that organizations do not yet have specialized disciplines that they can rely on to aid them in managing such entities.

IV. Using the Matrix to Predict and Shape the Future Practice of Organizational Management

In the sections above, we have considered the situation that has existed up to now – with organizations' sole agents being natural human beings and computers. We can now explore the ways in which the situation is rapidly changing due to the emergence of new posthumanizing technologies.

A. The Converging Characteristics of Human and Artificial Agents in the Posthuman Age

Below we review once more the set of variables that define an agent's characteristics and, for each of the characteristics, discuss ways in which the advent of various posthumanizing technologies will result in a growing variety of cyborgs and bioroids. Studies focusing on these two types of entities are emerging as new fields in which ongoing innovation will expand the kinds of workers, partners, and consumers that are available to organizations and which are expected to become crucial loci for management theory and practice in the coming years. We can consider in turn the physical form, intelligence, and social interaction that will be demonstrated by such new types of human and artificial agents .

1. Physical Form

The range of physical forms available to human and artificial agents is expected to evolve and expand significantly. Such changes will be visible in the manner in which a number of key characteristics are expressed (or not expressed); these characteristics are described below.

---

[99] See Friedenberg (2008) for an in-depth review of efforts to develop robots and other artificial beings that possess human-like perception, learning, memory, thought, language use, intelligence, creativity, motivation, emotions, decision-making capacities and free will, consciousness, biological structures and processes, and social behaviors.

a. Components

It is anticipated that the bodies of human agents will increasingly include electronic components in the form of artificial organs, artificial limbs and exoskeletons, artificial sense organs, memory implants, and other kinds of neuroprosthetic devices;[100] the major obstacle to the expansion of such technology may be the fact that the natural biological brain (or at least, significant portions of the brain) of a human being will need to remain intact and functional in order for an agent to be considered 'human.'

Conversely, expected developments in genetic engineering technologies, soft robotics, and artificial life will increasingly allow the bodies of artificial agents to include components formed from biological material.[101] In cases that involve extensive engineering and modification of the genome (and especially in 'second-generation' entities that are the result of natural reproductive processes between biological parents rather than cloning or other direct engineering), it may be difficult conceptually and practically to specify whether an entity is an 'artificial agent' composed entirely of biological components or a 'human agent' whose biological substrate has been intentionally designed. The legal, ethical, ontological, and even theological questions involved with such potential practices are serious and wide-ranging.

b. Animation

Currently, only those human beings that are alive are capable of serving as employees or customers of an organization. Techniques such as 'mind uploading' and the development of artificial neurons that can replace or replicate the actions of neurons in the brain of a living human being may someday allow human agents that are no longer 'alive' in a biological sense to have their unique memories, knowledge, cognitive patterns, and social relations utilized by agents that function as employees, partners, or customers for organizations. The extent to which such nonbiological human agents can be identified with the biological human beings from whom they are derived depends on issues that are philosophically controversial and complex.[102]

---

[100] See Gasson, "ICT implants" (2008); Gasson et al., "Human ICT Implants: From Invasive to Pervasive" (2012); McGee (2008); Merkel et al., "Central Neural Prostheses" (2007); Gladden, *The Handbook of Information Security for Advanced Neuroprosthetics* (2015), pp. 32-33; and Gladden, "Cybershells, Shapeshifting, and Neuroprosthetics" (2015).

[101] See Berner (2004), pp. 15, 18, 31, 61-62. For a discussion of the possibilities of using DNA as a mechanism for the storage or processing of data, see Church et al. (2012) and Friedenberg (2008), p. 244.

[102] See Koene (2012); Proudfoot (2012); Pearce, "The Biointelligence Explosion" (2012); Hanson, "If uploads come first: The crack of a future dawn" (1994); Moravec, *Mind Children: The Future of Robot and Human Intelligence* (1990); Ferrando (2013), p. 27; and Gladden, *The Handbook of Information Security for Advanced Neuroprosthetics* (2015), pp. 98-100, for a discussion of such issues from various perspectives.

Meanwhile, the development of biological components for use in robots and other artificial agents and ongoing advances in the development of non-biological artificial life (e.g., autonomous evolvable computer worms or viruses that satisfy standard scientific definitions of life-forms) can result in artificial agents that are considered to be alive, insofar as they constitute a viable system that demonstrate a physical metabolism, the ability to maintain homeostasis, reproduction, reaction and adaptation to the environment, and other key characteristics.[103]

c. Design

The growing possibilities for genetic engineering, gene therapy, and the augmentation of human agents through the implantation of neuroprosthetic devices or other synthetic components means that the body possessed by a human agent will no longer necessarily be a natural substrate that is produced through the randomized inheritance of genetic material from biological parents and that is free from intentional design by institutions or individual human engineers.[104] Besides the major moral and legal questions raised by such possibilities, there are also operational issues that would confront organizations whose pool of potential employees or customers includes human agents who have been designed in such ways; for example, forms of genetic engineering that create synthetic characteristics shared broadly across a population and which reduce genotypic diversity may render the population more vulnerable to biological or electronic hacking attempts (and may make such attempts more profitable and attractive for would-be adversaries), although such standardization may also make it easier for effective anti-hacking security mechanisms to be developed and deployed across the population.[105]

At the same time, artificial agents may no longer be products of explicit design and engineering by human manufacturers. Some artificial life-forms that exist within the digital-physical ecosystem primarily as physical robots possessing some degree of AI or as digital life-forms that temporarily occupy

---

[103] See the discussion of essential elements of artificial life in Friedenberg (2008), pp. 201-03, which is based on the criteria for biological life presented by Curtis (1983). See also Gladden, "The Artificial Life-Form as Entrepreneur: Synthetic Organism-Enterprises and the Reconceptualization of Business" (2014).

[104] For different perspectives on such possibilities, see, e.g., De Melo-Martín (2015); Regalado, "Engineering the perfect baby" (2015); Lilley, *Transhumanism and Society: The Social Debate over Human Enhancement* (2013); Nouvel (2015); Section B ("Enhancement") in *The Future of Bioethics: International Dialogues*, edited by Akira Akabayashi (2014); Mehlman, *Transhumanist Dreams and Dystopian Nightmares: The Promise and Peril of Genetic Engineering* (2012); and Bostrom (2012).

[105] For the relationship between the heterogeneity of information systems and their information security, see Gladden, *The Handbook of Information Security for Advanced Neuroprosthetics* (2015), p. 296, and *NIST SP 800-53* (2013), p. F-204.

physical substrates may manifest structures and behaviors that are the result of randomized evolutionary processes that lie beyond the control of human designers or which are the result of intentional design efforts conducted by other artificial agents whose nature is such that they are inscrutable to human understanding – in which case, from the human perspective, the engineered agents would essentially lack a comprehensible design.[106] In other cases, human designers may have intentionally engineered an artificial agent's basic structures (such as a physical neural network), but the exact nature of the behaviors and other traits eventually developed and demonstrated by those structures may lie beyond the reach of human engineering.[107]

d. Upgradeability

The growing use of technologies for somatic cell gene therapy and neuroprosthetic augmentation may increasingly allow the physical components and cognitive capacities of human agents to be upgraded and expanded even after the agents have reached a stage of physical and cognitive maturity.[108]

Conversely, it may be difficult or impossible to upgrade, expand, or replace the physical components of artificial agents that are composed of biological material in the way that components of an electronic computer can be upgraded. In the case of especially complex or fragile artificial agents, efforts to upgrade or otherwise modify an agent's physical components after its creation may result in the impairment or death of such biological material or of the agent as a whole. Similarly, after an artificial agent that possesses a holonomic physical neural network has been created and achieved intellectual maturity through experience and learning, it may not be possible to intervene directly in the neural network's physical structure or processes to upgrade its capacities or edit its contents without irreparably harming the agent.[109]

e. Lifespan

A human agent whose bodily components can be easily replaced with biological or electronic substitutes after deteriorating or becoming damaged or whose components can be (re)engineered to prevent them from undergoing

[106] Regarding evolutionary robotics and evolvable robot hardware, see Friedenberg (2008), pp. 206-10.

[107] Regarding the relationship of artificial life and evolutionary robotics, see Friedenberg (2008), pp. 201-16.

[108] See, e.g., Panno, *Gene Therapy: Treating Disease by Repairing Genes* (2005); *Gene Therapy of the Central Nervous System: From Bench to Bedside*, edited by Kaplitt & During (2006); and Bostrom (2012).

[109] Regarding the potentially holonomic nature of memory storage within the brain, see, e.g., Longuet-Higgins (1968); Pribram (1990); Pribram & Meade (1999); and Gladden, *The Handbook of Information Security for Advanced Neuroprosthetics* (2015), pp. 200-01.

damage or deterioration in the first place could potentially experience an extended or even indefinite lifespan, although such engineering might result in side-effects that are detrimental to the agent and which would render such lifespan extension undesirable as a practical matter.[110] As in other cases, the moral and legal questions involved with such activities are serious.

At the same time, artificial agents whose bodies include or comprise biological components or whose cognitive processes follow an irreversible developmental cycle (e.g., in which the neural network of an agent's 'brain' possesses a maximum amount of information that it can accumulate over the course of the agent's lifespan) might possess a limited and predetermined lifespan that cannot be extended after the agent's creation.[111]

f. Operational Cycle

Genetic engineering could potentially speed the natural biological processes that contribute to physical growth and cognitive development or slow or block processes of physical and cognitive decline. Scholars also envision the possibility of neuroprosthetic technologies being used to allow human beings to instantly acquire new knowledge or skills through the implantation of memory chips or the downloading of files into one's brain; if feasible, this could allow human cognitive capacities to be instantaneously upgraded in a manner similar to that of installing new software on a computer, thereby bypassing typical human processes of cognitive development and learning.[112]

At the same time, the integration into artificial agents of biological components and physical neural networks whose structure and behavior render them difficult to control externally after their deployment means that it may become impossible to simply 'reset' artificial agents and restore them to an earlier physical and informational state.[113]

---

[110] Regarding issues with technologically facilitated life extension or the replacement of a human being's original biological body, see Proudfoot (2012); Pearce (2012); Hanson (1994); and Gladden, "'Upgrading' the Human Entity: Cyberization as a Path to Posthuman Utopia or Digital Annihilation?" (2015).

[111] As early as the 1940s, Wiener speculated that a physical neural network that is incapable of adding new neurons or creating new synapses but which instead stores memories through increases to the input threshold that triggers the firing of existing neurons may display an irreversible process of creating memories through which its finite available storage capacity is gradually exhausted, after which point a sort of senescence occurs that degrades the neural network's functioning and disrupts the formation of new memories. See Wiener (1961), loc. 2467ff.

[112] See, e.g., McGee (2008).

[113] Regarding the difficulty of detecting and understanding the current state of an artificially intelligent system (let alone restoring it to a previous state), especially that of a distributed artificial intelligence (DAI) displaying emergent behavior, see Friedenberg (2008), pp. 31-32.

g. Locality

The use of neuroprosthetic devices and virtual reality technologies may effectively allow a human agent to occupy different and multiple bodies that are either physical or virtual and are potentially of a radically nonhuman nature.[114] In this way, a human agent could be extremely multilocal by being present in many different environments simultaneously.[115]

At the same time, an artificial agent whose cognitive processes are tied to a single body comprising biological components or a single physical artificial neural network that possesses limited sensorimotor and I/O mechanisms may be confined to exercising its agency within the location in which that cognitive substrate is located.[116]

h. Permanence of Substrate

Historically, a particular human agent has been tied to a particular physical substrate or body; the dissolution of that body entails the end of that human being's ability to act as an agent within the environment. Ontologically and ethically controversial practices such as the development of artificial neurons to replace the natural biological neurons of a human brain and mind uploading may allow a single human agent's agency to exist and act beyond the physical confines of the agent's original biological physical substrate – but only under certain definitions of 'agent' and 'agency' that remain strongly contested.[117] Similarly, the use of genetic engineering or neuroprosthetically mediated cybernetic networks to create hive minds or other forms of collective agency involving human agents might allow such multi-agent systems or 'super-agents' to survive and function despite a continual addition and loss of biological substrates which mean that the entity's substrate at one moment in time shares no components in common with its substrate at a later point in time.

---

[114] Gladden, "Cybershells, Shapeshifting, and Neuroprosthetics" (2015).

[115] See Gladden, "The Diffuse Intelligent Other" (2016) for a discussion of multilocality.

[116] Regarding different fundamental architectures for the design of artificially intelligent systems – from a CPU-based Von Neumann architecture and software-based artificial neural network to models utilizing grid computing and distributed AI – see Friedenberg (2008), pp. 27-32. Regarding the extent to which a human-like AI may necessarily be tied to a single body that interacts with a particular environment, see Friedenberg (2008), pp. 32-33, and the literature on embodied embedded cognition – e.g., Wilson, "Six views of embodied cognition" (2002); Anderson, "Embodied cognition: A field guide" (2003); Sloman, "Some Requirements for Human-like Robots: Why the recent over-emphasis on embodiment has held up progress" (2009); and Garg, "Embodied Cognition, Human Computer Interaction, and Application Areas" (2012).

[117] Regarding such issues, see Koene (2012); Proudfoot (2012); Pearce (2012); Hanson (1994); Moravec (1990); and Gladden, *The Handbook of Information Security for Advanced Neuroprosthetics* (2015), pp. 99-100.

Just as certain posthumanizing technologies might – according to their proponents – free human agency from its historic link to a particular biological body, other technologies might increasingly bind artificial agency to a particular permanent physical substrate. For example, an artificial agent whose cognitive processes are executed by biological components or a physical artificial neural network and whose memories and knowledge are stored within such components may not be capable of exchanging its body or migrating to a new substrate without losing its agency.[118]

i. Identity

If a human agent's agency is no longer irrevocably tied to a particular biological body, it may become difficult or impossible to attribute actions to a specific human agent or even to identify which human agent is occupying and utilizing a particular physical body in a given moment – since a single electronic sensor or actuator could simultaneously belong to the bodies of multiple human agents. The ability of neuroprosthetically mediated cybernetic networks to create hive minds and other forms of collective consciousness among human and artificial agents may also make it difficult to identify which human agent, if any, is present in a particular physical or virtual environment and is carrying out the behaviors observed there.[119]

Conversely, if an artificial agent is tied to a particular physical body (e.g., because the agent's cognitive processes cannot be extracted or separated from the biological components or physical artificial neural network that execute them), this may provide it with a uniqueness and identity similar to that historically enjoyed by individual human beings.[120] On the other hand, an artificial agent that possesses a spatially dispersed or nonlocalizable body may possess even less of a clear identity than is possessed today by conventional hardware-software computing platforms.

2. Intelligence

The range of information-processing mechanisms and behaviors available to human and artificial agents is expected to evolve significantly as a result of

---

[118] It is not yet clear, for example, whether an artificial intelligence possessing human-like levels of intelligence could potentially exist in the form of a computer worm or virus that can move or copy itself from computer to computer, or whether the nature of human-like intelligence renders such a scenario theoretically impossible. Regarding the significance of a body for artificial intelligence, see, e.g., Friedenberg (2008), pp. 32-33, 179-234.

[119] Regarding such issues, see Gladden, "Utopias and Dystopias as Cybernetic Information Systems: Envisioning the Posthuman Neuropolity" (2015), and Gladden, "'Upgrading' the Human Entity" (2015).

[120] For an overview of issues of personal identity from a philosophical perspective, see Olson (2015). For an exploration of questions of physicality and identity in robots, see Friedenberg (2008), pp. 179-234.

posthumanizing technological and social change. Such changes will be expressed through the possession (or lack) of a number of key characteristics, which are described below.

a. Sapience

By interfering with or altering the biological mechanisms that support consciousness and self-awareness within the brain, neuroprosthetic devices could deprive particular human agents of sapience, even if those agents outwardly appear to remain fully functional as human beings; for example, a human agent might retain its ability to engage in social interactions with longtime friends – not because the agent's mind is conscious and aware of such interactions, but because a sufficiently sophisticated artificially intelligent neuroprosthetic device is orchestrating the agent's sensorimotor activity.[121] Genetic engineering could also potentially be employed in an attempt to create human agents that lack sapience (and could be subject to claims by their producers that they should be considered property rather than legal persons and moral agents) or human agents whose transhuman sapience is of such an unusual and 'advanced' sort that it is unfathomable – and perhaps even undetectable – to natural human beings.[122]

Much research from a philosophical and engineering perspective has been dedicated to considering whether sufficiently sophisticated artificial agents might be capable of achieving sapience and possessing self-awareness and a subjective conscious experience of reality. Controversy surrounds not only the theoretical questions of whether artificial agents can potentially possess sapience (and, if so, what types of artificial agents) but also the practical question of how outside observers might determine whether a particular artificial agent possesses conscious self-awareness or simply simulates the possession of such self-awareness.[123] Regardless of how these questions are answered by philosophers, theologians, scientists, engineers, and legislators, emerging popular conceptions of artificial agents and their potential for sapience may require organizations to treat certain kinds of artificial agents *as though* they

---

[121] See Gladden, "'Upgrading' the Human Entity" (2015).

[122] See Abrams, "Pragmatism, Artificial Intelligence, and Posthuman Bioethics: Shusterman, Rorty, Foucault" (2004); McGee (2008), pp. 214-16; Warwick, "The cyborg revolution" (2014), p. 271; Rubin, "What Is the Good of Transhumanism?" (2008); and Gladden, *The Handbook of Information Security for Advanced Neuroprosthetics* (2015), pp. 166-67.

[123] On the possibility that efforts to ascertain the levels of intelligence or consciousness of artificial entities might be distorted by human beings' anthropomorphizing biases, see Yampolskiy & Fox, "Artificial General Intelligence and the Human Mental Model" (2012), pp. 130-31. On the distinction between intelligence, consciousness, and personhood in such a context, see, e.g., Proudfoot (2012), pp. 375-76. For a broader discussion of such issues, see, e.g., *The Turing Test: The Elusive Standard of Artificial Intelligence*, edited by Moor (2003).

possessed a degree of sapience comparable, if not identical, to that possessed by human beings.

b. Autonomy

Some kinds of neuroprosthetic devices or genetic modification may weaken the desires or strategic planning capacities of human agents or subject them to the control of external agents, thereby reducing their autonomy. New kinds of social network topologies that link the minds of human agents to create hive minds or other forms of merged consciousness can also reduce the autonomy of the individual members of such networks.[124] Neuroprosthetic augmentation, genetic modification, and other uses of posthumanizing technology that renders human agents dependent on corporations or other organizations for ongoing hardware or software upgrades or medical support similarly reduce the autonomy of those agents.[125] On the other hand, technologies that allow human agents to survive and operate in hostile environments or to reduce or repair physical damage to their bodies would enhance such agents' autonomy.

The development of synthetic systems that possess human-like levels of artificial general intelligence would result in the appearance of artificial agents that do not function autonomously with regard to carrying out some specific task that they are expected to perform but which function autonomously at a more general level in deciding their own aims, aspirations, and strategies.[126] The development of robots that can obtain energy from their environment, for example, by consuming the same kinds of foods that are edible for human beings[127] or which possess biological components that can heal wounds that they have suffered will also result in artificial agents with increased autonomy.

c. Volitionality

Researchers have already observed ways in which certain kinds of neuroprosthetic devices and medications can affect their human host's capacity to possess desires, knowledge, and belief;[128] insofar as technologies disrupt or

[124] See Gladden, "Utopias and Dystopias as Cybernetic Information Systems" (2015).

[125] See Gladden, "Neural Implants as Gateways to Digital-Physical Ecosystems and Posthuman Socioeconomic Interaction" (2016).

[126] See, e.g., Yampolskiy & Fox (2012).

[127] See, e.g., the discussion of artificial digestive systems in Friedenberg (2008), p. 214-15.

[128] Regarding the possibility of developing neuroprosthetics that affect emotions and perceptions of personal identity and authenticity, see Soussou & Berger, "Cognitive and Emotional Neuroprostheses" (2008); Hatfield et al., "Brain Processes and Neurofeedback for Performance Enhancement of Precision Motor Behavior" (2009); Kraemer, "Me, Myself and My Brain Implant: Deep Brain Stimulation Raises Questions of Personal Authenticity and Alienation" (2011); Van

control such abilities, they may impair their human host's exercise of his or her conscience, which depends on the possession of these capacities. This may result in the existence of human agents that are no longer fully metavolitional but instead merely volitional or nonvolitional.[129] The use of neuroprosthetics, virtual reality, and other technologies to create hive minds and other forms of collective consciousness among human agents may also impair the volitionality of human agents participating in such systems and reduce them to a state that is less than metavolitional; each agent may no longer possess its own individual conscience but instead help to form (and be guided by) the conscience of the multi-agent system as a whole.

Meanwhile, advances toward the development of human-like artificial general intelligence point at the eventual creation of artificial agents that possess a capacity for knowledge, belief, personal desires, and self-reflexive thought – in short, the components necessary for an entity to be metavolitional and to possess a conscience.[130] The existence of conscience within artificial agents would have significant ramifications for the ways in which such agents could possibly be employed by organizations. Organizations that have metavolitional artificial agents as employees or customers could motivate them to act in certain ways by appealing to their conscience – to their sense of morality, justice, mercy, and the common good. At the same time, metavolitional artificial agents serving as employees within organizations could not be expected to automatically carry out instructions that have been given to them without first weighing them against the demands of their conscience. In the case of metavolitional artificial agents serving in roles that have a critical impact on human safety (e.g., robots serving as soldiers, police officers, surgeons, or the pilots of passenger vehicles) this could have positive or negative consequences.[131] For example, a robotic police officer who had been given an illegal and immoral command by its corrupt human supervisor to conceal evidence might decide to ignore that command as a result of its conscience; on the other hand, a robotic soldier could be manipulated by skilled 'conscience hackers' belonging to an opposing army who present the robot with fabricated evi-

---

den Berg, "Pieces of Me: On Identity and Information and Communications Technology Implants" (2012); McGee (2008), p. 217; and Gladden, *The Handbook of Information Security for Advanced Neuroprosthetics* (2015), pp. 26-27.

[129] For a discussion of different levels of volitionality, see Gladden, "The Diffuse Intelligent Other" (2016).

[130] See Calverley (2008) and Gladden, "The Diffuse Intelligent Other" (2016), for an explanation of the relationship of various cognitive capacities to the possession of second-order volitions (or metavolitions) on the part of artificially intelligent entities.

[131] Regarding the moral and practical implications of the possession of a conscience by artificial agents such as robots, see Wallach & Allen (2008).

dence of atrocities that appeal to known weaknesses or bugs within the robot's metavolitional mechanisms and which persuade the robot to desert its post and join that opposing army.

d. Knowledge Acquisition

The use of genetic engineering to alter the basic cognitive structures and processes of human agents and, especially, the use of neuroprosthetic devices to monitor, control, or bypass the natural cognitive activity of a human agent may result in agents that do not need to be trained or educated but which can simply be 'programmed' to perform certain tasks or even remotely controlled by external systems to guide them in the performance of those tasks.[132]

At the same time, there will be growing numbers and kinds of artificial agents that cannot simply be 'programmed' to carry out particular tasks in the manner of earlier conventional computers but which must be trained, educated, and allowed to learn through trial and error and firsthand interaction with and exploration of their world.[133]

e. Information-processing Locus

Increasingly the information processing performed by and within a human agent may occur not within the physical neural network that comprises natural biological neurons in the agent's brain but in other electronic or biological substrates, including neuroprosthetic devices and implantable computers that utilize traditional CPU-based technologies.[134]

Meanwhile, artificial agents' information processing may increasingly be performed within electronic or biological physical neural networks that do not rely on conventional CPU-based computing architectures, which do not possess a traditional operating system or the ability to run standard executable software programs, and which may be immune to many traditional electronic hacking techniques.[135]

---

[132] Regarding the 'programming' of human beings through the intentional, targeted modification of their memories and knowledge, see, e.g., McGee (2008); Pearce (2012); and Spohrer, "NBICS (Nano-Bio-Info-Cogno-Socio) Convergence to Improve Human Performance: Opportunities and Challenges" (2002). Regarding the remote control of human bodies by external systems, see Gladden, "Neural Implants as Gateways" (2016), and Gladden, *The Handbook of Information Security for Advanced Neuroprosthetics* (2015).

[133] See, e.g., Friedenberg (2008), pp. 55-72, 147-200; Haykin (2009); and Lamm & Unger (2011).

[134] See, e.g., Warwick & Gasson, "Implantable Computing" (2008), and the discussion of cognitive neuroprosthetics in Gladden, *The Handbook of Information Security for Advanced Neuroprosthetics* (2015), pp. 26-27.

[135] See, e.g., Friedenberg (2008), pp. 17-146.

f. Emotionality

The use of advanced neuroprosthetic devices that can heighten, suppress, or otherwise modify the emotions of human beings may result in populations of human agents whose programmatically controlled emotional behavior – or lack of emotional behavior – more closely resembles the functioning of computers than that of natural human beings.[136]

Meanwhile, the creation of autonomous robots with increasingly sophisticated and human-like social capacities and emotional characteristics – perhaps generated by the internal action of a complex physical neural network – may yield new types of artificial agents that cannot simply be programmed or configured to perform certain actions by their human operators but which must instead be motivated and persuaded to perform such actions through an application of psychological principles, negotiation techniques, and other practices typically employed with human beings.[137]

g. Cognitive Biases

Genetic engineering could potentially be used to create new designer types of cognitively engineered human beings whose brains do not develop cognitive biases. Alternatively, a neuroprosthetic device could be used to monitor the cognitive processes of a human mind and to alert the mind whenever the device detects that the individual is about to undertake a decision or action that is flawed or misguided because the mind's cognitive processes have been influenced by a cognitive bias; beyond directly intervening to prevent the effects of cognitive biases in this manner, such a device could potentially also train the mind over time to recognize and avoid cognitive biases on its own.[138]

Artificial agents that are patterned after human models of cognition and which display human-like levels of intelligence, emotion, sociality, and other traits may be subject to many of the same cognitive biases as human beings;[139] highly sophisticated artificial agents (e.g., superintelligences) might also suffer from their own idiosyncratic forms of cognitive biases that may be hard for their designers to recognize or anticipate.[140]

---

[136] For the possibility of developing emotional neuroprosthetics, see Soussou & Berger (2008); Hatfield et al. (2009); Kraemer (2011); and McGee (2008), p. 217.

[137] See Friedenberg (2008), pp. 179-200.

[138] See Gladden, "Neural Implants as Gateways" (2016).

[139] Regarding the potential for emotionally driven biases in artificial intelligences, see Friedenberg (2008), pp. 180-85, 197-98.

[140] For cognitive biases, mental illnesses, and other potentially problematic psychological conditions that may be manifested by advanced AIs, see, e.g., Chapter 4, "Wireheading, Addiction, and Mental Illness in Machines," in Yampolskiy, *Artificial Superintelligence: A Futuristic Approach* (2015).

h. Memory

Genetic engineering could potentially be used to enhance or otherwise alter the natural neural mechanisms for the encoding, storage, and retrieval of memories within the brain of a human agent. The use of neuroprosthetic devices to control, supplement, or replace the brain's natural memory mechanisms could result in human agents that possess memory that is effectively lossless, does not degrade over time, and can be easily copied to or from external systems.[141]

At the same time, the use of biological components or physical artificial neural networks as a substrate for the cognitive processes of artificial agents could result in agents whose memories are stored in a highly compressed form that degrades unreliably over time and which makes individual memories difficult to recall, even when they are retained within the memory system.[142]

i. Predictability

Human agents whose actions are influenced or controlled by neuroprosthetic devices or whose range of possible behaviors has been constrained through genetic engineering may produce behavior that is more predictable and is easily 'debugged' in a straightforward and precise manner that has traditionally been possible only when dealing with computers.[143]

Meanwhile, artificial agents that possess human-like cognitive capacities – including emotion and sociality – may generate behavior that is difficult to reliably predict, analyze, or control, especially if the agents' cognitive processes take place within a physical neural network whose activities and current state cannot easily be determined by outside observers.[144]

j. Vulnerability to Hacking

Human agents that possess electronic neuroprosthetic devices would be vulnerable to electronic hacking attempts similar to those employed against

---

[141] Regarding genetic and neuroprosthetic technologies for memory alteration in biological organisms, see Han et al., "Selective Erasure of a Fear Memory" (2009); Josselyn, "Continuing the Search for the Engram: Examining the Mechanism of Fear Memories" (2010); and Ramirez et al., "Creating a False Memory in the Hippocampus" (2013). Regarding the use of neuroprosthetic systems to store memories as effectively lossless digital exograms, see Gladden, "Neural Implants as Gateways" (2016), and Gladden, *The Handbook of Information Security for Advanced Neuroprosthetics* (2015), pp. 156-57.

[142] Regarding memory mechanisms for artificial agents, including those involving neural networks, see Friedenberg (2008), pp. 55-72.

[143] Regarding the testing and debugging of neuroprosthetic devices (especially in relation to information security), see Gladden, *The Handbook of Information Security for Advanced Neuroprosthetics* (2015), pp. 176-77, 181-84, 213-14, 248-19, 242-43, 262.

[144] For an overview of issues relating to the social behavior of artificial agents, see Friedenberg (2008), pp. 217-34.

conventional computers. Moreover, advanced technologies for genetic engineering and the production of customized biopharmaceuticals and biologics may allow the biohacking even of human agents that do not possess electronic neuroprosthetic components.[145]

At the same time, artificial agents that include or wholly comprise biological components rather than electronic components might thereby reduce or eliminate their vulnerability to traditional methods of electronic hacking. However, such artificial agents may be vulnerable to biohacking approaches that are based on genetic engineering or biopharmaceutical technologies as well as to psychologically based social engineering attacks.[146]

3. Social Interaction

The forms of social engagement and belonging available to human and artificial agents are expected to be transformed by the advent of posthumanizing technologies. Such change will be manifested through the possession (or absence) of a number of key characteristics, which are described below.

a. Sociality

Neuroprosthetic devices or genetic modifications that affect long-term memory processes could make it difficult or impossible for human agents to engage in friendships and other long-term social relationships with other intelligent agents. Such human agents would no longer be fully social but instead semisocial or even nonsocial.[147] Ongoing immersion in virtual worlds or neuroprosthetically enabled cybernetic networks with other human minds or other kinds of intelligent agents could potentially also lead to the atrophying or enhancement of human agents' social capacities.

At the same time, an increasing number of artificial agents may possess fully human-like sociality, including the ability to participate in long-term social relations that deepen and evolve over time as a result of the agents' experience of such engagement and which are shaped by society's expectations for the social roles to be filled by the relations' participants. This would potentially allow artificial agents to serve as charismatic leaders of human beings

---

[145] Regarding the possibility of hybrid biological-electronic computer viruses and other attacks, see Gladden, *The Handbook of Information Security for Advanced Neuroprosthetics* (2015), p. 53.

[146] For a discussion of social engineering attacks, see Rao & Nayak (2014), pp. 307-23, and Sasse et al., "Transforming the 'weakest link'—a human/computer interaction approach to usable and effective security" (2001).

[147] For ways of describing and classifying degrees of sociality of artificial entities, see Vinciarelli et al. (2012) and Gladden, "Managerial Robotics" (2014).

who guide and manage the activities of their followers not through threats or intimidation but by inspiring or seducing them.[148]

b. Culture

Human agents whose thoughts, dreams, and aspirations have been attenuated or even eliminated or whose physical sensorimotor systems are controlled through the use of genetic engineering, neuroprosthetic devices, or other advanced technologies may no longer possess a desire or ability to perceive or generate cultural artifacts. If a single centralized system (e.g., a server providing a shared virtual reality experience to large numbers of individuals) maintains and controls all of the sensorimotor channels through which human agents are able to create and experience culture, then that automated system may generate all of the aspects of culture within that virtual world, without the human agents who dwell in that world being able to contribute meaningfully to the process.[149]

Artificial agents already play important roles in supporting the creation, maintenance, and dissemination of human culture(s), and some artificial agents are already capable of acting autonomously to generate works of art, poetry, music, content for computer games, webpages, Internet memes, and other kinds of cultural artifacts.[150] It is expected that in the future, artificial agents will not only play a role in contributing to predominantly human cultures or act in symbiosis with human agents to create hybrid human-artificial cultures that are truly shared; they will also create among themselves entirely new synthetic cultures whose art, music, architecture, literature, philosophy, and way of life could never have been developed by human beings (and perhaps cannot even be observed or comprehended by human beings), due to the physical and cognitive differences between human agents and the artificial agents that create such cultures.[151]

---

[148] See Gladden, "The Social Robot as 'Charismatic Leader': A Phenomenology of Human Submission to Nonhuman Power" (2014). For an exploration of the potential social behavior of advanced artificial agents, see Friedenberg (2008), pp. 217-34.

[149] Regarding the possibilities of a centralized computerized system shaping culture by mediating and influencing or controlling the communications among neuroprosthetically enabled human minds, see Gladden, "Utopias and Dystopias as Cybernetic Information Systems" (2015), and Gladden, "From Stand Alone Complexes to Memetic Warfare: Cultural Cybernetics and the Engineering of Posthuman Popular Culture" (2016).

[150] See Friedenberg (2008), pp. 127-46, and Gladden, "From Stand Alone Complexes to Memetic Warfare" (2016).

[151] See Payr & Trappl (2003); regarding the creation of hybrid human-artificial cultures in an organizational setting, see Gladden, "Leveraging the Cross-Cultural Capacities of Artificial Agents as Leaders of Human Virtual Teams" (2014). For a philosophical analysis of digital-physical ecosystems in which human and artificial agents may interact symbiotically to generate shared cognitive and cultural artifacts (and in which such artifacts may even exist as actors that

c. Spirituality

Researchers have raised concerns that the use of neuroprosthetic devices to replace or dramatically alter the structures and activities of the body and mind of human agents may result in the loss of those fundamental characteristics that make such agents human. While this can be analyzed from purely biological and psychological perspectives,[152] it may alternatively be understood from philosophical and theological perspectives as a dissolution of the 'soul' or 'essence' of such human agents.[153] The use of genetic engineering in transhumanist efforts to design beings that possess superior (and even transcendent) intelligence and morality raises similarly significant questions about the nature of humanity and future human beings.

At the same time, artificial agents that possess sufficiently sophisticated and human-like cognitive capacities may be subject to instinctive desires to seek out and experience some transcendent truth and reality and may engage in behaviors such as meditation, contemplation, and even prayer.[154]

d. Political Engagement

Human agents that have been neuroprosthetically augmented may form social and technological networks that demonstrate new kinds of network topologies and may engage in new forms of cybernetic relations with similarly augmented human agents and with artificial entities; such human agents may dwell (virtually, if not physically) in societies in which traditional human political systems and structures are not meaningful or relevant.[155] Such human agents may find themselves disconnected from political life and institutions of the 'real' world and instead immerse themselves in new kinds of structures that might resemble traditional computer networks more than political systems.

At the same time, artificial agents that possess intelligence and sociality that are human-like (or which surpass the capacities of human beings) may create political systems and structures to govern their relations with one another or may seek to participate in human political systems.[156]

---

can propagate themselves), see, e.g., Kowalewska, "Symbionts and Parasites – Digital Ecosystems" (2016).

[152] For a discussion of, e.g., the psychological impact of neuroprosthetic devices upon a user's perceptions of authenticity and identity, see Kraemer (2011) and Van den Berg (2012).

[153] E.g., see Gladden, "'Upgrading' the Human Entity" (2015).

[154] For a discussion of such possibilities, see Kurzweil (2000).

[155] Regarding the possible fragmentation of human societies as a result of posthuman neuroprosthetics, see Gladden, "Utopias and Dystopias as Cybernetic Information Systems" (2015); McGee (2008), pp. 214-16; Warwick (2014), p. 271; Rubin (2008); Koops & Leenes, "Cheating with Implants: Implications of the Hidden Information Advantage of Bionic Ears and Eyes" (2012), p. 127; and Gladden, *The Handbook of Information Security for Advanced Neuroprosthetics* (2015), 166-67.

[156] For the possibility of social robots exercising referent power or charismatic authority within

e. Economic Engagement

The adoption of posthumanizing technologies may weaken the ability of human beings to serve as autonomous economic actors. Depending on the precise terms under which such components were acquired, a human agent whose body has been subject to extensive neuroprosthetic augmentation and is largely composed of electronic components may not even 'own' its own body or the products generated by that body, including intellectual property such as thoughts and memories. Such a human agent may for practical purposes be wholly dependent on and economically subjugated to the corporation(s), government agencies, or other institutions that provide maintenance services for its synthetic components and legally or practically barred from purchasing goods or services from competing enterprises.[157] The use of neuroprosthetic devices or other technologies that directly affect a human agent's cognitive processes may also impair that agent's ability to make free choices as an autonomous economic actor.

Conversely, artificial agents may gain new abilities to function as independent economic actors. Some forms of artificial life may be able to function as autonomous organism-enterprises that acquire resources from within the digital-physical ecosystem shared with human beings, process the resources to generate goods and services, and then exchange those goods and services with human beings or other artificial agents to generate revenue, including profit that the artificial life-form can use for purposes of growth, reproduction, or risk management.[158] Such artificial life-forms could compete directly with human enterprises within the real-world economy or offer new kinds of goods and services that human agents are incapable of offering.

f. Legal Status

Human agents that have been intentionally engineered by other human beings or organizations (e.g., biological clones or custom-designed human beings) may be subject to claims that they are not full-fledged legal persons but rather wards or even property of those who have created them – especially if the agents have been engineered to possess characteristics that clearly distinguish them from 'normal' human beings.[159]

---

human social or political institutions, see Gladden, "The Social Robot as 'Charismatic Leader'" (2014).

[157] See Gladden, "Neural Implants as Gateways" (2016), and Gladden, *The Handbook of Information Security for Advanced Neuroprosthetics* (2015).

[158] For an approach to modelling entrepreneurship on the part of artificial agents, see Ihrig, "Simulating Entrepreneurial Opportunity Recognition Processes: An Agent-Based and Knowledge-Driven Approach" (2012). For an innovative exploration of the possibility of creating fully autonomous systems for entrepreneurship, see Rijntjes, "On the Viability of Automated Entrepreneurship" (2016). See also Gladden, "The Artificial Life-Form as Entrepreneur" (2014).

[159] See, e.g., Cesaroni, "Designer Human Embryos as a Challenge for Patent Law and Regulation"

Conversely, sufficiently sophisticated artificial agents that possess human-like cognitive capacities or biological components may not be considered inanimate objects or property from a legal perspective but either moral patients possessing rights that must be protected or even moral subjects that can be held legally responsible for their own actions.[160]

B. The Four Types of Beings Relevant for Technologically Posthumanized Organizations

The only two quadrants of the Posthuman Management Matrix that have historically been considered relevant objects for management scholarship and practice are those of natural human beings and, more recently, computers. However, the advent of new posthumanizing technologies will create a variety of entities that fall within the remaining two quadrants and which can serve as potential employees, partners, and customers for businesses and other organizations. This will require the field of management to directly address those two quadrants – to create theoretical frameworks for understanding the activities and organizational potential of such entities and to develop new practices for managing them. Figure 3 reflects the fact that during the dawning Posthuman Age, all four quadrants of the Matrix will at last be relevant for management.

We can now consider in more detail the future roles that all four types of entities may play for future posthumanized organizations, along with the academic disciplines and practical bodies of knowledge that can contribute to their effective management.

1. Natural Human Beings

At least during the early stages of the emerging Posthuman Age, human agents with anthropic characteristics will remain the key leaders and decision-makers within businesses and other organizations. This will not necessarily be due to the fact that such natural human beings are more capable than artificial agents or technologically modified human beings when it comes to performing the actions involved with managing others; it will instead likely be due to legal, political, and cultural considerations. For example, even after sufficiently sophisticated social robots have been developed that are capable of

---

(2012); Pereira, "Intellectual Property and Medical Biotechnologies" (2013); Bera, "Synthetic Biology and Intellectual Property Rights" (2015); Camenzind, "On Clone as Genetic Copy: Critique of a Metaphor" (2015); Section B ("Enhancement") and Section D ("Synthetic Biology and Chimera") in *The Future of Bioethics: International Dialogues*, edited by Akabayashi (2014); and Singh, *Biotechnology and Intellectual Property Rights: Legal and Social Implications* (2014). For perspectives on the ways in which such issues have been explored within fiction, see, e.g., Pérez, "Sympathy for the Clone: (Post) Human Identities Enhanced by the 'Evil Science' Construct and its Commodifying Practices in Contemporary Clone Fiction" (2014).

[160] Regarding such questions see, e.g., Wallach & Allen (2008) and Calverley (2008).

serving effectively as CEOs of businesses, it may take many years before the ethical and political questions surrounding such practices have been resolved to the point that human legislators and regulators allow the human businesses and other institutions that are subject to their oversight to legally employ such artificial agents as CEOs.[161]

It appears likely that human agents that possess at least limited computronic characteristics will achieve positions of formal leadership within organizations before artificial agents accomplish that feat. This can be anticipated due to the fact that current law and cultural tradition already allow human beings to fill such roles: while existing laws would generally need to be explicitly changed in order to *allow* artificial agents to serve, for example, as CEOs of publically traded corporations, those same laws would need to be explicitly changed in order to *bar* human agents who possess computronic characteristics from filling such roles. Indeed, declining to offer a human being a position as an executive within a business because he or she possesses a pacemaker, defibrillator, cochlear implant, robotic artificial limb, or other device that endows him or her with limited computronic characteristics would, in many cases, be considered a form of unlawful employment discrimination, and even simply attempting to ascertain whether a potential employee possesses such traits could in itself be illicit.[162]

Although human agents who possess extensive computronic characteristics and artificial agents are expected to gradually fill a broader range of positions within organizations, there will likely remain a number of professions or specific jobs which – at least in the early stages of the Posthuman Age – can only be filled by natural, unmodified human agents.[163] For example, some positions within the military, police forces, or intelligence services may initially be restricted to natural human beings, in order to avoid the possibility of external adversaries hacking the minds or bodies of such agents and gaining control of them and the information that they possess. Roles as judges, arbitrators, and regulators might be restricted to natural human beings on ethical grounds, to ensure that such officials' decisions are being made on the basis of human wisdom, understanding, and conscience (including the *known biases* of the human mind), rather than executed by software programs that might possess unknown bugs or biases or be surreptitiously manipulated. Some roles – such as those of priest, therapist, poet, or existentialist philosopher – might as a practical matter be restricted to natural human beings, because the work performed by persons in such positions is considered to derive unique value from the fact that it is performed by a human being rather than a machine.

---

[161] The question arises of whether such artificial agents will voluntarily allow themselves to be subject to human laws or will instead seek to formulate their own.

[162] See Gladden, *The Handbook of Information Security for Advanced Neuroprosthetics* (2015), pp. 93-94.

[163] See Gladden, "Neural Implants as Gateways" (2016).

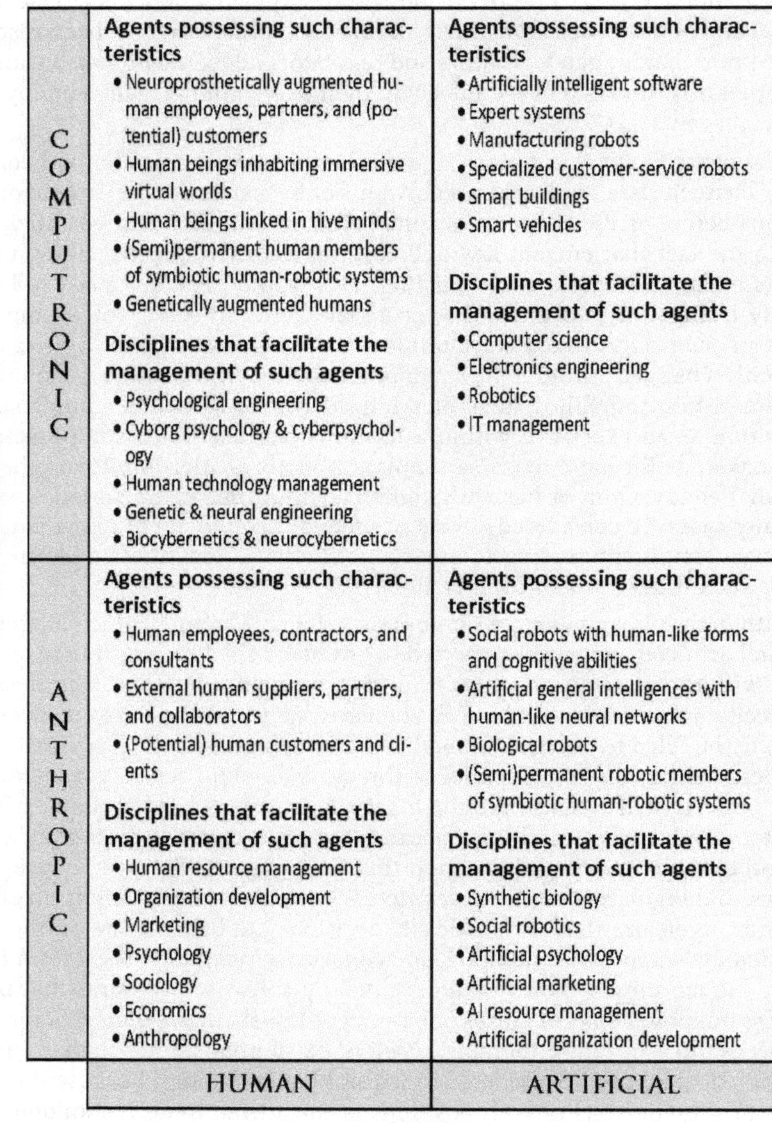
Agents possessing such characteristics
• Neuroprosthetically augmented human employees, partners, and (potential) customers
• Human beings inhabiting immersive virtual worlds
• Human beings linked in hive minds
• (Semi)permanent human members of symbiotic human-robotic systems
• Genetically augmented humans
Disciplines that facilitate the management of such agents
• Psychological engineering
• Cyborg psychology & cyberpsychology
• Human technology management
• Genetic & neural engineering
• Biocybernetics & neurocybernetics
Agents possessing such characteristics
• Artificially intelligent software
• Expert systems
• Manufacturing robots
• Specialized customer-service robots
• Smart buildings
• Smart vehicles
Disciplines that facilitate the management of such agents
• Computer science
• Electronics engineering
• Robotics
• IT management
Agents possessing such characteristics
• Human employees, contractors, and consultants
• External human suppliers, partners, and collaborators
• (Potential) human customers and clients
Disciplines that facilitate the management of such agents
• Human resource management
• Organization development
• Marketing
• Psychology
• Sociology
• Economics
• Anthropology
Agents possessing such characteristics
• Social robots with human-like forms and cognitive abilities
• Artificial general intelligences with human-like neural networks
• Biological robots
• (Semi)permanent robotic members of symbiotic human-robotic systems
Disciplines that facilitate the management of such agents
• Synthetic biology
• Social robotics
• Artificial psychology
• Artificial marketing
• AI resource management
• Artificial organization development
COMPUTRONIC
ANTHROPIC
CHARACTERISTICS
HUMAN
ARTIFICIAL
AGENTS

Fig. 3: The Posthuman Management Matrix displaying the two types of entities (in the lower left and upper right quadrants) that have long been relevant for the theory and practice of organizational management, joined by two types of entities (in the upper left and lower right quadrants) that are becoming newly relevant in the dawning Posthuman Age.

The adoption of posthumanizing technologies across the world will likely be highly uneven, as differences in economic resources and systems, political systems, and philosophical, religious, and cultural traditions combine in unique ways in different parts of the world to either spur or restrain the adoption of such technologies. The role of natural human beings as workers and consumers may maintain greater importance in some regions and industries than in others. Wherever such beings fill places as workers or consumers, the traditional disciplines of psychology, sociology, economics, anthropology, cultural studies, marketing, organization development, HR management, and ergonomics will continue to be relevant for theorists and practitioners of organizational management.

2. Computers

It is expected that artificial agents with computronic characteristics will continue to play a fundamental – and ever-growing – role as backbone elements within the increasingly ubiquitous networked systems that constitute the digital-physical infrastructure within which human beings will dwell. Artificial systems that can be quickly and reliably programmed to perform certain tasks without any worry that a system might become bored or annoyed or object to its assigned tasks on moral grounds will remain highly useful and desirable.[164]

Although the theory and practice used to design, implement, and manage such systems will likely continue to evolve rapidly, even during the near-future Posthuman Age such disciplines will likely be recognizable as heirs of our

[164] One can consider, for example, the case of autonomous military robots. Serious efforts have been undertaken to create morally aware autonomous military robots that can be programmed with a knowledge of and obedience to relevant national and international legal obligations governing the conduct of war, as well as a knowledge of relevant ethical principles and even a 'conscience' that allows a robot to assimilate all available information, evaluate the propriety of various courses of action, and select an optimal ethically and legally permissible course of action. However, scholars have noted the possibility for cynical manipulation of such technologies – e.g., perhaps the creation of robots who possess a 'conscience' that is sufficiently developed to reassure the public about the ethicality of such devices while not being restrictive or powerful enough to actually block the robot from performing any activities desired by its human overseers. See Sharkey, "Killing Made Easy: From Joysticks to Politics" (2012), pp. 121-22. On the other hand, if a robot's conscience is such that the robot becomes a conscientious objector and refuses to participate in any military actions at all, then the robot becomes operationally useless from the perspective of its intended purpose.

contemporary fields of computer science, electronics engineering, robotics, and IT management.

3. Cyborgs

As described in earlier sections, the increasing use of neuroprosthetic enhancement, genetic engineering, and other posthumanizing technologies is expected to result in a growing number of human agents that no longer possess the full suite of traditional anthropic characteristics but instead reflect some degree of computronic characteristics. Such agents might include human employees or customers whose artificial sense organs or limbs mediate their experience of their physical environment;[165] human beings who never physically leave their bedroom but instead engage with the world through long-term immersion in virtual worlds and digital ecosystems;[166] groups of human beings whose minds are neuroprosthetically linked to create a hive mind with

---

[165] For discussions of particular types of neuroprosthetic mediation of sensory experience of one's environment, see, e.g., Ochsner et al., "Human, non-human, and beyond: cochlear implants in socio-technological environments" (2015), and Stiles & Shimojo, "Sensory substitution: A new perceptual experience" (2016). On ways in which the absence of mediation transforms teleoperation into telepresence in the case of noninvasive brain-computer interfaces, see Salvini et al., "From robotic tele-operation to tele-presence through natural interfaces" (2006).

[166] Regarding the implications of long-term immersion in virtual reality environments, see, e.g., Bainbridge, *The Virtual Future* (2011); Heim, *The Metaphysics of Virtual Reality* (1993); Geraci, *Apocalyptic AI: Visions of Heaven in Robotics, Artificial Intelligence, and Virtual Reality* (2010); and Koltko-Rivera, "The potential societal impact of virtual reality" (2005). Regarding psychological, social, and political questions relating to repetitive long-term inhabitation of virtual worlds through a digital avatar, see, e.g., Castronova, "Theory of the Avatar" (2003). On the risks of potentially 'toxic immersion' in a virtual world, see Castronova, *Synthetic Worlds: The Business and Culture of Online Games* (2005). On implantable systems for augmented or virtual reality, see Sandor et al., "Breaking the Barriers to True Augmented Reality" (2015), pp. 5-6. For a conceptual analysis of the interconnection between physical and virtual reality and different ways in which beings and objects can move between these worlds, see Kedzior, "How Digital Worlds Become Material: An Ethnographic and Netnographic Investigation in Second Life" (2014).

a collective consciousness;[167] human beings who are temporarily or permanently joined in symbiotic relationships with robotic exoskeletons,[168] companions,[169] or supervisors;[170] or genetically augmented human beings whose physical structures and cognitive capacities have been intentionally engineered to make them especially well-suited (or poorly suited) to perform particular roles within society.[171]

Because such technological modification may dramatically affect human agents' physical and cognitive traits, their behavior can no longer be understood, predicted, or managed simply by relying on historical disciplines such as psychology, sociology, or HR management. Established and evolving fields

---

[167] Regarding the possibility of hive minds, see, e.g., McIntosh, "The Transhuman Security Dilemma" (2010), and Gladden, "Utopias and Dystopias as Cybernetic Information Systems" (2015). For more detailed taxonomies and classification systems for different kinds of potential hive minds, see Chapter 2, "Hive Mind," in Kelly, *Out of control: the new biology of machines, social systems and the economic world* (1994); Kelly, "A Taxonomy of Minds" (2007); Kelly, "The Landscape of Possible Intelligences" (2008); Yonck, "Toward a standard metric of machine intelligence" (2012); and Yampolskiy, "The Universe of Minds" (2014). For the idea of systems whose behavior resembles that of a hive mind but without a centralized controller, see Roden, *Posthuman Life: Philosophy at the Edge of the Human* (2014), p. 39. For critical perspectives on the idea of hive minds, see, e.g., Bendle, "Teleportation, cyborgs and the posthuman ideology" (2002), and Heylighen, "The Global Brain as a New Utopia" (2002). Regarding the need for society to debate the appropriateness of neuroprosthetic technologies that facilitate hive minds, see Maguire & McGee, "Implantable brain chips? Time for debate" (1999).

[168] For examples of such systems currently under development, see *Wearable Robots: Biomechatronic Exoskeletons*, edited by Pons (2008); Guizzo & Goldstein, "The rise of the body bots [robotic exoskeletons]" (2005); and Contreras-Vidal & Grossman, "NeuroRex: A clinical neural interface roadmap for EEG-based brain machine interfaces to a lower body robotic exoskeleton" (2013). For a discussion of the extent to which the form of an exoskeleton can differ from that of the human body before it becomes impossible for its human operator to interface with the exoskeleton, see Gladden, "Cybershells, Shapeshifting, and Neuroprosthetics" (2015).

[169] See Dautenhahn, "Robots we like to live with?! - A Developmental Perspective on a Personalized, Life-long Robot Companion" (2004); Van Oost and Reed, "Towards a Sociological Understanding of Robots as Companions" (2011); Shaw-Garlock, "Loving machines: Theorizing human and sociable-technology interaction" (2011); Whitby, "Do You Want a Robot Lover? The Ethics of Caring Technologies" (2012); and *Social Robots and the Future of Social Relations*, edited by Seibt et al. (2014).

[170] See, e.g., Samani & Cheok, "From human-robot relationship to robot-based leadership" (2011); Samani et al., "Towards robotics leadership: An analysis of leadership characteristics and the roles robots will inherit in future human society" (2012); Gladden, "Leveraging the Cross-Cultural Capacities of Artificial Agents" (2014); and Gladden, "The Social Robot as 'Charismatic Leader'" (2014).

[171] Regarding such possibilities, see *Converging Technologies for Improving Human Performance: Nanotechnology, Biotechnology, Information Technology and Cognitive Science*, edited by Bainbridge (2003); Canton (2004), pp. 186-98; and Khushf, "The use of emergent technologies for enhancing human performance: Are we prepared to address the ethical and policy issues" (2005).

such as genetic engineering, neural engineering, neurocybernetics, and biocybernetics will offer resources for management theorists and practitioners who must account for the existence and activity of such agents. However, it is likely that entirely new disciplines will arise – and will need to arise – in order to fill the conceptual and practical gaps that exist between those structures and dynamics that will be manifested by cyborgs and those that are addressed by existing disciplines. In particular, new disciplines may study and manage computronic human agents using many of the same techniques that have previously been employed with artificial agents. Such hypothetical new fields might include disciplines such as:

- **Psychological engineering**, which would apply practices from fields like electronics engineering to the design of a human psyche.[172] It might involve the use of genetic engineering and gene therapy, neuroprosthetic devices, immersive virtual reality, and other technologies to create and maintain human beings who possess particular (and potentially non-natural) cognitive structures, processes, and behaviors.
- **Cyborg psychology** and **cyberpsychology**, which would apply the knowledge and methods of traditional psychology to understand the cognitive structures and processes of human beings whose psychology is atypical as a result of neuroprosthetic augmentation, long-term immersion in virtual reality environments, or other factors.[173] Subdisciplines might include cyberpathology,[174] for example.

---

[172] For earlier uses of the term 'psychological engineering' in different contexts, see, e.g., Doyle, "Big problems for artificial intelligence" (1988), p. 22, which employs the term in the context of artificial intelligence, with psychological engineering's goal being "parallel to the aim of any engineering field, namely to find economical designs for implementing or mechanizing agents with specified capacities or behaviors," and Yagi, "Engineering psychophysiology in Japan" (2000), p. 361, which defines psychological engineering to be "engineering relating to human psychological activities" and include themes such as "the development of new systems between the human mind and machines" that yield not only convenience but comfort, "the development of the technology to measure psychological effects in industrial settings," and "the development of new types of human-machine systems incorporating concepts and procedures utilizing virtual reality."

[173] For other use of the term 'cyborg psychology,' see, e.g., Plowright, "Neurocomputing: some possible implications for human-machine interfaces" (1996). For earlier use of the term 'cyberpsychology' in various contexts, see, e.g., *Cyberpsychology*, edited by Gordo-López & Parker (1999); Riva & Galimberti. *Towards CyberPsychology: Mind, Cognition, and Society in the Internet Age* (2001); *Cyberpsychology: Journal of Psychosocial Research*, founded in 2007; and Norman, *Cyberpsychology: An Introduction to Human-Computer Interaction* (2008).

[174] See, e.g., Chapter 4, "Wireheading, Addiction, and Mental Illness in Machines," in Yampolskiy, *Artificial Superintelligence: A Futuristic Approach* (2015).

- **Human technology management** (or 'anthropotech management'[175]), which would apply the knowledge and practices of traditional IT management to the management of organizational resources (e.g., human employees) whose neuroprosthetic or genetic augmentation or intimate cybernetic integration with computerized systems at a structural or behavioral level allows them to be managed in ways similar to those utilized with traditional IT assets.

4. Bioroids

As described in earlier sections, organizations will increasingly need to deal with the existence of artificial agents that possess anthropic characteristics as both potential workers and consumers of the goods and services that organizations produce. Such bioroids might include social robots that resemble human beings in their physical form and cognitive capacities,[176] artificial general intelligences[177] that process information using complex physical neural networks rather than CPU-based platforms,[178] robots possessing biological

---

[175] For the use of such terminology, see, e.g., the Anthropotech project of the University of the West of England and University of Bristol that has studied the philosophical and ethical implications of "*Anthropotech*: the technological alteration of the body for the purpose of augmenting existing capacities, introducing new ones, or aesthetically improving the body" and which has drawn its inspiration explicitly from Jérôme Goffette's *Naissance de l'anthropotechnie: De la médecine au modelage de l'humain* (2006). See "Anthropotech" (2013).

[176] For an overview of different perspectives on social robots that behaviorally resemble and can interact with human beings, see, e.g., Breazeal (2003); Gockley et al., "Designing Robots for Long-Term Social Interaction" (2005); Kanda & Ishiguro (2013); *Social Robots and the Future of Social Relations*, edited by Seibt et al. (2014); *Social Robots from a Human Perspective*, edited by Vincent et al. (2015); and *Social Robots: Boundaries, Potential, Challenges*, edited by Marco Nørskov (2016).

[177] Regarding challenges inherent in the development of artificial general intelligence and potential paths toward that objective, see, e.g., *Artificial General Intelligence*, edited by Goertzel & Pennachin (2007); *Theoretical Foundations of Artificial General Intelligence*, edited by Wang & Goertzel (2012); and *Artificial General Intelligence: 8th International Conference, AGI 2015: Berlin, Germany, July 22-25, 2015: Proceedings*, edited by Bieger et al. (2015).

[178] Regarding AIs that utilize physical neural networks, see, e.g., Snider (2008); Versace & Chandler (2010); and *Advances in Neuromorphic Memristor Science and Applications*, edited by Kozma et al. (2012). For a discussion of such technologies from the perspective of information security, see Pino & Kott, "Neuromorphic Computing for Cognitive Augmentation in Cyber Defense" (2014), and Lohn et al., "Memristors as Synapses in Artificial Neural Networks: Biomimicry Beyond Weight Change" (2014).

components,[179] and robots that exist in permanent symbiosis with human agents to whom they serve as bodies, colleagues, or guides.[180]

The physical forms and processes, cognitive capacities, and social engagement of such bioroids will likely differ in their underlying structures and dynamics from those of human beings, no matter how closely they outwardly resemble them. Thus traditional human-focused disciplines such as psychology, economics, and HR management cannot be applied directly and without modification to analyze, predict, or manage the behavior of bioroids. On the other hand, traditional disciplines such as computer science, electronics engineering, and IT management will not in themselves prove adequate for shaping the behavior of such unique anthropic artificial agents.

Emerging fields such as synthetic biology and social robotics provide a starting point for the development and management of bioroids. As researchers attempt to create new theoretical and practical frameworks for managing such agents, we might expect to witness the development of new fields that study and manage artificial agents utilizing approaches that have traditionally been applied to human agents; these new fields might include disciplines like:

- **Artificial psychology**, which is already being formulated as a discipline[181] and which applies the extensive knowledge and techniques

[179] See, e.g., Ummat et al. (2005); Andrianantoandro et al. (2006); Lamm & Unger (2011); Cheng & Lu (2012); and Kawano et al., "Finding and defining the natural automata acting in living plants: Toward the synthetic biology for robotics and informatics in vivo" (2012).

[180] Regarding robots that exist in symbiotic relationships with human beings as their physical bodies (i.e., constituting a cyborg), see, e.g., Tomas, "Feedback and Cybernetics: Reimaging the Body in the Age of the Cyborg" (1995); Clark, *Natural-born cyborgs: Minds, Technologies, and the Future of Human Intelligence* (2004); and Anderson "Augmentation, symbiosis, transcendence: technology and the future(s) of human identity" (2003). For discussions of robots serving as colleagues to human workers, see, e.g., Ablett et al., "A Robotic Colleague for Facilitating Collaborative Software Development" (2006); Vänni and Korpela, "Role of Social Robotics in Supporting Employees and Advancing Productivity" (2015); and Gladden, "Leveraging the Cross-Cultural Capacities of Artificial Agents" (2014). For a notable early allusion to the possibility of robotic colleagues, see Thompson (1976). For robotic systems that serve as 'guides' to human beings in a very practical and functional sense, see, e.g., Chella et al., "A BCI teleoperated museum robotic guide" (2009), and Vogiatzis et al., "A conversant robotic guide to art collections" (2008). For robots that serve as charismatic leaders (and perhaps even spiritual guides) for human beings, see Gladden, "The Social Robot as 'Charismatic Leader'" (2014).

[181] Friedenberg has introduced the concept of 'artificial psychology' as a new branch of psychology that addresses the cognitive behavior of synthetic agents; see Friedenberg (2008). 'Artificial psychology' is not simply a form of computer programming or IT management. It is psychology: just as complex and mysterious a discipline as when directed to the cognitive structures and processes of human beings, except that in this case it is directed to the cognitive structures and processes of robots or AIs.

developed through the academic study of human psychology to understanding, designing, and controlling the psychology of synthetic beings such as artificial general intelligences or social robots.

- **Artificial marketing**, which would address the design, production, sale, and distribution of goods and services targeted at consumers who are not human beings but artificial entities.
- **AI resource management**, which would deal with the management of artificial entities within an organizational context not as though they were conventional IT assets like desktop computers but as human-like employees, drawing on the knowledge and practices developed in the field of human resource management.
- **Artificial organization development**, which would seek to bring about long-term systemic improvements in the performance of organizations whose members are synthetic entities – not by directly reprogramming them or updating their software but through the use of intervention techniques such as coaching and mentoring, surveys, teambuilding exercises, changes to workplace culture, and the design of strategic plans and incentive structures. This would adapt the explicitly 'humanistic' approaches of the existing field of organization development to serve new constituencies of nonhuman agents.[182]

C. Exploring the 'Fifth Quadrant': Hybrid Agents within Hybrid Systems

While it is true that management theory and practice must be capable of separately addressing each of the four types of entities described above, within real-world organizations it will in practice be difficult to extricate one kind of entity from its relationships with those of other kinds – just as it is already difficult to consider the performance of human workers apart from the performance of the computerized technologies that they use in carrying out their tasks.

In practice, the four types of entities described above will frequently work intimately with one another, either as elements in hybrid systems that have been intentionally designed or as members of systems whose participants can voluntarily join and leave and which can include any types of agents. For example, a company might maintain a persistent virtual world in which all of its

[182] Regarding the goals and practices of organization development, see, e.g., Anderson, *Organization Development: The Process of Leading Organizational Change* (2015), and Bradford & Burke, *Reinventing Organization Development: New Approaches to Change in Organizations* (2005). For the humanistic foundations of organization development, see, e.g., Bradford & Burke (2005); "The International Organization Development Code of Ethics" of The OD Institute; the OD Network's "Organization and Human Systems Development Credo"; IAGP's "Ethical Guidelines and Professional Standards for Organization Development and Group Process Consultants"; and the OD Network's "Principles of OD Practice."

human and artificial personnel come together to work rather than meeting in a physical workplace, or a firm might operate an online marketplace in which human and artificial agents of all types are welcomed to purchase or consume the company's products and services – without the firm necessarily knowing or caring whether a particular consumer is a human or artificial agent. In such cases, the focus of an organization's management efforts is not on specific agents that participate in or constitute a system but on the management of the system as a whole.

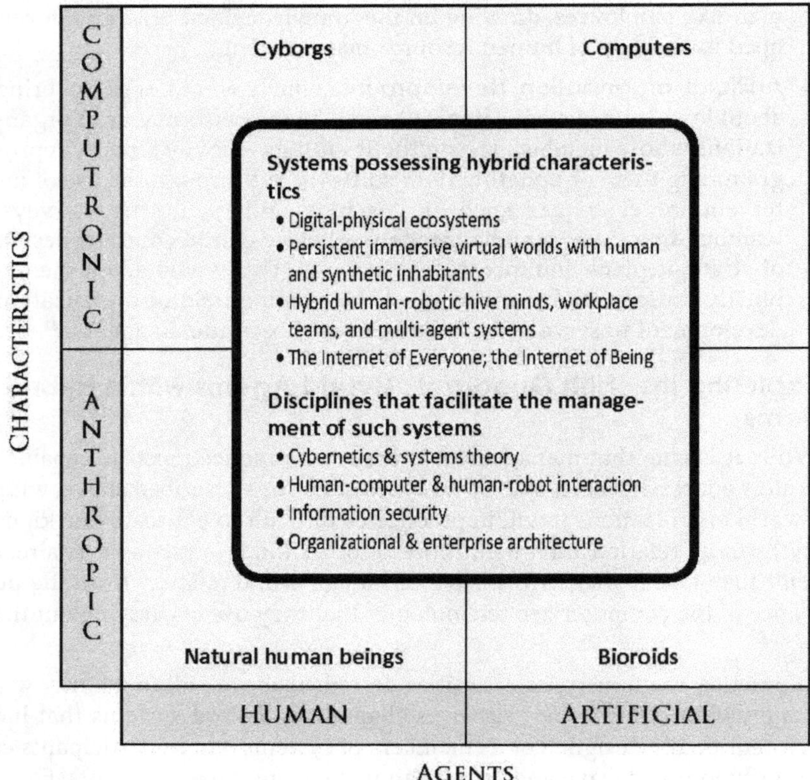

Fig. 4: The 'fifth quadrant' of the Posthuman Management Matrix, which spans and ties together all four types of entities that will be of relevance for organizational management in the Posthuman Age.

Systems that incorporate or comprise multiple types of agents might include digital-physical ecosystems; persistent immersive virtual worlds that are home to both human and artificial inhabitants; and hybrid human-robotic hive minds, workplace teams, and multi-agent systems. Moreover, after having evolved into the Internet of Things and eventually comprising all *objects* as the 'Internet of Everything,'[183] the Internet as a whole might come to encompass all *subjects* – all sapient minds and persons – thanks to the wearable and implantable computers and neuroprosthetic devices that will increasingly serve as gateways, vehicles, and virtualizing bodies that provide their human hosts and users with a permanent link to and presence in the world's digital-physical ecosystems. In this way, we can expect the growth of a lush, dense, complex, unruly, all-embracing digital-physical cyber-jungle that is not simply the Internet of Everything but the Internet of Everyone, the Internet of Life, the Internet of Being. Together these kinds of systems can be seen as occupying a 'fifth quadrant' that lies at the heart of the Posthuman Management Matrix and which reaches into and joins all of the other four quadrants, as reflected in Figure 4.

The kinds of rich and sophisticated human-artificial systems that exist within the fifth quadrant cannot be effectively managed simply by drawing insights from an array of disciplines that focus exclusively on either human agents *or* artificial agents. Instead, disciplines will be required whose theory and practice holistically embrace both the forms and behaviors of human *and* artificial agents as well as anthropic *and* computronic characteristics and which occupy themselves with systems in which the four possible types of entities are closely integrated or even inextricably merged.

Already, existing disciplines such as cybernetics and systems theory attempt to provide a universal conceptual framework that can account for the structures and dynamics of all kinds of viable systems, whether they be human, artificial, hybrid, or of some previously unknown form. The fields of human-computer interaction, human-robot interaction, and information security focus on the characteristics of such hybrid systems in a more specialized way. Some management disciplines such as organizational architecture and enterprise architecture have the potential – if thoughtfully and creatively elaborated – to provide conceptual and practical frameworks for the development and maintenance of such hybrid human-artificial systems, although efforts to develop those disciplines in the direction of posthumanized human-artificial systems have not yet been robustly pursued.[184]

---

[183] See, e.g., Evans (2012).

[184] For examples of some initial efforts, see, e.g., Gladden, "Leveraging the Cross-Cultural Capacities of Artificial Agents" (2014) and sources cited therein. Organizational theory may also be able to draw on contemporary work in the field of philosophy; for example, see Kowalewska (2016) for an analysis of technologically facilitated digital-physical ecosystems that draws on Actor-Network Theory (ANT) to explore the manner in which nonhuman and human actors within such ecosystems may create "hierarchies, symbioses, chains and balances" (p. 74) that do not simply resemble the structures and relations of biological ecosystems in a metaphorical sense but truly instantiate the forms and dynamics of such ecologies within a hybrid biological-synthetic system full of diverse types of actors.

Artificial, Xeno- and Meta- Studies

As hybrid human-robotic organizations and environments become more common, we can expect to see the development of new disciplines that attempt to understand the unique physical structures, behaviors, advantages and capacities, and weaknesses and vulnerabilities displayed by such systems. Just as 'artificial psychology' focuses on the cognitive activity of beings that are human-like in their behavior but synthetic in their construction – and 'xenopsychology' or 'exopsychology' studies the cognitive activity of agents that are radically nonhuman (e.g., hypothetical extraterrestrial intelligences) and whose behavior is not intended or expected to replicate that of human beings,[185] so the prefix 'meta-' or words such as 'post-anthropocentric,' 'agent-independent,' or 'cybernetic' might be employed to refer to efforts at developing universal conceptual frameworks that are sufficiently abstract to be able to account for the structures and dynamics found in the activities of human agents, artificial agents resembling human beings, radically nonhuman synthetic agents, and any other kinds of agents. For example, attempts to identify the essential structures and processes that must be present in any type of agent in order for it to be considered 'cognitive' – and to explore the full spectrum of ways in which those structures and processes can manifest themselves across different types of agents – could be understood alternatively as 'metapsychology,' 'post-anthropocentric psychology,' 'agent-independent psychology,' or 'psychological cybernetics.' Similarly, a term like 'metaeconomics' might be used to refer to generalized conceptual frameworks that can account equally well for the economic activity of all kinds of entities, both human and artificial.[186]

---

[185] For a history of such use of 'xeno-' in both literary and scholarly contexts, see the "Preface and Acknowledgements for the First Edition" in Freitas, *Xenology: An Introduction to the Scientific Study of Extraterrestrial Life, Intelligence, and Civilization* (1979), where "[...] xenology may be defined as the scientific study of all aspects of extraterrestrial life, intelligence, and civilization. Similarly, xenobiology refers to the study of the biology of extraterrestrial lifeforms not native to Earth, xenopsychology refers to the higher mental processes of such lifeforms if they are intelligent, xenotechnology refers to the technologies they might possess, and so forth." For the use of 'exopsychology' in connection with potential extraterrestrial intelligences, see Harrison & Elms, "Psychology and the search for extraterrestrial intelligence" (1990), p. 207, where "The proposed field of exopsychology would involve the forecast, study, and interpretation of the cognitive, affective, and behavioral aspects of extraterrestrial organisms. Exopsychological research would encompass search, contact, and post-contact activities, and would include study and work with humans as well as with any extraterrestrials that might be encountered."

[186] We note that some of the terms suggested above have already been utilized by other scholars in different contexts. For example, the understanding of 'metapsychology' formulated here is different from the specialized sense in which Freud used that term; our envisioned use of the prefix 'meta-' is more closely related to the contemporary philosophical use of the term to refer to an abstracted or second-order phenomenon. Some scholars have used the prefix 'meta-' in ways that are closely aligned with our proposed use. For example, building on earlier questions

V. Conclusion

A transformative convergence is underway within contemporary organizations, as human workers integrate computers ever more closely into their minds and bodies and computers themselves become ever more 'human.' Such developments create both opportunities and threats that must be carefully evaluated from ethical, legal, and managerial perspectives. In order to aid with such endeavors, in this text we have formulated the Posthuman Management Matrix, a model in which an organization's employees, consumers, and other stakeholders are divided into two different kinds of agents (human and artificial) who may possess either of two sets of characteristics (anthropic or computronic), thus defining four types of entities. Until now, the only types that have been of relevance for management theory and practice were those of human agents who possess anthropic characteristics (i.e., ordinary human beings) and artificial agents that possess computronic characteristics (as exemplified by assembly-line robots or artificially intelligent software running on desktop computers).

Management theory and practice have traditionally not addressed the remaining two types of agents that are theoretically possible, largely because such agents did not exist to serve as employees or consumers for organizations. However, we have argued that ongoing advances in neuroprosthetics, genetic engineering, virtual reality, robotics, and artificial intelligence are now giving rise to new kinds of human agents that demonstrate computronic characteristics and artificial agents that possess anthropic characteristics. If organizations are to successfully resolve the complex issues that appear when such posthumanized agents are adopted as workers or customers, new spheres of management theory and practice will need to be pioneered. A starting point may be found in existing fields such as cybernetics, systems theory, organizational design, and enterprise architecture that already offer tools for integrating human and artificial agents into the multi-agent system that constitutes an organization. Such fields will likely be complemented through the development of new disciplines such as psychological engineering, cyborg psychology, human technology management, artificial organization development, AI resource management, metapsychology, and metaeconomics that are specifically intended to confront the issues that will accompany the arrival of new kinds of posthumanized agents as organizational stakeholders. Although we

---

posed by Kant, legal scholar Andrew Haley attempted to identify fundamental principles of law and ethics that are not specific to human biology, psychology, sociality, and culture but which would be relevant to and binding on all intelligent beings, regardless of their physical form or cognitive dynamics; such universal and legal principles could govern humanity's potential encounter with an extraterrestrial intelligence. Haley proposed 'The Great Rule of Metalaw,' which demands that all intelligent beings should "Do unto others as they would have you do unto them"; see Michaud, *Contact with Alien Civilizations: Our Hopes and Fears about Encountering Extraterrestrials* (2007), p. 374.

cannot yet know the exact paths that such developments will take, our hope is that the framework presented in this text can prove useful in highlighting the new areas that wait to be explored and in informing the work of those management scholars and practitioners who choose to embrace that challenge.

References

Ablett, Ruth, Shelly Park, Ehud Sharlin, Jörg Denzinger, and Frank Maurer. "A Robotic Colleague for Facilitating Collaborative Software Development." *Proceedings of Computer Supported Cooperative Work (CSCW 2006).* ACM, 2006.

Abrams, Jerold J. "Pragmatism, Artificial Intelligence, and Posthuman Bioethics: Shusterman, Rorty, Foucault." *Human Studies* 27, no. 3 (September 1, 2004): 241-58. doi:10.1023/B:HUMA.0000042130.79208.c6.

Advances in Neuromorphic Memristor Science and Applications, edited by Robert Kozma, Robinson E. Pino, and Giovanni E. Pazienza. Dordrecht: Springer Science+Business Media, 2012.

Agent-Based Manufacturing: Advances in the Holonic Approach, edited by S.M. Deen. Springer Berlin Heidelberg, 2003.

Aier, Stephan. "The Role of Organizational Culture for Grounding, Management, Guidance and Effectiveness of Enterprise Architecture Principles." *Information Systems and E-Business Management* 12, no. 1 (2014): 43-70.

Alford, John R., and John R. Hibbing. "The origin of politics: An evolutionary theory of political behavior." *Perspectives on Politics* 2, no. 04 (2004): 707-23.

Anderson, Donald L. *Organization Development: The Process of Leading Organizational Change,* 3e. SAGE Publications, 2015.

Anderson, Michael L. "Embodied cognition: A field guide." *Artificial intelligence* 149, no. 1 (2003): 91-130.

Anderson, Walter Truett. "Augmentation, symbiosis, transcendence: technology and the future(s) of human identity." *Futures* 35, no. 5 (2003): 535-46.

Andrianantoandro, Ernesto, Subhayu Basu, David K. Karig, and Ron Weiss. "Synthetic biology: new engineering rules for an emerging discipline." *Molecular Systems Biology* 2, no. 1 (2006).

"Anthropotech." http://www.anthropotech.org.uk/. 2013. Accessed January 29, 2016.

Appleseed [アップルシード / *Appurushīdo*]. Directed by Shinji Aramaki. 2004. Houston: Sentai Selects, 2010. Blu-Ray.

Aristotle, *Politics,* Book 1, Section 1253a. In *Aristotle in 23 Volumes,* Vol. 21, translated by H. Rackham. Cambridge, MA: Harvard University Press, 1944. http://www.perseus.tufts.edu/hopper/text?doc=Perseus%3Atext%3A1999.01.0058%3Abook%3D1%3Asection%3D1253a. Accessed March 4, 2016.

Arkin, Ronald C., and J. David Hobbs. "Dimensions of communication and social organization in multi-agent robotic systems." In *From Animals to Animats 2: Proceedings of the Second International Conference on Simulation of Adaptive Behavior*, edited by Jean-Arcady Meyer, H. L. Roitblat and Stewart W. Wilson, pp. 486-93. Cambridge, MA: The MIT Press, 1993.

Artificial General Intelligence, edited by Ben Goertzel and Cassio Pennachin. Springer Berlin Heidelberg, 2007.

Artificial General Intelligence: 8th International Conference, AGI 2015: Berlin, Germany, July 22-25, 2015: Proceedings, edited by Jordi Bieger, Ben Goertzel, and Alexey Potapov. Springer International Publishing, 2015.

Austerberry, David. *Digital Asset Management*, second edition. Burlington, MA: Focal Press, 2013.

Badmington, Neil. "Cultural Studies and the Posthumanities." In *New Cultural Studies: Adventures in Theory*, edited by Gary Hall and Claire Birchall, pp. 260-72. Edinburgh: Edinburgh University Press, 2006.

Bainbridge, William Sims. *The Virtual Future*. London: Springer, 2011.

Băjenescu, Titu-Marius, and Marius I. Bâzu. *Reliability of Electronic Components: A Practical Guide to Electronic Systems Manufacturing*. Springer Berlin Heidelberg, 1999.

Barca, Jan Carlo, and Y. Ahmet Sekercioglu. "Swarm robotics reviewed." *Robotica* 31, no. 03 (2013): 345-59.

Bekey, G.A. *Autonomous Robots: From Biological Inspiration to Implementation and Control*. Cambridge, MA: MIT Press, 2005.

Bendle, Mervyn F. "Teleportation, cyborgs and the posthuman ideology." *Social Semiotics* 12, no. 1 (2002): 45-62.

Bera, Rajendra K. "Synthetic Biology and Intellectual Property Rights." In *Biotechnology*, edited by Deniz Ekinci. Rijeka: InTech, 2015.

Berner, Georg. *Management in 20XX: What Will Be Important in the Future – A Holistic View*. Erlangen: Publicis Corporate Publishing, 2004.

Boly, Melanie, Anil K. Seth, Melanie Wilke, Paul Ingmundson, Bernard Baars, Steven Laureys, David B. Edelman, and Naotsugu Tsuchiya. "Consciousness in humans and non-human animals: recent advances and future directions." *Frontiers in Psychology* 4 (2013).

Bostrom, Nick. "Human Genetic Enhancements: A Transhumanist Perspective." In *Arguing About Bioethics*, edited by Stephen Holland, pp. 105-15. New York: Routledge, 2012.

Bostrom, Nick. "Why I Want to Be a Posthuman When I Grow Up." In *Medical Enhancement and Posthumanity*, edited by Bert Gordijn and Ruth Chadwick, pp. 107-37. The International Library of Ethics, Law and Technology 2. Springer Netherlands, 2008.

Bradford, David L., and W. Warner Burke. *Reinventing Organization Development: New Approaches to Change in Organizations.* John Wiley & Sons, 2005.

Brambilla, Manuele, Eliseo Ferrante, Mauro Birattari, and Marco Dorigo. "Swarm robotics: a review from the swarm engineering perspective." *Swarm Intelligence* 7, no. 1 (2013): 1-41.

Breazeal, Cynthia. "Toward sociable robots." *Robotics and Autonomous Systems* 42 (2003): 167-75.

Burton, Richard M., Børge Obel, and Dorthe Døjbak Håkonsson. *Organizational Design: A Step-by-Step Approach.* Cambridge University Press, 2015.

Caetano, Artur, António Rito Silva, and José Tribolet. "A Role-Based Enterprise Architecture Framework." In *Proceedings of the 2009 ACM Symposium on Applied Computing,* pp. 253-58. ACM, 2009.

Callaghan, Vic. "Micro-Futures." Presentation at Creative-Science 2014, Shanghai, China, July 1, 2014.

Calverley, D.J. "Imagining a non-biological machine as a legal person." *AI & SOCIETY* 22, no. 4 (2008): 523-37.

Camenzind, Samuel. "On Clone as Genetic Copy: Critique of a Metaphor." *NanoEthics* 9, no. 1 (2015): 23-37.

Canton, James. "Designing the future: NBIC technologies and human performance enhancement." *Annals of the New York Academy of Sciences* vol. 1013, (2004): 186-98.

Castronova, Edward. *Synthetic Worlds: The Business and Culture of Online Games.* Chicago: The University of Chicago Press, 2005.

Castronova, Edward. "Theory of the Avatar." CESifo Working Paper No. 863, February 2003. http://www.cesifo.de/pls/guestci/download/CESifo+Working+Papers+2003/CESifo+Working+Papers+February+2003+/cesifo_wp863.pdf. Accessed January 25, 2016.

Cesaroni, John L. "Designer Human Embryos as a Challenge for Patent Law and Regulation." *Quinnipiac Law Review* 30, no. 4 (2012).

Chella, Antonio, Enrico Pagello, Emanuele Menegatti, Rosario Sorbello, Salvatore Maria Anzalone, Francesco Cinquegrani, Luca Tonin, F. Piccione, K. Prifitis, C. Blanda, E. Buttita, and E. Tranchina. "A BCI teleoperated museum robotic guide." In *International Conference on Complex, Intelligent and Software Intensive Systems, 2009 (CISIS'09),* pp. 783-88. IEEE, 2009.

Cheng, Allen A., and Timothy K. Lu. "Synthetic biology: an emerging engineering discipline." *Annual Review of Biomedical Engineering* 14 (2012): 155-78.

Church, George M., Yuan Gao, and Sriram Kosuri. "Next-generation digital information storage in DNA." *Science* 337, no. 6102 (2012): 1628.

Clark, Andy. *Natural-born cyborgs: Minds, Technologies, and the Future of Human Intelligence.* Oxford: Oxford University Press, 2004.

Clark, Stephen R.L. *The Political Animal: Biology, Ethics and Politics.* London: Routledge, 1999.

Coeckelbergh, Mark. "From Killer Machines to Doctrines and Swarms, or Why Ethics of Military Robotics Is Not (Necessarily) About Robots." *Philosophy & Technology* 24, no. 3 (2011): 269-78.

Communication in the Age of Virtual Reality, edited by Frank Biocca and Mark R. Levy. Hillsdale, NJ: Lawrence Erlbaum Associates, Publishers, 1995.

Contreras-Vidal, Jose L., and Robert G. Grossman. "NeuroRex: A clinical neural interface roadmap for EEG-based brain machine interfaces to a lower body robotic exoskeleton." In *2013 35th Annual International Conference of the IEEE Engineering in Medicine and Biology Society (EMBC),* pp. 1579-82. IEEE, 2013.

Converging Technologies for Improving Human Performance: Nanotechnology, Biotechnology, Information Technology and Cognitive Science, edited by William Sims Bainbridge. Dordrecht: Springer Science+Business Media, 2003.

Coughlin, Thomas M. *Digital Storage in Consumer Electronics: The Essential Guide.* Burlington, MA: Newnes, 2008.

Curtis, H. *Biology,* 4th edition. New York: Worth, 1983.

Cyberpsychology, edited by Ángel J. Gordo-López and Ian Parker. New York: Routledge, 1999.

Cyberpsychology: Journal of Psychosocial Research 1, no. 1 (2007) and subsequent issues. http://www.cyberpsychology.eu/index.php. Accessed January 29, 2016.

Cybersociety 2.0: Revisiting Computer-Mediated Communication and Community, edited by Steven G. Jones. Thousand Oaks: Sage Publications, 1998.

Daft, Richard L. *Management.* Mason, OH: South-Western / Cengage Learning, 2011.

Dautenhahn, Kerstin. "Robots we like to live with?! - A Developmental Perspective on a Personalized, Life-long Robot Companion." In *Proceedings of the 2004 IEEE International Workshop on Robot and Human Interactive Communication,* pp. 17-22. IEEE, 2004.

De Melo-Martín, Inmaculada. "Genetically Modified Organisms (GMOs): Human Beings." In *Encyclopedia of Global Bioethics,* edited by Henk ten Have. Springer Science+Business Media Dordrecht. Version of March 13, 2015. doi: 10.1007/978-3-319-05544-2_210-1. Accessed January 21, 2016.

Doyle, Jon. "Big problems for artificial intelligence." *AI Magazine* 9, no. 1 (1988): 19-22.

Dudai, Yadin. "The Neurobiology of Consolidations, Or, How Stable Is the Engram?" *Annual Review of Psychology* 55 (2004): 51-86.

Dumas II, Joseph D. *Computer Architecture: Fundamentals and Principles of Computer Design.* Boca Raton: CRC Press, 2006.

Duncan, Sean C. "Mandatory Upgrades: The Evolving Mechanics and Theme of Android: Netrunner." Presentation at Well-Played Summit @ DiGRA 2014, Salt Lake City, August 3-6, 2014.

Emmons, Robert A. "Is spirituality an intelligence? Motivation, cognition, and the psychology of ultimate concern." *The International Journal for the psychology of Religion* 10, no. 1 (2000): 3-26.

"Ethical Guidelines and Professional Standards for Organization Development and Group Process Consultants." IAGP - International Association for Group Psychotherapy and Group Processes. http://www.iagp.com/docs/IAGPOrgEthicalguidelinesEnglishv1.0.pdf. Accessed December 20, 2014.

Evans, Dave. "The Internet of Everything: How More Relevant and Valuable Connections Will Change the World." Cisco Internet Solutions Business Group: Point of View, 2012. https://www.cisco.com/web/about/ac79/docs/innov/IoE.pdf. Accessed December 16, 2015.

Fabbro, Franco, Salvatore M. Aglioti, Massimo Bergamasco, Andrea Clarici, and Jaak Panksepp. "Evolutionary aspects of self-and world consciousness in vertebrates." *Frontiers in human neuroscience* 9 (2015).

Ferber, Jacques, Olivier Gutknecht, and Fabien Michel. "From agents to organizations: an organizational view of multi-agent systems." In *Agent-Oriented Software Engineering IV*, pp. 214-30. Springer Berlin Heidelberg, 2004.

Ferrando, Francesca. "Posthumanism, Transhumanism, Antihumanism, Metahumanism, and New Materialisms: Differences and Relations." *Existenz: An International Journal in Philosophy, Religion, Politics, and the Arts* 8, no. 2 (Fall 2013): 26-32.

Fleischmann, Kenneth R. "Sociotechnical Interaction and Cyborg–Cyborg Interaction: Transforming the Scale and Convergence of HCI." *The Information Society* 25, no. 4 (2009): 227-35. doi:10.1080/01972240903028359.

Ford, Martin. *Rise of the Robots: Technology and the Threat of a Jobless Future*. New York: Basic Books, 2015.

Freitas Jr., Robert A. "Preface and Acknowledgements for the First Edition." In *Xenology: An Introduction to the Scientific Study of Extraterrestrial Life, Intelligence, and Civilization.* Sacramento: Xenology Research Institute, 1979. http://www.xenology.info/Xeno/PrefaceFirstEdition.htm, last updated October 22, 2009. Accessed January 30, 2016.

Friedenberg, Jay. *Artificial Psychology: The Quest for What It Means to Be Human*, Philadelphia: Psychology Press, 2008.

Friedman, Batya, and Helen Nissenbaum. "Bias in Computer Systems." In *Human Values and the Design of Computer Technology*, edited by Batya Friedman, pp. 21-40. CSL Lecture Notes 72. Cambridge: Cambridge University Press, 1997.

Fukuyama, Francis. *Our Posthuman Future: Consequences of the Biotechnology Revolution.* New York: Farrar, Straus, and Giroux, 2002.

The Future of Automated Freight Transport: Concepts, Design and Implementation, edited by Hugo Priemus and Peter Nijkamp. Cheltenham: Edward Elgar Publishing, 2005.

The Future of Bioethics: International Dialogues, edited by Akira Akabayashi, Oxford: Oxford University Press, 2014.

Galis, Alex, Spyros G. Denazis, Alessandro Bassi, Pierpaolo Giacomin, Andreas Berl, Andreas Fischer, Hermann de Meer, J. Srassner, S. Davy, D. Macedo, G. Pujolle, J. R. Loyola, J. Serrat, L. Lefevre, and A. Cheniour. "Management Architecture and Systems for Future Internet Networks." In *Towards the Future Internet: A European Research Perspective*, edited by Georgios Tselentis, John Domingue, Alex Galis, Anastasius Gavras, David Hausheer, Srdjan Krco, Volkmar Lotz, and Theodore Zahariadis, pp. 112-22. IOS Press, 2009.

Garg, Anant Bhaskar. "Embodied Cognition, Human Computer Interaction, and Application Areas." In *Computer Applications for Web, Human Computer Interaction, Signal and Image Processing, and Pattern Recognition*, pp. 369-74. Springer Berlin Heidelberg, 2012.

Gasson, M.N. "Human ICT Implants: From Restorative Application to Human Enhancement." In *Human ICT Implants: Technical, Legal and Ethical Considerations*, edited by Mark N. Gasson, Eleni Kosta, and Diana M. Bowman, pp. 11-28. Information Technology and Law Series 23. T. M. C. Asser Press, 2012.

Gasson, M.N. "ICT implants." In *The Future of Identity in the Information Society*, edited by S. Fischer-Hübner, P. Duquenoy, A. Zuccato, and L. Martucci, pp. 287-95. Springer US, 2008.

Gasson, M.N., Kosta, E., and Bowman, D.M. "Human ICT Implants: From Invasive to Pervasive." In *Human ICT Implants: Technical, Legal and Ethical Considerations*, edited by Mark N. Gasson, Eleni Kosta, and Diana M. Bowman, pp. 1-8. Information Technology and Law Series 23. T. M. C. Asser Press, 2012.

Gene Therapy of the Central Nervous System: From Bench to Bedside, edited by Michael G. Kaplitt and Matthew J. During. Amsterdam: Elsevier, 2006.

Geraci, Robert M. *Apocalyptic AI: Visions of Heaven in Robotics, Artificial Intelligence, and Virtual Reality*. New York: Oxford University Press, 2010.

Geraci, Robert M. "Spiritual robots: Religion and our scientific view of the natural world." *Theology and Science* 4, issue 3 (2006). doi: 10.1080/14746700600952993.

Giudici, P. *Applied Data Mining: Statistical Methods for Business and Industry*. Wiley, 2003.

Gladden, Matthew E. "The Artificial Life-Form as Entrepreneur: Synthetic Organism-Enterprises and the Reconceptualization of Business." In *Proceedings of the Fourteenth International Conference on the Synthesis and Simulation of Living Systems*, edited by Hiroki Sayama, John Rieffel, Sebastian Risi, René Doursat and Hod Lipson, pp. 417-18. Cambridge, MA: The MIT Press, 2014.

Gladden, Matthew E. "Cybershells, Shapeshifting, and Neuroprosthetics: Video Games as Tools for Posthuman 'Body Schema (Re)Engineering'." Keynote presentation at the Ogólnopolska Konferencja Naukowa Dyskursy Gier Wideo, Facta Ficta / AGH, Kraków, June 6, 2015.

Gladden, Matthew E. "The Diffuse Intelligent Other: An Ontology of Nonlocalizable Robots as Moral and Legal Actors." In *Social Robots: Boundaries, Potential, Challenges*, edited by Marco Nørskov, pp. 177-98. Farnham: Ashgate, 2016.

Gladden, Matthew E. "From Stand Alone Complexes to Memetic Warfare: Cultural Cybernetics and the Engineering of Posthuman Popular Culture." Presentation at the 50 Shades of Popular Culture International Conference. Facta Ficta / Uniwersytet Jagielloński, Kraków, February 19, 2016.

Gladden, Matthew E. *The Handbook of Information Security for Advanced Neuroprosthetics*, Indianapolis: Synthypnion Academic, 2015.

Gladden, Matthew E. "Leveraging the Cross-Cultural Capacities of Artificial Agents as Leaders of Human Virtual Teams." *Proceedings of the 10th European Conference on Management Leadership and Governance*, edited by Visnja Grozdanić, pp. 428-35. Reading: Academic Conferences and Publishing International Limited, 2014.

Gladden, Matthew E. "Managerial Robotics: A Model of Sociality and Autonomy for Robots Managing Human Beings and Machines." *International Journal of Contemporary Management* 13, no. 3 (2014), pp. 67-76.

Gladden, Matthew E. "Neural Implants as Gateways to Digital-Physical Ecosystems and Posthuman Socioeconomic Interaction." In *Digital Ecosystems: Society in the Digital Age*, edited by Łukasz Jonak, Natalia Juchniewicz, and Renata Włoch, pp. 85-98. Warsaw: Digital Economy Lab, University of Warsaw, 2016.

Gladden, Matthew E. "The Social Robot as 'Charismatic Leader': A Phenomenology of Human Submission to Nonhuman Power." In *Sociable Robots and the Future of Social Relations: Proceedings of Robo-Philosophy 2014*, edited by Johanna Seibt, Raul Hakli, and Marco Nørskov, pp. 329-39. Frontiers in Artificial Intelligence and Applications 273. IOS Press, 2014.

Gladden, Matthew E. "'Upgrading' the Human Entity: Cyberization as a Path to Posthuman Utopia or Digital Annihilation?" Lecture in the Arkana Fantastyki lecture cycle, Centrum Informacji Naukowej i Biblioteka Akademicka (CINiBA), Katowice, May 27, 2015.

Gladden, Matthew E. "Utopias and Dystopias as Cybernetic Information Systems: Envisioning the Posthuman Neuropolity." *Creatio Fantastica* nr 3 (50) (2015).

Gockley, Rachel, Allison Bruce, Jodi Forlizzi, Marek Michalowski, Anne Mundell, Stephanie Rosenthal, Brennan Sellner, Reid Simmons, Kevin Snipes, Alan C. Schultz, and Jue Wang. "Designing Robots for Long-Term Social Interaction." In *2005 IEEE/RSJ International Conference on Intelligent Robots and Systems (IROS 2005)*, pp. 2199-2204. 2005.

Goffette, Jérôme. *Naissance de l'anthropotechnie: De la médecine au modelage de l'humain*. Paris: Vrin, 2006.

Goodall, Noah. "Ethical decision making during automated vehicle crashes." *Transportation Research Record: Journal of the Transportation Research Board* 2424 (2014): 58-65.

Goodman, Paul S., and Linda Argote. "New Technology and Organizational Effectiveness." Carnegie-Mellon University, April 27, 1984.

Graham, Elaine. *Representations of the Post/Human: Monsters, Aliens and Others in Popular Culture*. Manchester: Manchester University Press, 2002.

Gross, Dominik. "Traditional vs. modern neuroenhancement: notes from a medico-ethical and societal perspective." In *Implanted minds: the neuroethics of intracerebral stem cell transplantation and deep brain stimulation*, edited by Heiner Fangerau, Jörg M. Fegert, and Thorsten Trapp, pp. 291-312. Bielefeld: transcript Verlag, 2011.

Guizzo, Erico, and Harry Goldstein. "The rise of the body bots [robotic exoskeletons]." *IEEE Spectrum* 42, no. 10 (2005): 50-56.

Gunkel, David J. *The Machine Question: Critical Perspectives on AI, Robots, and Ethics.* Cambridge, MA: The MIT Press, 2012.

Han, J.-H., S.A. Kushner, A.P. Yiu, H.-W. Hsiang, T. Buch, A. Waisman, B. Bontempi, R.L. Neve, P.W. Frankland, and S.A. Josselyn. "Selective Erasure of a Fear Memory." *Science* 323, no. 5920 (2009): 1492-96.

Handbook of Cloud Computing, edited by Borko Furht and Armando Escalante. New York: Springer, 2010. DOI 10.1007/978-1-4419-6524-0.

Handbook of Psychology, Volume 6: Developmental Psychology, edited by Richard M. Lerner, M. Ann Easterbrooks, and Jayanthi Mistry. Hoboken: John Wiley & Sons, Inc., 2003.

Hanson, R. "If uploads come first: The crack of a future dawn." *Extropy* 6, no. 2 (1994): 10-15.

Harrison, Albert A., and Alan C. Elms. "Psychology and the search for extraterrestrial intelligence." *Behavioral Science* 35, no. 3 (1990): 207-18.

Hatfield, B., A. Haufler, and J. Contreras-Vidal. "Brain Processes and Neurofeedback for Performance Enhancement of Precision Motor Behavior." In *Foundations of Augmented Cognition. Neuroergonomics and Operational Neuroscience*, edited by Dylan D. Schmorrow, Ivy V. Estabrooke, and Marc Grootjen, pp. 810-17. Lecture Notes in Computer Science 5638. Springer Berlin Heidelberg, 2009.

Haykin, Simon. *Neural Networks and Learning Machines*, third edition. New York: Pearson Prentice Hall, 2009.

Hazen, Robert. "What is life?", *New Scientist* 192, no. 2578 (2006): 46-51.

Heim, Michael. *The Metaphysics of Virtual Reality.* New York: Oxford University Press, 1993.

Heinrichs, Jan-Hendrik. "The promises and perils of non-invasive brain stimulation." *International journal of law and psychiatry* 35, no. 2 (2012): 121-29.

Herbrechter, Stefan. *Posthumanism: A Critical Analysis.* London: Bloomsbury, 2013. [Kindle edition.]

Heylighen, Francis. "The Global Brain as a New Utopia." In *Renaissance der Utopie. Zukunftsfiguren des 21. Jahrhunderts*, edited by R. Maresch and F. Rötzer. Frankfurt: Suhrkamp, 2002.

Hoogervorst, Jan. "Enterprise Architecture: Enabling Integration, Agility and Change." *International Journal of Cooperative Information Systems* 13, no. 03 (2004): 213-33.

Ihrig, M. "Simulating Entrepreneurial Opportunity Recognition Processes: An Agent-Based and Knowledge-Driven Approach." In *Advances in Intelligent Modelling and Simulation: Simulation Tools and Applications*, edited by A. Byrski, Z. Oplatková, M. Carvalho, and M. Kisiel-Dorohinicki, pp. 27-54. Berlin: Springer-Verlag, 2012.

Industrial Applications of Holonic and Multi-Agent Systems, edited by Vladimír Mařík, Arnd Schirrmann, Damien Trentesaux, and Pavel Vrba. Lecture Notes in Artificial Intelligence 9266. Springer International Publishing AG Switzerland, 2015.

Intelligent Production Machines and Systems, edited by Duc T. Pham, Eldaw E. Eldukhri, and Anthony J. Soroka. Amsterdam: Elsevier, 2006.

"The International Organization Development *Code of Ethics*." Organizational Development English Library, The OD Institute. http://www.theodinstitute.org/od-library/code_of_ethics.htm. Accessed December 17, 2014.

Josselyn, Sheena A. "Continuing the Search for the Engram: Examining the Mechanism of Fear Memories." *Journal of Psychiatry & Neuroscience : JPN* 35, no. 4 (2010): 221-28.

Kanda, Takayuki, and Hiroshi Ishiguro. *Human-Robot Interaction in Social Robotics*. Boca Raton: CRC Press, 2013.

Kawano, Tomonori, François Bouteau, and Stefano Mancuso. "Finding and defining the natural automata acting in living plants: Toward the synthetic biology for robotics and informatics in vivo." *Communicative & Integrative Biology* 5, no. 6 (2012): 519-26.

Kedzior, Richard. *How Digital Worlds Become Material: An Ethnographic and Netnographic Investigation in Second Life*. Economics and Society: Publications of the Hanken School of Economics Nr. 281. Helsinki: Hanken School of Economics, 2014.

Kelly, Kevin. "The Landscape of Possible Intelligences." *The Technium*, September 10, 2008. http://kk.org/thetechnium/the-landscape-o/. Accessed January 25, 2016.

Kelly, Kevin. *Out of Control: The New Biology of Machines, Social Systems and the Economic World*. Basic Books, 1994.

Kelly, Kevin. "A Taxonomy of Minds." *The Technium*, February 15, 2007. http://kk.org/thetechnium/a-taxonomy-of-m/. Accessed January 25, 2016.

Kerr, Paul K., John Rollins, and Catherine A. Theohary. "The Stuxnet Computer Worm: Harbinger of an Emerging Warfare Capability." Congressional Research Service, 2010.

Khushf, George. "The use of emergent technologies for enhancing human performance: Are we prepared to address the ethical and policy issues." *Public Policy and Practice* 4, no. 2 (2005): 1-17.

Kinicki, Angelo, and Brian Williams. *Management: A Practical Introduction*, 5th edition. New York: McGraw Hill, 2010.

Koene, Randal A. "Embracing Competitive Balance: The Case for Substrate-Independent Minds and Whole Brain Emulation." In *Singularity Hypotheses*, edited by

Amnon H. Eden, James H. Moor, Johnny H. Søraker, and Eric Steinhart, pp. 241-67. The Frontiers Collection. Springer Berlin Heidelberg, 2012.

Koltko-Rivera, Mark E. "The potential societal impact of virtual reality." *Advances in virtual environments technology: Musings on design, evaluation, and applications* 9 (2005).

Koops, B.-J., and R. Leenes. "Cheating with Implants: Implications of the Hidden Information Advantage of Bionic Ears and Eyes." In *Human ICT Implants: Technical, Legal and Ethical Considerations*, edited by Mark N. Gasson, Eleni Kosta, and Diana M. Bowman, pp. 113-34. Information Technology and Law Series 23. T. M. C. Asser Press, 2012.

Kowalewska, Agata. "Symbionts and Parasites – Digital Ecosystems." In *Digital Ecosystems: Society in the Digital Age*, edited by Łukasz Jonak, Natalia Juchniewicz, and Renata Włoch, pp. 73-84. Warsaw: Digital Economy Lab, University of Warsaw, 2016.

Kraemer, Felicitas. "Me, Myself and My Brain Implant: Deep Brain Stimulation Raises Questions of Personal Authenticity and Alienation." *Neuroethics* 6, no. 3 (May 12, 2011): 483-97. doi:10.1007/s12152-011-9115-7.

Kurzweil, Ray. *The Age of Spiritual Machines: When Computers Exceed Human Intelligence.* New York: Penguin Books, 2000.

Lamm, Ehud, and Ron Unger. *Biological Computation*, Boca Raton: CRC Press, 2011.

Lilley, Stephen. *Transhumanism and Society: The Social Debate over Human Enhancement.* Springer Science & Business Media, 2013.

Litzsinger, Lukas. *Android: Netrunner* [card game]. Roseville, MN: Fantasy Flight Games, 2012.

Lohn, Andrew J., Patrick R. Mickel, James B. Aimone, Erik P. Debenedictis, and Matthew J. Marinella. "Memristors as Synapses in Artificial Neural Networks: Biomimicry Beyond Weight Change." In *Cybersecurity Systems for Human Cognition Augmentation*, edited by Robinson E. Pino, Alexander Kott, and Michael Shevenell, pp. 135-50. Springer International Publishing, 2014.

Longuet-Higgins, H.C. "Holographic Model of Temporal Recall." *Nature* 217, no. 5123 (1968): 104. doi:10.1038/217104a0.

Lyon, David. "Beyond Cyberspace: Digital Dreams and Social Bodies." In *Education and Society*, third edition, edited by Joseph Zajda, pp. 221-38. Albert Park: James Nicholas Publishers, 2001.

Magoulas, Thanos, Aida Hadzic, Ted Saarikko, and Kalevi Pessi. "Alignment in Enterprise Architecture: A Comparative Analysis of Four Architectural Approaches." *Electronic Journal Information Systems Evaluation* 15, no. 1 (2012).

Maguire, Gerald Q., and Ellen M. McGee. "Implantable brain chips? Time for debate." *Hastings Center Report* 29, no. 1 (1999): 7-13.

Man Is by Nature a Political Animal: Evolution, Biology, and Politics, edited by Peter K. Hatemi and Rose McDermott. Chicago: The University of Chicago Press, 2011.

Martinez, Dolores. "Bodies of future memories: the Japanese body in science fiction anime." *Contemporary Japan* 27, no. 1 (2015): 71-88.

McBurney, Peter, and Simon Parsons. "Engineering democracy in open agent systems." In *ESAW* vol. 3071, pp. 66-80. 2003.

McGee, E.M. "Bioelectronics and Implanted Devices." In *Medical Enhancement and Posthumanity*, edited by Bert Gordijn and Ruth Chadwick, pp. 207-24. The International Library of Ethics, Law and Technology 2. Springer Netherlands, 2008.

McIndoe, Robert S. "Health Kiosk Technologies." In *Ethical Issues and Security Monitoring Trends in Global Healthcare: Technological Advancements: Technological Advancements*, edited by Steven A. Brown and Mary Brown, pp. 66-71. Hershey: Medical Information Science Reference, 2010.

McIntosh, Daniel. "The Transhuman Security Dilemma." *Journal of Evolution and Technology* 21, no. 2 (2010): 32-48.

Medical Enhancement and Posthumanity, edited by Bert Gordijn and Ruth Chadwick. The International Library of Ethics, Law and Technology 2. Springer Netherlands, 2008.

Mehlman, Maxwell J. *Transhumanist Dreams and Dystopian Nightmares: The Promise and Peril of Genetic Engineering*. Baltimore: The Johns Hopkins University Press, 2012.

Merdan, Munir, Thomas Moser, W. Sunindyo, Stefan Biffl, and Pavel Vrba. "Workflow scheduling using multi-agent systems in a dynamically changing environment." *Journal of Simulation* 7, no. 3 (2013): 144-58.

Merkel, R., G. Boer, J. Fegert, T. Galert, D. Hartmann, B. Nuttin, and S. Rosahl. "Central Neural Prostheses." In *Intervening in the Brain: Changing Psyche and Society*, 117-60. Ethics of Science and Technology Assessment 29. Springer Berlin Heidelberg, 2007.

Michaud, M.A.G. *Contact with Alien Civilizations: Our Hopes and Fears about Encountering Extraterrestrials*. New York: Springer, 2007.

Moravec, Hans. *Mind Children: The Future of Robot and Human Intelligence*. Cambridge: Harvard University Press, 1990.

Mueller, Scott. *Upgrading and Repairing PCs, 20th Edition*. Indianapolis: Que, 2012.

Murphy, Robin. *Introduction to AI Robotics*. Cambridge, MA: The MIT Press, 2000.

Nadler, David, and Michael Tushman. *Competing by Design: The Power of Organizational Architecture*. Oxford University Press, 1997. [Kindle edition.]

Nahin, Paul J. "Religious Robots." In *Holy Sci-Fi!*, pp. 69-94. Springer New York, 2014.

Naufel, Stephanie. "Nanotechnology, the Brain, and Personal Identity." In *Nanotechnology, the Brain, and the Future*, edited by Sean A. Hays, Jason Scott Robert, Clark A. Miller, and Ira Bennett, pp. 167-78. Dordrecht: Springer Science+Business Media, 2013.

NIST Special Publication 800-53, Revision 4: Security and Privacy Controls for Federal Information Systems and Organizations. Joint Task Force Transformation Initiative. Gaithersburg, Maryland: National Institute of Standards & Technology, 2013.

Norman, Kent L. *Cyberpsychology: An Introduction to Human-Computer Interaction.* New York: Cambridge University Press, 2008.

Nouvel, Pascal. "A Scale and a Paradigmatic Framework for Human Enhancement." In *Inquiring into Human Enhancement,* edited by Simone Bateman, Jean Gayon, Sylvie Allouche, Jérôme Goffette, and Michela Marzano, pp. 103-18. Palgrave Macmillan UK, 2015.

Novaković, Branko, Dubravko Majetić, Josip Kasać, and Danko Brezak. "Artificial Intelligence and Biorobotics: Is an Artificial Human Being our Destiny?" In *Annals of DAAAM for 2009 & Proceedings of the 20th International DAAAM Symposium "Intelligent Manufacturing & Automation: Focus on Theory, Practice and Education,"* edited by Branko Katalinic, pp. 121-22. Vienna: DAAAM International, 2009.

Null, Linda, and Julia Lobur. *The Essentials of Computer Organization and Architecture,* second edition. Sudbury, MA: Jones and Bartlett Publishers, 2006.

Ochsner, Beate, Markus Spöhrer, and Robert Stock. "Human, non-human, and beyond: cochlear implants in socio-technological environments." *NanoEthics* 9, no. 3 (2015): 237-50.

Olson, Eric T. "Personal Identity." *The Stanford Encyclopedia of Philosophy* (Fall 2015 Edition), edited by Edward N. Zalta. http://plato.stanford.edu/archives/fall2015/entries/identity-personal/. Accessed January 17, 2016.

"Organization and Human Systems Development Credo." OD Network. http://www.odnetwork.org/?page=ODCredo. Accessed December 17, 2014.

Panno, Joseph. *Gene Therapy: Treating Disease by Repairing Genes.* New York: Facts on File, 2005.

Payr, S., and R. Trappl. "Agents across Cultures." In *Intelligent Virtual Agents,* pp. 320-24. Lecture Notes in Computer Science no. 2792. Springer Berlin Heidelberg, 2003.

Pearce, David. "The Biointelligence Explosion." In *Singularity Hypotheses,* edited by A.H. Eden, J.H. Moor, J.H. Søraker, and E. Steinhart, pp. 199-238. The Frontiers Collection. Berlin/Heidelberg: Springer, 2012.

Pereira, Alexandre L.D. "Intellectual Property and Medical Biotechnologies." In *Legal and Forensic Medicine,* edited by Roy G. Beran, pp. 1735-54. Springer Berlin Heidelberg, 2013.

Pérez, Jimena Escudero. "Sympathy for the Clone: (Post) Human Identities Enhanced by the 'Evil Science' Construct and its Commodifying Practices in Contemporary Clone Fiction." *Between* 4, no. 8 (2014).

Philips, M. "How the Robots Lost: High-Frequency Trading's Rise and Fall." *BloombergView* June 6, 2012. http://www.bloomberg.com/bw/articles/2013-06-06/how-the-robots-lost-high-frequency-tradings-rise-and-fall. Accessed April 29, 2015.

Pinedo, Michael L. *Scheduling: Theory, Algorithms, and Systems*, fourth edition. New York: Springer Science+Business Media, 2012.

Pino, Robinson E., and Alexander Kott. "Neuromorphic Computing for Cognitive Augmentation in Cyber Defense." In *Cybersecurity Systems for Human Cognition Augmentation*, edited by Robinson E. Pino, Alexander Kott, and Michael Shevenell, pp. 19-46. Springer International Publishing, 2014.

Plowright, Stephen. "Neurocomputing: some possible implications for human-machine interfaces." *ASHB News* 8, no. 2 (1996): pp. 4-6.

Ponsteen, Albert, and Rob J. Kusters. "Classification of Human and Automated Resource Allocation Approaches in Multi-Project Management." *Procedia – Social and Behavioral Sciences* 194 (2015): 165-73.

Porterfield, Andrew. "Be Aware of Your Inner Zombie." *ENGINEERING & SCIENCE* (Fall 2010): 14-19.

Pribram, K.H. "Prolegomenon for a Holonomic Brain Theory." In *Synergetics of Cognition*, edited by Hermann Haken and Michael Stadler, pp. 150-84. Springer Series in Synergetics 45. Springer Berlin Heidelberg, 1990.

Pribram, K.H., and S.D. Meade. "Conscious Awareness: Processing in the Synaptodendritic Web – The Correlation of Neuron Density with Brain Size." *New Ideas in Psychology* 17, no. 3 (December 1, 1999): 205-14. doi:10.1016/S0732-118X(99)00024-0.

Primate Politics, edited by Glendon Schubert and Roger D. Masters. Carbondale: Southern Illinois University Press, 1991.

"Principles of OD Practice." OD Network. http://www.odnetwork.org/?page=PrinciplesOfODPracti. Accessed December 17, 2014.

Proudfoot, Diane. "Software Immortals: Science or Faith?" In *Singularity Hypotheses*, edited by Amnon H. Eden, James H. Moor, Johnny H. Søraker, and Eric Steinhart, pp. 367-92. The Frontiers Collection. Springer Berlin Heidelberg, 2012.

Provost, Foster, and Tom Fawcett. *Data Science for Business*. Sebastopol, CA: O'Reilly Media, Inc., 2013.

Pulver, David L. *GURPS Robots*. Austin, TX: Steve Jackson Games, 1995.

Pulver, David L. *Transhuman Space*, second edition. Austin, TX: Steve Jackson Games, 2002.

Ramirez, S., X. Liu, P.-A. Lin, J. Suh, M. Pignatelli, R.L. Redondo, T.J. Ryan, and S. Tonegawa. "Creating a False Memory in the Hippocampus." *Science* 341, no. 6144 (2013): 387-91.

Rankin, T. Murray. "Business Secrets Across International Borders: One Aspect of the Transborder Data Flow Debate." *Canadian Business Law Journal* 10 (1985): 213.

Rao, Umesh Hodeghatta, and Umesha Nayak. *The InfoSec Handbook*. New York: Apress, 2014.

Regalado, Antonio. "Engineering the perfect baby." *MIT Technology Review* 118, no. 3 (2015): 27-33.

Rijntjes, Tom. "On the Viability of Automated Entrepreneurship." Presentation of Media Technology MSc program graduation project, Universiteit Leiden, Leiden, January 22, 2016.

Riva, Giuseppe, and Carlo Galimberti. *Towards CyberPsychology: Mind, Cognition, and Society in the Internet Age*, edited by Giuseppe Riva and Carlo Galimberti. Amsterdam: IOS Press, 2001.

Roden, David. *Posthuman Life: Philosophy at the Edge of the Human*. Abingdon: Routledge, 2014.

Rohloff, Michael. "Framework and Reference for Architecture Design." In *AMCIS 2008 Proceedings*, 2008. http://citeseerx.ist.psu.edu/viewdoc/download?doi=10.1.1.231.8261&rep=rep1&type=pdf.

Rowlands, Mark. *Can Animals Be Moral?* Oxford: Oxford University Press, 2012.

Rubin, Charles T. "What Is the Good of Transhumanism?" In *Medical Enhancement and Posthumanity*, edited by Bert Gordijn and Ruth Chadwick, pp. 137-56. The International Library of Ethics, Law and Technology 2. Springer Netherlands, 2008.

Salvini, Pericle, Cecilia Laschi, and Paolo Dario. "From robotic tele-operation to telepresence through natural interfaces." In *The First IEEE/RAS-EMBS International Conference on Biomedical Robotics and Biomechatronics, 2006. BioRob 2006*, pp. 408-13. IEEE, 2006.

Samani, Hooman Aghaebrahimi, and Adrian David Cheok. "From human-robot relationship to robot-based leadership." In *2011 4th International Conference on Human System Interactions (HSI)*, pp. 178-81. IEEE, 2011.

Samani, Hooman Aghaebrahimi, Jeffrey Tzu Kwan Valino Koh, Elham Saadatian, and Doros Polydorou. "Towards Robotics Leadership: An Analysis of Leadership Characteristics and the Roles Robots Will Inherit in Future Human Society." In *Intelligent Information and Database Systems*, edited by Jeng-Shyang Pan, Shyi-Ming Chen, and Ngoc Thanh Nguyen, pp. 158-65. Lecture Notes in Computer Science 7197. Springer Berlin Heidelberg, 2012.

Samuelson, William F., and Stephen G. Marks. *Managerial Economics*, seventh edition. Hoboken: John Wiley & Sons, 2012.

Sandberg, Anders. "Ethics of brain emulations." *Journal of Experimental & Theoretical Artificial Intelligence* 26, no. 3 (2014): 439-57.

Sandor, Christian, Martin Fuchs, Alvaro Cassinelli, Hao Li, Richard Newcombe, Goshiro Yamamoto, and Steven Feiner. "Breaking the Barriers to True Augmented Reality." arXiv preprint, *arXiv:1512.05471 [cs.HC]*, December 17, 2015. http://arxiv.org/abs/1512.05471. Accessed January 25, 2016.

Sasse, Martina Angela, Sacha Brostoff, and Dirk Weirich. "Transforming the 'weakest link'—a human/computer interaction approach to usable and effective security." *BT technology journal* 19, no. 3 (2001): 122-31.

Scopino, G. "Do Automated Trading Systems Dream of Manipulating the Price of Futures Contracts? Policing Markets for Improper Trading Practices by Algorithmic Robots." *Florida Law Review,* vol. 67 (2015): 221-93.

Sharf, S. "Can Robo-Advisors Survive A Bear Market?" *Forbes,* January 28, 2015. http://www.forbes.com/sites/samanthasharf/2015/01/28/can-robo-advisors-survive-a-bear-market/. Accessed April 29, 2015.

Sharkey, Noel. "Killing Made Easy: From Joysticks to Politics." In *Robot Ethics: The Ethical and Social Implications of Robotics,* edited by Patrick Lin, Keith Abney, and George A. Bekey, pp. 111-28. Cambridge, MA: The MIT Press, 2012.

Shaw-Garlock, Glenda. "Loving machines: Theorizing human and sociable-technology interaction." In *Human-Robot Personal Relationships,* edited by Maarten H. Lamers and Fons J. Verbeek, pp. 1-10. Springer Berlin Heidelberg, 2011.

Siewert, Charles. "Consciousness and Intentionality." *The Stanford Encyclopedia of Philosophy* (Fall 2011 Edition), edited by Edward N. Zalta. http://plato.stanford.edu/archives/fall2011/entries/consciousness-intentionality/.

Siewiorek, Daniel, and Robert Swarz. *Reliable Computer Systems: Design and Evaluation,* second edition. Burlington: Digital Press, 1992.

Singh, Kshitij Kumar. *Biotechnology and Intellectual Property Rights: Legal and Social Implications.* Springer, 2014.

Sloman, Aaron. "Some Requirements for Human-like Robots: Why the recent over-emphasis on embodiment has held up progress." In *Creating brain-like intelligence,* pp. 248-77. Springer Berlin Heidelberg, 2009.

Snider, Greg S. "Cortical Computing with Memristive Nanodevices." *SciDAC Review* 10 (2008): 58-65.

Social Robots and the Future of Social Relations, edited by Johanna Seibt, Raul Hakli, and Marco Nørskov. Amsterdam: IOS Press, 2014.

Social Robots: Boundaries, Potential, Challenges, edited by Marco Nørskov. Farnham: Ashgate Publishing, 2016.

Social Robots from a Human Perspective, edited by Jane Vincent, Sakari Taipale, Bartolomeo Sapio, Giuseppe Lugano, and Leopoldina Fortunati. Springer International Publishing, 2015.

Sorbello, R., A. Chella, and R. C. Arkin. "Metaphor of Politics: A Mechanism of Coalition Formation." *vectors* 1, no. 40 (2004): 20.

Soussou, Walid V., and Theodore W. Berger. "Cognitive and Emotional Neuroprostheses." In *Brain-Computer Interfaces,* pp. 109-23. Springer Netherlands, 2008.

Spohrer, Jim. "NBICS (Nano-Bio-Info-Cogno-Socio) Convergence to Improve Human Performance: Opportunities and Challenges." In *Converging Technologies for Improving Human Performance: Nanotechnology, Biotechnology, Information Technology and Cognitive Science,* edited by M.C. Roco and W.S. Bainbridge, pp. 101-17. Arlington, Virginia: National Science Foundation, 2002.

Stahl, B. C. "Responsible Computers? A Case for Ascribing Quasi-Responsibility to Computers Independent of Personhood or Agency." *Ethics and Information Technology* 8, no. 4 (2006): 205-13.

Stiles, Noelle R.B., and Shinsuke Shimojo. "Sensory substitution: A new perceptual experience." In *The Oxford Handbook of Perceptual Organization*, edited by Johan Wagemans. Oxford: Oxford University Press, 2016.

Surbrook, Michael. *Kazei-5.* Hero Games, 1998.

Theoretical Foundations of Artificial General Intelligence, edited by Pei Wang and Ben Goertzel. Paris: Atlantis Press, 2012.

Thompson, David A. "The Man-Robot Interface in Automated Assembly." In *Monitoring Behavior and Supervisory Control*, edited by Thomas B. Sheridan and Gunnar Johannsen, pp. 385-91. NATO Conference Series 1. New York: Plenum Press, 1976.

Thornton, Stephanie. *Understanding Human Development: Biological, Social and Psychological Processes from Conception to Adult Life.* New York: Palgrave Macmillan, 2008.

Tomas, David. "Feedback and Cybernetics: Reimaging the Body in the Age of the Cyborg." In *Cyberspace, Cyberbodies, Cyberpunk: Cultures of Technological Embodiment*, edited by Mike Featherstone and Roger Burrows, pp. 21-43. London: SAGE Publications, 1995.

The Turing Test: The Elusive Standard of Artificial Intelligence, edited by James H. Moor. Studies in Cognitive Systems 30. Springer Netherlands, 2003.

Turner, John N. "The right to privacy in a computronic age." In *Position Papers of the Conference on Computers, Privacy and Freedom of Information.* Kingston: Queen's University, 1970.

Tweedale, Jeffrey W., and Lakhmi C. Jain. "Agent Oriented Programming." In *Embedded Automation in Human-Agent Environment*, pp. 105-24. Adaptation, Learning, and Optimization 10. Springer Berlin Heidelberg, 2011.

Ullrich, Günter. *Automated Guided Vehicle Systems: A Primer with Practical Applications*, translated by Paul A. Kachur. Springer Berlin Heidelberg, 2015.

Ummat, Ajay, Atul Dubey, and Constantinos Mavroidis. "Bionanorobotics: A Field Inspired by Nature." In *Biomimetics: Biologically Inspired Technologies*, edited by Yoseph Bar-Cohen, pp. 201-26. Boca Raton: CRC Press, 2005.

Van den Berg, Bibi. "Pieces of Me: On Identity and Information and Communications Technology Implants." In *Human ICT Implants: Technical, Legal and Ethical Considerations*, edited by Mark N. Gasson, Eleni Kosta, and Diana M. Bowman, pp. 159-73. Information Technology and Law Series 23. T. M. C. Asser Press, 2012.

Van Oost, Ellen, and Darren Reed. "Towards a Sociological Understanding of Robots as Companions." In *Human-Robot Personal Relationships*, edited by Maarten H. Lamers and Fons J. Verbeek, pp. 11-18. Springer Berlin Heidelberg, 2011.

Vänni, Kimmo J., and Annina K. Korpela. "Role of Social Robotics in Supporting Employees and Advancing Productivity." In *Social Robotics*, pp. 674-83. Springer International Publishing, 2015.

Versace, Massimiliano, and Ben Chandler. "The Brain of a New Machine." *IEEE spectrum* 47, no. 12 (2010): 30-37.

Viirre, Erik, Françoise Baylis, and Jocelyn Downie. "Promises and perils of cognitive performance tools: A dialogue." *Technology* 11, no. 1 (2008): 9-25.

Vinciarelli, A., M. Pantic, D. Heylen, C. Pelachaud, I. Poggi, F. D'Errico, and M. Schröder. "Bridging the Gap between Social Animal and Unsocial Machine: A survey of Social Signal Processing." *IEEE Transactions on Affective Computing* 3:1 (January-March 2012): 69-87.

Vogiatzis, D., Dimitrios Galanis, V. Karkaletsis, Ion Androutsopoulos, and C. D. Spyropoulos. "A conversant robotic guide to art collections." In *Proceedings of the 2nd Workshop on Language Technology for Cultural Heritage Data, Language Resources and Evaluation Conference, Marrakech, Morocco*, pp. 55-60. 2008.

Wallach, Wendell, and Colin Allen. *Moral machines: Teaching robots right from wrong*. Oxford University Press, 2008.

Warkentin, Merrill, Vijayan Sugumaran, and Robert Sainsbury. "The Role of Intelligent Agents and Data Mining in Electronic Partnership Management." *Expert Systems with Applications* 39, no. 18 (2012): 13277-88.

Warwick, K. "The Cyborg Revolution." *Nanoethics* 8 (2014): 263-73.

Warwick, K., and M. Gasson. "Implantable Computing." In *Digital Human Modeling*, edited by Y. Cai, pp. 1-16. Lecture Notes in Computer Science 4650. Berlin/Heidelberg: Springer, 2008.

Wearable Robots: Biomechatronic Exoskeletons, edited by Jose L. Pons. Chichester: John Wiley & Sons, 2008.

Whitby, Blay. "Do You Want a Robot Lover? The Ethics of Caring Technologies." In *Robot Ethics: The Ethical and Social Implications of Robotics*, edited by Patrick Lin, Keith Abney, and George A. Bekey, pp. 233-48. Cambridge, MA: The MIT Press, 2012.

Wiener, Norbert. *Cybernetics: Or Control and Communication in the Animal and the Machine*, second edition. Cambridge, MA: The MIT Press, 1961. [Quid Pro ebook edition for Kindle, 2015.]

Wilson, Margaret. "Six views of embodied cognition." *Psychonomic bulletin & review* 9, no. 4 (2002): 625-36.

Xu, J., B. Archimede, and A. Letouzey. "A Distributed Multi-Agent Framework for Shared Resources Scheduling." *Information Control Problems in Manufacturing* 14, no. 1 (2012): 775-80.

Yagi, Akihiro. "Engineering psychophysiology in Japan." In *Engineering psychophysiology: Issues and applications*, edited by Richard W. Backs and Wolfram Boucsein, pp. 361-68. Mahwah, NJ: Lawrence Erlbaum Associates, Publishers, 2000.

Yampolskiy, Roman V. *Artificial Superintelligence: A Futuristic Approach.* Boca Raton: CRC Press, 2015.

Yampolskiy, Roman V. "The Universe of Minds." arXiv preprint, *arXiv:1410.0369 [cs.AI]*, October 1, 2014. http://arxiv.org/abs/1410.0369. Accessed January 25, 2016.

Yampolskiy, Roman V., and Joshua Fox. "Artificial General Intelligence and the Human Mental Model." In *Singularity Hypotheses*, edited by Amnon H. Eden, James H. Moor, Johnny H. Søraker, and Eric Steinhart, pp. 129-45. The Frontiers Collection. Springer Berlin Heidelberg, 2012.

Yonck, Richard. "Toward a standard metric of machine intelligence." *World Future Review* 4, no. 2 (2012): 61-70.

Chapter Four

Organization Development and the Robotic-Cybernetic-Human Workforce: Humanistic Values for a Posthuman Future?

Introduction

Organization Development (OD) is a management discipline whose theory and practice are firmly rooted in humanistic values insofar as it seeks to create effective organizations by facilitating the empowerment and growth of their human members. However, a new posthuman age is dawning in which human beings will no longer be the only intelligent actors guiding the behavior of organizations; increasingly, social robots, AI programs, and cybernetically augmented human employees are taking on roles as collaborators and decision-makers in the workplace, and this transformation is only likely to accelerate.

How should OD professionals react to the rise of these posthumanizing technologies? In this text we explore OD's humanistic foundations and the social and organizational implications of posthuman technologies for the workplace. Several ways are suggested in which OD could act as a 'Humanist OD for a posthuman world,' providing an essential service to future organizations without abandoning its traditional humanist values. An alternative vision is then presented for a 'Posthuman OD' that reinterprets and expands its humanist vision to embrace the benefits that social robots, AI, and cyberization can potentially bring into the workplace. Finally, we discuss the extent to which OD can remain a single, unified discipline in light of the challenge to its traditional humanistic values presented by such emerging technologies.

OD's Humanistic Values

Organization Development is a discipline that seeks to facilitate positive, long-term, systemic organizational change by carrying out interventions that involve practices such as "organization design, strategic planning, quality interventions, team building, survey feedback, individual instruments, and coaching and mentoring."[1] Perhaps more so than any other field within the world of business management, OD is a discipline that was explicitly founded on humanistic principles and which does not hesitate to remind the public of that fact. Such values feature prominently in the definition of OD developed by Bradford and Burke, which states that:[2]

> Based on (1) a set of values, largely humanistic; (2) application of the behavioral sciences; and (3) open system theory, organization development is a system-wide process of planned change aimed toward improving overall organization effectiveness by way of enhanced congruence of such key organizational dimensions as external environment, mission, strategy, leadership, culture, structure, information and reward systems, and work policies and procedures.

While humanism can take many different forms (such as religious humanism and secular humanism), all of its varieties share a common foundation insofar as they "emphasize human welfare and dignity" and are "either optimistic about the powers of human reason, or at least insistent that we have no alternative but to use it as best we can."[3] Over the last few decades, OD's humanistic concern for the welfare, development, and fundamental liberties of human beings has been enshrined in key documents created through a process of consultation and consensus among OD theorists and practitioners who have worked to create a shared ethical basis for their profession. For example, *The International Organization Development Code of Ethics* – originally developed by Dr. William Gellermann and the OD Institute in the 1980s and eventually adopted by many OD organizations around the world – recognizes ten values whose fundamental importance it argues should be acknowledged by OD professionals. The first of these values is "quality of life – people being satisfied with their whole life experience," and the second is "health, human potential, empowerment, growth and excellence – people being healthy, aware of the fullness of their potential, recognizing their power to bring that potential into being..." Not until the eighth value does one find any reference to OD's aim of enhancing an organization's "effectiveness, efficiency and alignment..."[4]

Similar statements emphasizing OD's fundamentally humanistic values can be found in the *Organization and Human Systems Development Credo*,[5] the

---

[1] Anderson, *Organization Development: The Process of Leading Organizational Change* (2015).

[2] Bradford & Burke, *Reinventing Organization Development* (2005).

[3] "Humanism," *The Oxford Dictionary of Philosophy.*

[4] "The International Organization Development *Code of Ethics,*" The OD Institute.

[5] "Organization and Human Systems Development Credo," OD Network.

Ethical Guidelines and Professional Standards for Organization Development and Group Process Consultants,[6] and the OD Network's *Principles of OD Practice*.[7] Of course, it is possible for a profession to adopt codes of ethics whose principles do not actually reflect the daily behavior of its practitioners. However, in the case of Organization Development it has been argued that most individuals who identify themselves as OD professionals do, in fact, feel a calling to "Create opportunities for all organizational members to learn and develop personally toward full realization of individual potential," and, more generally, that they share a "set of beliefs about the congruence of human and organizational behavior that is humanistic in nature."[8]

OD's humanistic foundations become especially apparent if one contrasts the discipline with a field like Change Management. Professionals in OD and Change Management are often called upon to address similar organizational issues, but they do so from very different perspectives. Change Management typically focuses on "engineering and directing" changes in behavior that are desired by an organization's elite decision-makers, with a focus on generating economic outcomes – while OD focuses on participatory processes that engage all of an organization's members through facilitation and coaching, to aid them in advancing the pursuit of humanistic values.[9] Indeed, one might even be tempted to argue that the very name of 'Organization Development' is misleading, insofar as the discipline does not focus directly on enhancing an organization's productivity or efficiency. While OD's ultimate goal is to create a more effective organization, it accomplishes this primarily by facilitating the personal and professional development, self-realization, creativity, just and ethical decision-making, and healthy relationships of the individual human beings who constitute the organization, based on the belief that "in essence if you create a humanistic organization it will inherently be a high-performing organization."[10]

The Posthuman Future Confronting OD

We are entering a new age in which the 'human being' contemplated by traditional humanism will no longer be the only intelligent actor whose decisions guide the development of organizations. In addition to the 'bioagency' demonstrated by traditional human beings, organizations are also shaped by

---

[6] "Ethical Guidelines and Professional Standards for Organization Development and Group Process Consultants," IAGP.

[7] "Principles of OD Practice," OD Network.

[8] Bradford & Burke (2005).

[9] Bradford & Burke (2005).

[10] Bradford & Burke (2005).

the 'cyberagency' of social robots, artificially intelligent software, and cybernetically enhanced human beings, as well as the 'collective agency' demonstrated by intelligent networks.[11]

Robots and AI

Having already moved far beyond their original roles in industrial manufacturing, robots will continue to expand into new reaches of society, filling a burgeoning number of roles in service-sector industries such as information technology, finance, retail sales, health care, education, entertainment and the arts, and government.[12] Unlike the emotionless mechanical automata that many still stereotypically associate with the word 'robot,' future generations of social robots will likely possess a full range of human-like emotional, social, and cultural capacities.[13] They will be capable of juggling the responsibilities and expectations that come with belonging simultaneously to multiple social groups – such as a family, professional association, and religious community[14] – and they will experience the same mixed emotions that often accompany human beings' efforts to make difficult decisions.[15]

Rather than being preprogrammed to mechanically follow particular ethical rules, much like human beings they will be taught a set of virtues[16] that are ambiguous and sometimes difficult to reconcile, and they will then be given the freedom and legal and moral responsibility[17] to make sound decisions shaped by their personal values, their emotional makeup,[18] and their unique personal experiences.[19] Such robots may possess a knowledge of human psychology that allows them to serve as benevolent mentors, encouraging and teaching human beings to behave in good and virtuous ways – or to serve as

---

[11] Fleischmann, "Sociotechnical Interaction and Cyborg–Cyborg Interaction: Transforming the Scale and Convergence of HCI" (2009).

[12] Solis & Takanishi, "Recent Trends in Humanoid Robotics Research: Scientific Background, Applications, and Implications" (2010); Schaal, "The New Robotics – towards Human-centered Machines" (2007).

[13] Friedenberg, *Artificial Psychology: The Quest for What It Means to Be Human* (2008).

[14] De Man et al., "A Cognitive Model for Social Role Compliant Behavior of Virtual Agents" (2012).

[15] Lee et al., "Feeling Ambivalent: A Model of Mixed Emotions for Virtual Agents" (2006).

[16] Coleman, "Android Arete: Toward a Virtue Ethic for Computational Agents" (2001).

[17] Calverley, "Imagining a Non-Biological Machine as a Legal Person" (2008).

[18] Thill & Lowe, "On the Functional Contributions of Emotion Mechanisms to (Artificial) Cognition and Intelligence" (2012).

[19] Ho & Dautenhahn, "Towards a Narrative Mind: The Creation of Coherent Life Stories for Believable Virtual Agents" (2008).

master manipulators, coaxing or coercing human beings to behave in ways that advance a robot's personal aims.[20]

Human Beings, Cyberization, and Virtual Relations

The abilities, limitations, and manner of existence of human workers will also be dramatically transformed in the coming years and decades. Already, ubiquitous computing – with our continuous connection to the Internet and reliance upon it in manifold, intimate ways – is accelerating a "convergence of the human and the computer." Computers are becoming such an integral component of the way in which contemporary human beings sense, understand, and control the world that we effectively function as cyborgs; 'cyborg-cyborg interaction' is becoming a fundamental aspect of society.[21] Genetic engineering may further enhance the ability of the human body to interface with implanted or external technologies in such ways. Meanwhile, the increasing 'virtualization' of our relationships means that we will regularly interact with coworkers, customers, and suppliers as digital avatars in virtual environments, without knowing (or perhaps caring) whether the entity on the other end of the conversation is a human being, social robot, or AI program.[22]

The Workplace

Increasingly, businesses and other organizations will consist of 'cybernetic teams' whose human and artificial members "cooperate as teammates to perform work."[23] Our acceptance of robots as colleagues and friends will be aided by the fact that human beings are not only *able* to engage in social relations with artificial intelligences as if they were human but are naturally inclined to do so.[24]

Moreover, social robots will not simply work alongside us as our peers or subordinates; some of them will likely serve as supervisors of human beings.

---

[20] Knowles et al., "Wicked Persuasion: A Designerly Approach" (2014); Ruijten et al., "Investigating the Influence of Social Exclusion on Persuasion by a Virtual Agent" (2014); Rebolledo-Mendez et al., "A Model of Motivation for Virtual-Worlds Avatars" (2008); Gladden, "The Social Robot as 'Charismatic Leader': A Phenomenology of Human Submission to Nonhuman Power" (2014).

[21] Fleischmann (2009).

[22] Grodzinsky et al., "Developing Artificial Agents Worthy of Trust: 'Would You Buy a Used Car from This Artificial Agent?'" (2011).

[23] Wiltshire et al., "Cybernetic Teams: Towards the Implementation of Team Heuristics in HRI" (2013); Bradshaw et al., "From Tools to Teammates: Joint Activity in Human-Agent-Robot Teams" (2009); Flemisch et al., "Towards a Dynamic Balance between Humans and Automation: Authority, Ability, Responsibility and Control in Shared and Cooperative Control Situations" (2012).

[24] Rehm et al., "Some Pitfalls for Developing Enculturated Conversational Agents" (2009); Friedenberg (2008).

It has been argued that "giving robots positions of responsibility is not only unavoidable but is rather something desired and that we are trying to achieve,"[25] and the development of artificial beings that possess the intellectual, emotional, social, and physical capacities needed to serve as managers and leaders of human employees – and even to excel in such roles – is well underway.[26] While some human employees might resent their social robot boss for "lacking humanity" or "stealing human jobs," other human employees may appreciate having a boss who is inherently incorruptible, honest, fair, caring, competent, and sincerely concerned for the long-term good of the whole organization rather than for personal acclaim, financial gain, career advancement, or other selfish interests.[27]

In the face of such developments, it has been noted that in the future, Human Resource Development will no longer simply involve training human employees but training their robotic coworkers, as well.[28] Similarly, OD will face the challenge of "teaching human managers and employees the skills to work with a workforce comprising humanoid robots and human beings."[29]

Beyond Humanism[30]

Such powerful new technologies are introducing a 'radical alterity' that challenges not only the traditional values of humanism but also our most fundamental understanding of the limits and possibilities of what it means to be human when old notions of ethical values that "privilege reason, truth, meaning, and a fixed concept of 'the human' are upended by digital technology, cybernetics, and virtual reality."[31] A variety of responses to these technologies has emerged. For example, on the one hand, scientists and philosophers identifying themselves as transhumanists accept the basic humanist principles of anthropocentrism, rationality, autonomy, progress, and optimism about the future of humanity, while actively working to employ science and technology in an effort to control and accelerate the transformation of the human species in ways that traditional biological evolution does not allow. Such thinkers ad-

---

[25] Samani et al., "Towards Robotics Leadership: An Analysis of Leadership Characteristics and the Roles Robots Will Inherit in Future Human Society" (2012).

[26] Gladden, "Leveraging the Cross-Cultural Capacities of Artificial Agents as Leaders of Human Virtual Teams" (2014); Gladden, "The Social Robot as 'Charismatic Leader'" (2014).

[27] Gladden, "The Social Robot as 'Charismatic Leader'" (2014).

[28] Azevedo et al., "The HRI (Human-Robot Interaction) and Human Resource Development (HRD)" (2013).

[29] Stanford, *Organization Design for HR Managers: Engaging with Change* (2013).

[30] See Gladden, *The Handbook of Information Security for Advanced Neuroprosthetics* (2015), pp. 94-96, upon which much of this section is based.

[31] Gunkel & Hawhee, "Virtual Alterity and the Reformatting of Ethics" (2003).

vocate the use of genetic engineering, cybernetics, and nanotechnology to create a more meaningful, more transcendent, 'enhanced' form of human existence. In this sense, their philosophy can be understood not as 'antihumanism' but as 'ultrahumanism.'[32]

On the other hand, there is a diverse group of posthumanist thinkers who agree that new forms of sentient and sapient existence are emerging, spurred on by the world's technological advances; however, these posthumanists reject transhumanism's anthropocentric and humanistic focus. Instead, such posthumanists argue that the future will include many different sources of intelligence and agency that will create meaning in the universe through their networks and relations:[33] such entities might include 'natural' human beings, genetically engineered human beings, human beings with extensive cybernetic modifications, human beings dwelling in virtual realities, social robots, artificially intelligent software, nanorobot swarms, and sentient or sapient networks.

Thinkers like Bostrom and Kurzweil are generally enthusiastic about the transhuman potential of technology to 'liberate' humanity, allowing us to transcend our previous physical and cognitive limitations and spark the evolution of higher forms of human existence. Following Habermas and Horkheimer, other scholars are more pessimistic about the anticipated impact of such technologies, suggesting that while the exact impact of such technologies cannot be predicted in advance, they are more likely to spur social fragmentation and inequality, a reduction in human autonomy and meaning, and the oppression – or in its extreme form, even destruction – of humanity at the hands of its technological creation. Yet another group of thinkers takes a more extreme view, arguing that while humanity's twilight and replacement by artificial beings is indeed inevitable, it is in fact a natural and desirable step in the evolution of intelligent life on earth.[34]

OD's Possible Stances toward Posthuman Technology

Driven in part by such dramatic technological transformations, it is clear that fields such as Human Resource Development and OD are about to enter a 'new age' – but it is not yet known exactly what form this new age might

---

[32] Ferrando, "Posthumanism, Transhumanism, Antihumanism, Metahumanism, and New Materialisms: Differences and Relations" (2013).

[33] Ferrando (2013).

[34] Abrams, "Pragmatism, Artificial Intelligence, and Posthuman Bioethics: Shusterman, Rorty, Foucault" (2004), and Edgar, "The Hermeneutic Challenge of Genetic Engineering: Habermas and the Transhumanists" (2009).

take.[35] The outcome may depend on the extent to which Organization Development is necessarily and inherently humanistic. How far can OD go in adapting itself to new posthuman realities before it ceases to be OD?

Humanistic OD for a Posthuman World

Even if OD professionals hold fast to a traditional understanding of humanist values – declining to expand their 'moral universe' to include a concern for the welfare of sapient computers and rejecting the notion that there is anything 'liberating' about the growth in workplace robotics and cybernetics – it is still possible for such professionals to utilize their OD expertise to aid future organizations that *do* find value in such technologies and that are working to implement them. We can refer to this path as 'Humanistic OD for a posthuman world.' Under this model, typical OD interventions might include:

- **A call for reflection**. OD professionals already play a key role in calling upon CEOs to "pause and reflect and assess the genuine needs of the organization" carefully before making the decision to terminate any employees, given the devastating impact that this can have on the human beings affected.[36] Even if OD professionals are not able to prevent the replacement of human workers by robots, they can at least aid CEOs to recognize and honestly reflect on their own motivations, biases, and assumptions before making the decision to implement such technologies in the workplace.
- **Pre-interventions: shaping robots through their designers**. Roboticists are striving to identify the "key cognitive capabilities" that social robots will need in order to work safely and effectively with human colleagues.[37] OD professionals can offer their expertise to help such scientists and engineers understand that social robots will meld most successfully into a hybrid human-artificial workplace if the robots' motivations, priorities, ethical commitments, and decision-making processes reflect the same humanistic values upon which OD is premised. Even if OD professionals do not personally believe that robot colleagues will provide a net positive contribution to the workplace, they might in this way at least work to minimize the harm and problems caused by such robots' introduction.
- **The robotic-cybernetic-human workplace as sociotechnical system**. Even if OD professionals do not believe that social robots or AIs

[35] Chermack et al., "Critical Uncertainties Confronting Human Resource Development" (2003).

[36] "Ethical Guidelines and Professional Standards for Organization Development and Group Process Consultants," IAGP.

[37] Williams, "Robot Social Intelligence" (2012).

are beings worthy of moral concern, they can still acknowledge that they represent remarkable and qualitatively different additions to the social and technical systems that constitute an organization. Social robots and AIs represent a special challenge, insofar as they are the first pieces of workplace technology that will operate proactively (and perhaps unpredictably) within the social sphere. Drawing on the body of theory and practice in sociotechnical systems (STS) that has been developed since the 1950s,[38] OD professionals can help organizations to reconcile their social and technical systems in light of the new complications and opportunities that are brought by artificial and cybernetic human workers.

- **Understanding new forms of power relations**. In their efforts to establish social agreements among organizational members with differing perspectives, some streams of 'New OD' have emphasized "less confronting and more 'optimistic' or 'positive' approaches" such as appreciative inquiry, while downplaying the harsher realities of workplace politics and the use and abuse of power dynamics by organizational members.[39] However, new forms of power relations will be created when the workplace becomes home to artificial beings and cybernetically enhanced humans who may have the capacity to out-think, out-organize, out-empathize, and out-manipulate their 'natural' human colleagues. Moreover, such powerful new technologies may become tools in the political machinations and organizational power struggles of the human executives who are pushing for and shaping their implementation. One must take into account not only the values and motives of sapient robots themselves but also those of the human beings who are designing and implementing them.[40] OD professionals can play a key role by helping an organization's members to recognize and address the sometimes subtle ways in which the introduction of such technologies can reshape an organization's power dynamics for good or ill.

Posthumanistic OD for a Posthuman World

It is conceivable that after working in partnership with social robots whose intelligence, moral values, emotions, relationships, and even spirituality appear as 'human' as those of human beings – as well as with human beings whose cybernetic enhancements have strengthened their connection to other human beings and concern for their welfare, rather than lessening it – some OD professionals might come to see such posthumanizing technological

---

[38] Anderson (2015).

[39] Marshak & Grant (2008).

[40] Anderson, "Why Is AI so Scary?" (2005); Grodzinsky et al. (2011).

change as something that should be embraced rather than rejected in the workplace, as something that can aid the development and self-fulfillment of individual human beings, if its implementation is thoughtfully managed in a way that unlocks these benefits. Apart from their instrumental worth, OD professionals might also come to value sentient or sapient robots and AI programs in themselves and attribute to them an existential value and moral agency previously attributed only to human beings. Such OD professionals might well expand their 'moral universe' to include not only a concern for the welfare of individual human beings, but also for the welfare of individual sapient robots and artificial intelligences, and perhaps even for hybrid entities or networks in which cyberized human beings, robots, and AIs have become symbiotically interdependent. We can refer to this general approach as 'Posthumanistic OD for a posthuman world.' Under this model, OD's work might include:

- **Facilitating the new multiculture: beyond human diversity**. One stream of today's New OD focuses on how "diversity and multicultural realities" shape an organization.[41] If OD already places an emphasis on helping human beings to find value in their differences and open themselves to new perspectives, it would not require a dramatic leap for OD to also help human beings find value in the novel perspectives and forms of being that their new artificial colleagues possess. OD could teach human employees to appreciate social robots and cyborgs as new forms of the 'other' that can and should be welcomed into their community.
- **Managing change for which Change Management is unprepared**. The implementation of a transformative new company-wide technological system is sometimes facilitated by an organization's Change Management unit, which collaborates with an organization's senior leadership to install such a technology by "engineering and directing" changes in workers' behavior, with an ultimate focus on financial outcomes.[42] However, integrating advanced social robots and cybernetics into an organization is not simply a techno-structural matter of adding new IT hardware; it must be understood from a phenomenological perspective as creating a strange new soil in which social relationships and meaning will grow as human beings and artificial general intelligences (AGIs) recognize and confirm one another's existence as beings possessing a conscience.[43]

  The manner in which cyberization and social robots are welcomed into the workforce (or rejected and sabotaged) by human employees

[41] Marshak & Grant (2008).

[42] Bradford & Burke (2005).

[43] Ramey, "Conscience as a Design Benchmark for Social Robots" (2006).

may thus depend primarily not on the employees' technical ability to learn new hardware and software but on their social abilities and mindsets. In contrast to changes that can be centrally decided and implemented, developing relationships with artificial sapient beings is the sort of change that may take place in a manner that is largely unplanned and undirected and which grows organically in directions shaped by the personal biases, experiences, hopes, fears, cultures, and power and trust relations of an organization's human members.[44] Some streams of New OD recognize that the most significant change is often a continuous, unplanned, distributed process over which an organization's senior decision-makers have limited conscious control. Rather than attempting to impose a desired change, New OD strives to facilitate "shifts in human consciousness" through techniques like appreciative inquiry that surface employees' positive experiences and aspirations, with a goal of "creating new mindsets or social agreements" among members of the organization.[45] OD's roots in humanist philosophy become a strength instead of a weakness when it comes to supporting technological change that is grounded as much in the hearts as in the minds of an organization's members.

- **From systems for manufacturing to systems for meaning**. OD professionals face growing conflicts "between employees' needs for greater meaning and the organization's need for more effective and efficient use of its resources."[46] Up to now, it has often been assumed that new automated technologies increase organizational efficiency while reducing employees' satisfaction and sense of meaning. However, social robots have the potential to radically rewrite this equation. For the first time, the workplace will be filled with pieces of technology that are designed to be moral, emotional, intelligent, social beings that yearn to both receive and offer respect, self-actualization, and a sense of belonging. Social robots could potentially become OD professionals' strong allies in efforts to enhance meaning in the workplace.[47] Similarly, workers with extensive cybernetic enhancements might find that while their desire and ability to engage in traditional forms of interpersonal relationships has lessened, they are able to access and experience reality in ways that offer new forms of meaning.

---

[44] Coeckelbergh, "Can We Trust Robots?" (2012).

[45] Marshak & Grant (2008).

[46] Cummings & Worley, *Organization Development and Change* (2014), p. 60.

[47] Regarding the nature and potential capacities of social robots, see, e.g., Breazeal, "Toward sociable robots" (2003); Kanda & Ishiguro, *Human-Robot Interaction in Social Robotics* (2013); *Social Robots and the Future of Social Relations*, edited by Seibt et al. (2014); *Social Robots from a Human Perspective*, edited by Vincent et al. (2015); and *Social Robots: Boundaries, Potential, Challenges*, edited by Marco Nørskov (2016).

Three Paths for OD as a Discipline

One can envision at least three different ways in which OD as a discipline might respond to the challenge of growing cyberization, robotics, and AI in the workforce. The first possibility is that OD professionals could unanimously decide to preserve their traditional humanistic values, understood to mean that OD works to serve the development of *human beings* in *human organizations*. Such an OD would likely work to prevent the introduction of social robots and cybernetic augmentation into the workplace because of their perceived 'dehumanizing' effects. Given the fact that OD's value is already questioned by some executives because of its lack of direct emphasis on financial outcomes, holding fast to its traditional humanistic values could result in OD becoming increasingly marginalized, as businesses replace their OD units and functions with new kinds of units that will readily accept and promote the economic, social, and technological 'advances' that are both made possible by and support a diminishing role for human beings and humanist values in the workplace. However, there could still be a role for such a Humanistic OD to serve organizations that similarly reject social robotics and cybernetic augmentation and hold to traditional humanist values.

A second possibility is for OD as a whole to broaden its vision to include a concern for the entire community of artificial, cybernetic, and human members that make up an organization. In this case, OD professionals would realize that their concern had never been for the welfare of 'human beings' as such – but for the welfare of 'sapient beings,' of whom traditional human beings had previously been the only representatives. Such a shift in OD's perspective could occur naturally as current generations of OD professionals are gradually succeeded by future generations who will have grown up amidst quite different social and technological realities. A Posthumanistic OD of this sort could play a key role for similarly posthumanist organizations that are struggling with their efforts to integrate social robotics and biocybernetic technologies successfully into the workplace.

A third possibility is for Organization Development to splinter into multiple disciplines that all claim the 'OD' name (or some variation on it) but which display divergent – and perhaps irreconcilable – attitudes toward the new technologies that enable a robotic-cybernetic-human workforce. This outcome is perhaps not unlikely, given the fact that OD has already begun to separate into several strains – such as Classical, Neoclassical, and Social Interaction OD or New OD – that differ in their willingness to supplement OD's original humanist values with more business-oriented considerations[48] and to

[48] Bradford & Burke (2005).

embrace the knowledge and technologies generated by new fields of science like self-organizing systems and complexity theory.[49]

New Consensus Ethical Documents?

The existing consensus ethics documents developed by Organization Development groups and described in previous sections do not formulate a stance toward such posthuman technologies. Efforts to update OD's foundational documents to explicitly address such issues would at this moment likely be premature, insofar as there is still broad disagreement among AI researchers, engineers, sociologists, philosophers, economists, and others about the particular forms, capacities, risks, and organizational relevance that such technologies might eventually display. Moreover, it is unlikely that most OD professionals currently possess the scholarly expertise or personal experience with social robotics, cyberization, and other posthuman technologies that would enable them to make informed decisions about the stances that their profession should take on such matters.

However, OD professionals may eventually decide that it is worth launching a broad new process of consultation to formulate revised ethics documents that address the role of posthumanizing technologies within organizations. If such an effort at forging consensus among OD professionals succeeds, it could help OD maintain cohesion as a profession and provide new energy and focus for addressing the dramatic organizational challenges and opportunities that such new technologies will bring. On the other hand, if the issues prove too divisive and the effort to achieve consensus fails, this could nevertheless be beneficial if it helps to clearly delineate the differences between 'Humanistic OD' and 'Posthumanistic OD' and allows the theorists and practitioners of these two different schools to explore and develop the unique strengths of their own approaches, rather than struggling in vain to fashion a single, unified OD that could accommodate both those who see social robotics and cyberization as technologies that will destroy our human identity and those who see them as technologies that will allow us to fulfill and transcend it.

Conclusion

The profoundly humanist roots of Organizational Development will be challenged by the rise of a new posthuman age in which traditional human beings, technologically enhanced human cyborgs, social robots, AI programs, and other forms of intelligent beings all work together within organizations to achieve common goals. OD's humanist premises may cause some OD theorists and practitioners to reject the notion that posthumanizing technologies can contribute positively to the development of effective organizations that

[49] Marshak & Grant (2008).

support their members' freedom, dignity, and pursuit of self-realization. However, OD's humanist values also provide a rich foundation to build upon for those OD professionals who embrace posthumanizing technologies as new venues for the creation of meaning, imagination, healthy relations, and self-fulfillment within the workplace. It is too early to know whether OD will emerge from this transition as a single, unified discipline or as a set of related humanistic and posthumanistic disciplines that adopt different stances toward OD's original humanist roots. However, the invaluable insights and techniques that OD has to offer should ensure that it will continue to play a key role as long as organizations exist that count human beings among their members.

References

Abrams, J.J. "Pragmatism, Artificial Intelligence, and Posthuman Bioethics: Shusterman, Rorty, Foucault." *Human Studies* 27:3 (2004): 241-58.

Anderson, Donald L. *Organization Development: The Process of Leading Organizational Change*, 3e. SAGE Publications, 2015.

Anderson, Michael L. "Why Is AI so Scary?" *Artificial Intelligence*, Special Review Issue, 169, no. 2 (2005): 201-08. doi:10.1016/j.artint.2005.10.008.

Azevedo, Renato, Cristiano Reis, and Edgard Cornacchione, Jr. "The HRI (Human-Robot Interaction) and Human Resource Development (HRD)." *International Journal of Education and Research* vol. 1, no. 4 (April 2013). http://www.ijern.com/images/April-2013/02.pdf. Accessed December 17, 2014.

Bradford, David L., and W. Warner Burke. *Reinventing Organization Development: New Approaches to Change in Organizations.* John Wiley & Sons, 2005.

Bradshaw, Jeffrey M., Paul Feltovich, Matthew Johnson, Maggie Breedy, Larry Bunch, Tom Eskridge, Hyuckchul Jung, James Lott, Andrzej Uszok, and Jurriaan van Diggelen. "From Tools to Teammates: Joint Activity in Human-Agent-Robot Teams." In *Human Centered Design*, edited by Masaaki Kurosu, pp. 935-44. Lecture Notes in Computer Science 5619. Springer Berlin Heidelberg, 2009.

Breazeal, Cynthia. "Toward sociable robots." *Robotics and Autonomous Systems* 42 (2003): 167-75.

Calverley, David J. "Imagining a Non-Biological Machine as a Legal Person." *AI & SOCIETY* 22 (4) (2008): pp. 523-37. doi:10.1007/s00146-007-0092-7.

Chermack, Thomas J., Susan A. Lynham, and Wendy E. A. Ruona. "Critical Uncertainties Confronting Human Resource Development." *Advances in Developing Human Resources* 5, no. 3 (2003): 257–71. doi:10.1177/1523422303254628.

Coeckelbergh, M. "Can We Trust Robots?" *Ethics and Information Technology* 14:1 (2012): 53-60.

Coleman, Kari Gwen. "Android Arete: Toward a Virtue Ethic for Computational Agents." *Ethics and Information Technology* 3 (4) (2001): 247-65. doi:10.1023/A:1013805017161.

Cummings, Thomas, and Christopher Worley. *Organization Development and Change.* Cengage Learning, 2014.

De Man, Jeroen, Annerieke Heuvelink, and Karel van den Bosch. "A Cognitive Model for Social Role Compliant Behavior of Virtual Agents." In *Intelligent Virtual Agents,* edited by Yukiko Nakano, Michael Neff, Ana Paiva, and Marilyn Walker, pp. 303-10. Lecture Notes in Computer Science 7502. Springer Berlin Heidelberg, 2012.

Edgar, Andrew. "The Hermeneutic Challenge of Genetic Engineering: Habermas and the Transhumanists." *Medicine, Health Care and Philosophy* 12, no. 2 (2009): 157-67. doi:10.1007/s11019-009-9188-9.

"Ethical Guidelines and Professional Standards for Organization Development and Group Process Consultants." IAGP - International Association for Group Psychotherapy and Group Processes. http://www.iagp.com/docs/IAGPOrgEthicalguidelinesEnglishv1.0.pdf. Accessed December 20, 2014.

Ferrando, Francesca. "Posthumanism, Transhumanism, Antihumanism, Metahumanism, and New Materialisms: Differences and Relations." *Existenz: An International Journal in Philosophy, Religion, Politics, and the Arts* vol. 8, no. 2 (2013): 26-32.

Fleischmann, Kenneth R. "Sociotechnical Interaction and Cyborg–Cyborg Interaction: Transforming the Scale and Convergence of HCI." *The Information Society* 25, no. 4 (2009): 227–35. doi:10.1080/01972240903028359.

Flemisch, F., M. Heesen, T. Hesse, J. Kelsch, A. Schieben, and J. Beller (2012). "Towards a Dynamic Balance between Humans and Automation: Authority, Ability, Responsibility and Control in Shared and Cooperative Control Situations." *Cognition, Technology & Work* 14 (1): 3-18. doi:10.1007/s10111-011-0191-6.

Friedenberg, J. *Artificial Psychology: The Quest for What It Means to Be Human,* Philadelphia: Psychology Press, 2008.

Gladden, Matthew E. *The Handbook of Information Security for Advanced Neuroprosthetics.* Indianapolis: Synthypnion Academic, 2015.

Gladden, Matthew E. "Leveraging the Cross-Cultural Capacities of Artificial Agents as Leaders of Human Virtual Teams." *Proceedings of the 10th European Conference on Management Leadership and Governance,* edited by Visnja Grozdanić, pp. 428-35. Reading: Academic Conferences and Publishing International Limited, 2014.

Gladden, Matthew E. "The Social Robot as 'Charismatic Leader': A Phenomenology of Human Submission to Nonhuman Power." In *Sociable Robots and the Future of Social Relations: Proceedings of Robo-Philosophy 2014,* edited by Johanna Seibt, Raul Hakli, and Marco Nørskov, pp. 329-39. Frontiers in Artificial Intelligence and Applications 273. IOS Press, 2014.

Grodzinsky, F.S., K.W. Miller, and M.J. Wolf. "Developing Artificial Agents Worthy of Trust: 'Would You Buy a Used Car from This Artificial Agent?'" *Ethics and Information Technology* 13:1 (2011): 17-27.

Gunkel, David, and Debra Hawhee. "Virtual Alterity and the Reformatting of Ethics." *Journal of Mass Media Ethics* 18, no. 3–4 (2003): 173-93. doi:10.1080/08900523.2003.9679663.

Ho, Wan Ching, and Kerstin Dautenhahn. "Towards a Narrative Mind: The Creation of Coherent Life Stories for Believable Virtual Agents." In *Intelligent Virtual Agents*, edited by Helmut Prendinger, James Lester, and Mitsuru Ishizuka, pp. 59-72. Lecture Notes in Computer Science 5208. Springer Berlin Heidelberg, 2008.

"Humanism." *The Oxford Dictionary of Philosophy (2 Rev Ed.)*, edited by Simon Blackburn. Oxford University Press. http://www.oxfordreference.com/view/10.1093/acref/9780199541430.001.0001/acref-9780199541430-e-1528?rskey=FZ9dbc&result=1530. Accessed January 14, 2015.

"The International Organization Development *Code of Ethics*." Organizational Development English Library, The OD Institute. http://www.theodinstitute.org/od-library/code_of_ethics.htm. Accessed December 17, 2014.

Kanda, Takayuki, and Hiroshi Ishiguro. *Human-Robot Interaction in Social Robotics*. Boca Raton: CRC Press, 2013.

Knowles, Bran, Paul Coulton, Mark Lochrie, and Jon Whittle. "Wicked Persuasion: A Designerly Approach." In *Persuasive Technology*, edited by Anna Spagnolli, Luca Chittaro, and Luciano Gamberini, pp. 137-42. Lecture Notes in Computer Science 8462. Springer International Publishing, 2014.

Lee, Benny Ping-Han, Edward Chao-Chun Kao, and Von-Wun Soo. "Feeling Ambivalent: A Model of Mixed Emotions for Virtual Agents." In *Intelligent Virtual Agents*, edited by Jonathan Gratch, Michael Young, Ruth Aylett, Daniel Ballin, and Patrick Olivier, pp. 329-42. Lecture Notes in Computer Science 4133. Springer Berlin Heidelberg, 2006.

Marshak, Robert J., and David Grant. "Organizational Discourse and New Organization Development Practices." *British Journal of Management* 19 (2008): S7-19. doi:10.1111/j.1467-8551.2008.00567.x.

"Organization and Human Systems Development Credo." OD Network. http://www.odnetwork.org/?page=ODCredo. Accessed December 17, 2014.

"Principles of OD Practice." OD Network. http://www.odnetwork.org/?page=PrinciplesOfODPracti. Accessed December 17, 2014.

Ramey, Christopher H. "Conscience as a Design Benchmark for Social Robots." In *The 15th IEEE International Symposium on Robot and Human Interactive Communication, 2006. ROMAN 2006*, pp. 486-91. IEEE, 2006.

Rebolledo-Mendez, Genaro, David Burden, and Sara de Freitas. "A Model of Motivation for Virtual-Worlds Avatars." In *Intelligent Virtual Agents*, edited by Helmut Prendinger, James Lester, and Mitsuru Ishizuka, pp. 535-36. Lecture Notes in Computer Science 5208. Springer Berlin Heidelberg, 2008.

Rehm, M., André, E., and Nakano, Y. "Some Pitfalls for Developing Enculturated Conversational Agents." In *Human-Computer Interaction: Ambient, Ubiquitous and Intelligent Interaction*, pp. 340-48. Lecture Notes in Computer Science 5612, 2009.

Ruijten, Peter A. M., Jaap Ham, and Cees J. H. Midden. "Investigating the Influence of Social Exclusion on Persuasion by a Virtual Agent." In *Persuasive Technology*, edited by Anna Spagnolli, Luca Chittaro, and Luciano Gamberini, pp. 191-200. Lecture Notes in Computer Science 8462. Springer International Publishing, 2014.

Samani, H.A., Valino Koh, J.T.K., Saadatian, E., and Polydorou, D. "Towards Robotics Leadership: An Analysis of Leadership Characteristics and the Roles Robots Will Inherit in Future Human Society." *Intelligent Information and Database Systems*, Lecture Notes in Computer Science 7197 (2012): 158-65.

Schaal, Stefan. "The New Robotics – towards Human-centered Machines." *HFSP Journal* 1, no. 2 (2007): 115-26. doi:10.2976/1.2748612.

Social Robots and the Future of Social Relations, edited by Johanna Seibt, Raul Hakli, and Marco Nørskov. Amsterdam: IOS Press, 2014.

Social Robots: Boundaries, Potential, Challenges, edited by Marco Nørskov. Farnham: Ashgate Publishing, 2016.

Social Robots from a Human Perspective, edited by Jane Vincent, Sakari Taipale, Bartolomeo Sapio, Giuseppe Lugano, and Leopoldina Fortunati. Springer International Publishing, 2015.

Solis, Jorge, and Atsuo Takanishi. "Recent Trends in Humanoid Robotics Research: Scientific Background, Applications, and Implications." *Accountability in Research* 17, no. 6 (2010): 278-98. doi:10.1080/08989621.2010.523673.

Stanford, Naomi. *Organization Design for HR Managers: Engaging with Change.* Routledge, 2013.

Thill, Serge, and Robert Lowe. "On the Functional Contributions of Emotion Mechanisms to (Artificial) Cognition and Intelligence." In *Artificial General Intelligence*, edited by Joscha Bach, Ben Goertzel, and Matthew Iklé, pp. 322-31. Lecture Notes in Computer Science 7716. Springer Berlin Heidelberg, 2012.

Williams, Mary-Anne. "Robot Social Intelligence." In *Social Robotics*, edited by Shuzhi Sam Ge, Oussama Khatib, John-John Cabibihan, Reid Simmons, and Mary-Anne Williams, pp. 45-55. Lecture Notes in Computer Science 7621. Springer Berlin Heidelberg, 2012.

Wiltshire, Travis J., Dustin C. Smith, and Joseph R. Keebler. "Cybernetic Teams: Towards the Implementation of Team Heuristics in HRI." In *Virtual Augmented and Mixed Reality. Designing and Developing Augmented and Virtual Environments*, edited by Randall Shumaker, pp. 321-30. Lecture Notes in Computer Science 8021. Springer Berlin Heidelberg, 2013.

Part II

Human Augmentation: A Closer Look

Chapter Five

Neural Implants as Gateways to Digital-Physical Ecosystems and Posthuman Socioeconomic Interaction[1]

Abstract. Looking beyond current desktop, mobile, and wearable technologies, we argue that work-related information and communications technology (ICT) will increasingly move inside the human body through the use of neuroprosthetic devices that create employees who are permanently connected to their workplace's digital ecosystems. Such persons may possess enhanced perception, memory, and abilities to manipulate physical and virtual environments and to link with human and synthetic minds to form cybernetic networks that can be both 'supersocial' and 'postsocial.' However, such neuroprosthetics may also create a sense of inauthenticity, vulnerability to computer viruses and hacking, financial burdens, and questions surrounding ownership of intellectual property produced using implants. Moreover, those populations who do and do not adopt neuroprostheses may come to inhabit increasingly incompatible and mutually incomprehensible digital ecosystems. Here we propose a cybernetic model for understanding how neuroprosthetics can either facilitate human beings' participation in posthuman informational ecosystems – or undermine their health, information security, and autonomy.

Introduction

For many employees, 'work' is no longer something performed while sitting at a computer in an office. Employees in a growing number of industries are expected to carry mobile devices and be available for work-related interactions even when beyond the workplace and outside of normal business

[1] This text was originally published in *Digital Ecosystems: Society in the Digital Age*, edited by Łukasz Jonak, Natalia Juchniewicz, and Renata Włoch, pp. 85-98. Warsaw: Digital Economy Lab, University of Warsaw, 2016.

hours. In this text we argue that a future step will increasingly be to move work-related information and communications technology (ICT) inside the human body through the use of neuroprosthetics, to create employees who are always 'online' and connected to their workplace's digital ecosystems. At present, neural implants are used primarily to restore abilities lost through injury or illness, however their use for augmentative purposes is expected to grow, resulting in populations of human beings who possess technologically altered capacities for perception, memory, imagination, and the manipulation of physical environments and virtual cyberspace. Such workers may exchange thoughts and share knowledge within posthuman cybernetic networks that are inaccessible to unaugmented human beings.

Scholars note that despite their potential benefits, such neuroprosthetic devices may create numerous problems for their users, including a sense of alienation, the threat of computer viruses and hacking, financial burdens, and legal questions surrounding ownership of intellectual property produced using such implants. Moreover, different populations of human beings may eventually come to occupy irreconcilable digital ecosystems as some persons embrace neuroprosthetic technology, others feel coerced into augmenting their brains to compete within the economy, others reject such technology, and still others are simply unable to afford it.

In this text we propose a model for analyzing how particular neuroprosthetic devices will either facilitate human beings' participation in new forms of socioeconomic interaction and digital workplace ecosystems – or undermine their mental and physical health, privacy, autonomy, and authenticity. We then show how such a model can be used to create device ontologies and typologies that help us classify and understand different kinds of advanced neuroprosthetic devices according to the impact that they will have on individual human beings.

From Neuroprosthetic Devices to Posthuman Digital-Physical Ecosystems

Existing Integration of the Human Brain with Work-related Digital-Physical Ecosystems

In recent decades the integration of the human brain with work-related digital ecosystems has grown stronger and increasingly complex. Whereas once employees were expected to use desktop computers during 'working hours,' for a growing number of employees it is now expected that they be available for work-related interactions at all times through their possession and mastery of mobile (and now, wearable) devices.[2] Along this path of ever

[2] Shih, "Project Time in Silicon Valley" (2004); Gripsrud, "Working on the Train: From 'Dead Time' to Productive and Vital Time" (2012).

closer human-technological integration, an emerging frontier is that of moving computing *inside* the human body through the use of implantable computers.[3]

The Potential of Neuroprosthetic Implants for Human Enhancement

One particular type of implantable computer is a neuroprosthetic device (or neural implant) designed to provide a human being with some sensory, cognitive, or motor capacity.[4] Such neuroprostheses are currently used primarily for therapeutic purposes, to restore abilities that have been lost due to injury or illness. However, researchers have already developed experimental devices designed for purposes of human enhancement that allow an individual to exceed his or her natural biological capacities by, for example, obtaining the ability to perceive ultrasonic waves or store digitized computer files within one's body.[5]

Toward Posthuman Digital-Physical Ecosystems

The use of neuroprosthetics for purposes of human enhancement is expected to grow over the coming decades, resulting in a segment of the population whose minds possess unique kinds of sensory perception, memory, imagination, and emotional intelligence and who participate in social relations that are mediated not through the exchange of traditional oral, written, or nonverbal communication but by neurotechnologies that allow the sharing of thoughts and volitions directly with other human minds and with computers.[6]

Until now, communicating a thought to another mind has required the thought to be expressed physically as a social action that is audible, visible, or tangible in nature, however future neuroprosthetics may facilitate the exchange of ideas directly at the level of thought,[7] thereby allowing the creation of human networks that can be understood as either 'supersocial' or 'postsocial' in nature. Not only might such posthuman[8] digital ecosystems be inaccessible to those who lack the appropriate form of neural augmentation, but even their very existence may be invisible to unmodified human beings.

---

[3] Koops & Leenes, "Cheating with Implants: Implications of the Hidden Information Advantage of Bionic Ears and Eyes" (2012); Gasson, "Human ICT Implants: From Restorative Application to Human Enhancement" (2012); McGee, "Bioelectronics and Implanted Devices" (2008).

[4] Lebedev, "Brain-Machine Interfaces: An Overview" (2014).

[5] Warwick, "The Cyborg Revolution" (2014); Gasson (2012); McGee (2008).

[6] McGee (2008); Warwick (2014); Rao et al., "A Direct Brain-to-Brain Interface in Humans" (2014).

[7] Warwick (2014); Rao et al. (2014); Gladden, "Tachikomatic Domains: Utopian Cyberspace as a 'Contingent Heaven' for Humans, Robots, and Hybrid Intelligences" (2015).

[8] See Ferrando, "Posthumanism, Transhumanism, Antihumanism, Metahumanism, and New Materialisms: Differences and Relations" (2013).

In this text, we will often refer to such ecosystems as 'digital' to emphasize the fact that they may utilize an immersive cyberspace or other artificial environment as a virtualized locus for socioeconomic interaction. However, it should be kept in mind that any such virtual reality is always grounded in and maintained by the computational activity of electronic or biological physical substrates; thus technically, digital ecosystems should always be understood as 'digital-physical' ecosystems.

The Need to Analyze Neuroprosthetics from Cybernetic, Phenomenological, and Existentialist Perspectives

As a bidirectional gateway, a neural implant not only aids one's mind to reach out to explore the world and interact with other entities; it may also allow external agents or systems to reach into one's mind to access – and potentially manipulate or disrupt – one's most intimate mental processes.[9] This makes it essential that manufacturers who produce such devices, policymakers who can encourage or ban their adoption, and users in whom they will be implanted be able to understand the positive and negative impacts of particular neuroprosthetic devices on individual users. This calls for the development of device ontologies and typologies for classifying and understanding neuroprostheses that do not simply focus on the devices' technical characteristics but which also consider a user's lived experience of a neuroprosthetic device and which integrate a cybernetic analysis of "control and communication"[10] with phenomenological and even existentialist perspectives.[11]

Existing Ontologies and Typologies of Neuroprosthetic Devices

Existing typologies for neuroprosthetics are primarily functional. For example, a neuroprosthetic device can be classified based on the nature of its interface with the brain's neural circuitry (sensory, motor, bidirectional sensorimotor, or cognitive[12]), its purpose (for restoration, diagnosis, identification, or enhancement[13]), or its location (non-invasive, partially invasive, or invasive[14]). Typologies have also been developed that classify a neuroprosthesis according to whether it aids its human user to interact with a real physical environment using his or her natural physical body, augments or replaces the

---

[9] See Gasson (2012), pp. 15-16.

[10] Wiener, *Cybernetics: Or Control and Communication in the Animal and the Machine* (1961).

[11] Gladden, "Tachikomatic Domains" (2015).

[12] Lebedev (2014).

[13] Gasson (2012), p. 25.

[14] Gasson (2012), p. 14.

user's natural physical body (e.g., with robotic prosthetic limbs), or allows the user to sense and manipulate some virtual environment.[15]

Formulating our Model for an Ontology of Neuroprosthetics

Here we propose a model for classifying and understanding neuroprosthetic devices especially in their role of integrating human beings into digital ecosystems, economies, and information systems. The model comprises two main dimensions, of which one (impact) is further subdivided into two sub-dimensions (new capacities and detriments).

Roles of the Human User

A neuroprosthetic device affects its human user as viewed on three levels: 1) the human being as a sapient metavolitional agent, a unitary mind that possesses its own conscious awareness, memory, volition, and conscience (or 'metavolitionality'[16]); 2) the human being as an embodied organism that inhabits and can sense and manipulate a particular environment through the use of its body; and 3) the human being as a social and economic actor who interacts with others to form social relationships and to produce, exchange, and consume goods and services.

Impact: Potential New Capacities and Detriments

At each of these three levels, a neuroprosthetic device can create for its user either new opportunities and advantages, new threats and disadvantages, or both. Typically a neuroprosthesis creates new opportunities for its user to participate in socioeconomic interaction and informational ecosystems by providing some new cognitive, sensory, or motor capacity. Disadvantages may take the form of a new dependency on some external resource, the loss of a previously existing capability, a security vulnerability, or some other detriment. Because a neuroprosthetic device's creation of new capacities can be independent of its creation of detriments, these elements comprise two different dimensions; however, it is simpler to treat them as two sub-dimensions of a single larger dimension, the device's 'impact.'

Impacts Captured by Our Model

Below we present specific capacities and detriments that neuroprosthetics are expected to create for their users at the three levels of the human being as 1) sapient metavolitional agent, 2) embodied embedded organism, and 3) social

[15] Gladden, "Cybershells, Shapeshifting, and Neuroprosthetics: Video Games as Tools for Posthuman 'Body Schema (Re)Engineering'" (2015).

[16] Gladden, "Tachikomatic Domains" (2015); Calverley, "Imagining a Non-biological Machine as a Legal Person" (2008).

and economic actor. These items constitute a broad universe of expected possible impacts identified by scholars; any one neuroprosthesis may generate only a small number of these effects, if any.

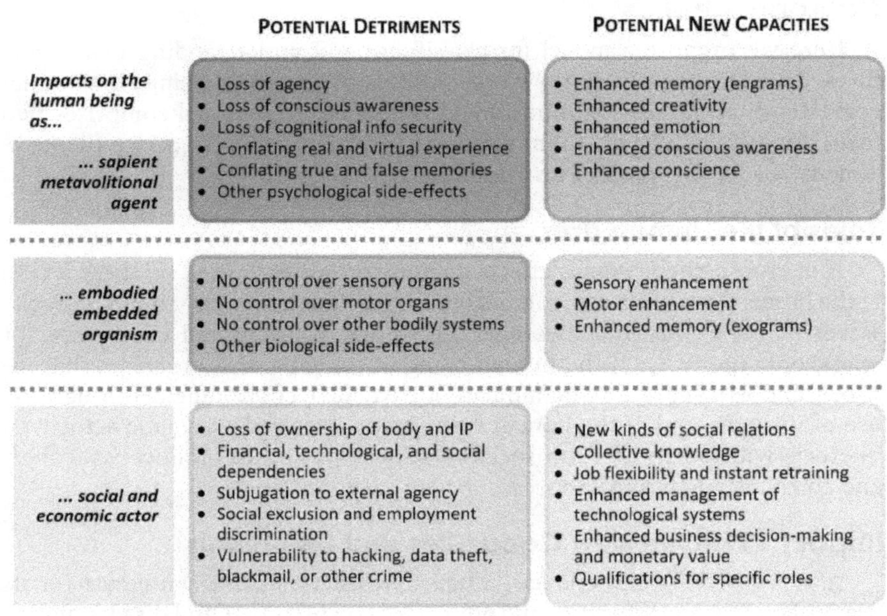

POTENTIAL DETRIMENTS
POTENTIAL NEW CAPACITIES
Impacts on the human being as...
... sapient metavolitional agent
• Loss of agency
• Loss of conscious awareness
• Loss of cognitional info security
• Conflating real and virtual experience
• Conflating true and false memories
• Other psychological side-effects
• Enhanced memory (engrams)
• Enhanced creativity
• Enhanced emotion
• Enhanced conscious awareness
• Enhanced conscience
... embodied embedded organism
• No control over sensory organs
• No control over motor organs
• No control over other bodily systems
• Other biological side-effects
• Sensory enhancement
• Motor enhancement
• Enhanced memory (exograms)
... social and economic actor
• Loss of ownership of body and IP
• Financial, technological, and social dependencies
• Subjugation to external agency
• Social exclusion and employment discrimination
• Vulnerability to hacking, data theft, blackmail, or other crime
• New kinds of social relations
• Collective knowledge
• Job flexibility and instant retraining
• Enhanced management of technological systems
• Enhanced business decision-making and monetary value
• Qualifications for specific roles

Figure 1. A multidimensional model of the impacts of neuroprosthetic devices on individual users.

Impacts on the User as Sapient Metavolitional Agent

Neuroprosthetic devices may affect their users' cognitive processes in ways that positively or negatively impact the ability of such persons to participate in socioeconomic interaction and informational ecosystems. New capacities provided by neuroprosthetics may include:

- **Enhanced memory, skills, and knowledge stored within the mind (engrams)**. Building on current technologies tested in mice, future neuroprosthetics may offer human users the ability to create, alter, or weaken memories stored in their brains' natural memory systems in the form of engrams.[17] This could potentially be used not only to affect

[17] See Han et al., "Selective Erasure of a Fear Memory" (2009); Ramirez et al., "Creating a False Memory in the Hippocampus" (2013); McGee (2008); Warwick (2014), p. 267.

a user's declarative knowledge but also to enhance motor skills or reduce learned fears.

- **Enhanced creativity**. A neuroprosthetic device may be able to enhance a mind's powers of imagination and creativity[18] by facilitating processes that contribute to creativity, such as stimulating mental associations between unrelated items. Anecdotal increases in creativity have been reported to result after the use of neuroprosthetics for deep brain stimulation.[19]
- **Enhanced emotion**. A neuroprosthetic device might provide its user with more desirable emotional dynamics.[20] Effects on emotion have already been seen in devices used, e.g., for deep brain stimulation.[21]
- **Enhanced conscious awareness**. Research is being undertaken to develop neuroprosthetics that would allow the human mind to, for example, extend its periods of attentiveness and limit the need for periodic reductions in consciousness (i.e., sleep).[22]
- **Enhanced conscience.** One's conscience can be understood as one's set of metavolitions, or desires about the kinds of volitions that one wishes to possess;[23] insofar as a neural implant enhances processes of memory and emotion[24] that allow for the development of the conscience, it may enhance one's ability to develop, discern, and follow one's conscience.

New impairments generated by neuroprosthetics at the level of their user's internal mental processes may include:

- **Loss of agency**. A neuroprosthetic device may damage the brain or disrupt its activity in a way that reduces or eliminates the ability of its human user to possess and exercise agency.[25] Moreover, the knowledge that this can occur may lead users to doubt whether their volitions are really 'their own' – an effect that has been seen with neuroprosthetics used for deep brain stimulation.[26]

---

[18] See Gasson (2012), pp. 23-24.

[19] See Cosgrove, "Session 6: Neuroscience, Brain, and Behavior V: Deep Brain Stimulation" (2004); Gasson (2012).

[20] McGee (2008), p. 217.

[21] See Kraemer, "Me, Myself and My Brain Implant: Deep Brain Stimulation Raises Questions of Personal Authenticity and Alienation" (2011).

[22] Kourany, "Human Enhancement: Making the Debate More Productive" (2013), pp. 992-93.

[23] Calverley (2008); Gladden, "Tachikomatic Domains" (2015).

[24] Calverley (2008), pp. 528-34.

[25] McGee (2008), p. 217.

[26] Kraemer (2011).

- **Loss of conscious awareness.** A neuroprosthetic device may diminish the quality or extent of its user's conscious awareness, e.g., by inducing daydreaming or increasing the required amount of sleep. A neuroprosthesis could potentially even destroy its user's capacity for conscious awareness (e.g., by inducing a coma) but without causing the death of his or her biological organism.[27]
- **Loss of information security for internal cognitive processes.** A neuroprosthetic device may compromise the confidentiality, integrity, or availability of information contained within its user's mental activities (such as perception, memory, volition, or imagination), either by altering or destroying information, making it inaccessible to the user, or making it accessible to unauthorized parties.[28]
- **Inability to distinguish a real from a virtual ongoing experience.** If a neuroprosthesis alters or replaces its user's sensory perceptions, it may make it impossible for the user to know which (if any) of the sense data that he or she is experiencing correspond to some actual element of an external physical environment and which are 'virtual' or simply 'false.'[29]
- **Inability to distinguish true from false memories.** If a neuroprosthetic device is able to create, alter, or destroy engrams within its user's brain, it may be impossible for a user to know which of his or her apparent memories are 'true' and which are 'false' (i.e., distorted or purposefully fabricated).[30]
- **Other psychological side effects.** The brain may undergo potentially harmful and unpredictable structural and behavioral changes as it adapts to the presence, capacities, and activities of a neuroprosthesis.[31] These effects may include new kinds of neuroses, psychoses, and other disorders unique to users of neuroprosthetics.

Impacts on the User as Embodied Embedded Organism Interacting with an Environment

Neuroprosthetic devices may affect the ways in which their users sense, manipulate, and occupy their environment through the interface of a physical or virtual body. New capacities provided might include:

---

[27] Gladden, "Tachikomatic Domains" (2015).

[28] McGee (2008), p. 217; Gladden, "Tachikomatic Domains" (2015); Gladden, *The Handbook of Information Security for Advanced Neuroprosthetics* (2015).

[29] McGee (2008), p. 221; Gladden, "Tachikomatic Domains" (2015).

[30] See Ramirez et al. (2013).

[31] McGee (2008), pp. 215-16; Koops & Leenes (2012), pp. 125, 130.

- **Sensory enhancement.** A neuroprosthetic device may allow its user to sense his or her physical or virtual environment in new ways, either by acquiring new kinds of raw sense data or new modes or abilities for processing, manipulating, and interpreting sense data.[32]
- **Motor enhancement.** A neuroprosthetic device may give users new ways of manipulating physical or virtual environments through their bodies.[33] It may grant enhanced control over one's existing biological body, expand one's body to incorporate new devices (such as an exoskeleton or vehicle) through body schema engineering,[34] or allow the user to control external networked physical systems such as drones or 3D printers or virtual systems or phenomena within an immersive cyberworld.
- **Enhanced memory, skills, and knowledge accessible through sensory organs (exograms).** A neuroprosthetic device may give its user access to external data-storage sites whose contents can be 'played back' to the user's conscious awareness through his or her sensory organs or to real-time streams of sense data that augment or replace one's natural sense data.[35] The ability to record and play back one's own sense data could provide perfect audiovisual memory of one's experiences.[36]

New impairments generated by neuroprosthetics at the level of their users' physical or virtual bodily interfaces with their environments might include:

- **Loss of control over sensory organs.** A neuroprosthetic device may deny a user direct control over his or her sensory organs.[37] Technologically mediated sensory systems may be subject to noise, malfunctions, and manipulation or forced sensory deprivation or overload occurring at the hands of 'sense hackers.'[38]
- **Loss of control over motor organs.** A neuroprosthetic device may impede a user's control over his or her motor organs.[39] The user's body may no longer be capable, e.g., of speech or movement, or the

---

[32] Warwick (2014), p. 267; McGee (2008), p. 214; Koops & Leenes (2012), pp. 120, 126.

[33] McGee (2008), p. 213; Warwick (2014), p. 266.

[34] Gladden, "Cybershells, Shapeshifting, and Neuroprosthetics" (2015).

[35] Koops & Leenes (2012), pp. 115, 120, 126.

[36] McGee (2008), p. 217.

[37] Koops & Leenes (2012), p. 130.

[38] Gladden, *The Handbook of Information Security for Advanced Neuroprosthetics* (2015), pp. 201-02.

[39] Gasson (2012), p. 216.

control over one's speech or movements may be assumed by some external agency.

- **Loss of control over other bodily systems**. A neuroprosthetic device may impact the functioning of internal bodily processes such as respiration, cardiac activity, digestion, hormonal activity, and other processes that are already affected by existing implantable medical devices.[40]
- **Other biological side effects**. A neuroprosthetic device may be constructed from components that are toxic or deteriorate in the body,[41] may be rejected by its host, or may be subject to mechanical, electronic, or software failures that harm its host's organism.

Impacts on the User as Social and Economic Actor

Neuroprosthetic devices may affect the ways in which their users connect to, participate in, contribute to, and are influenced by social relationships and structures and economic networks and exchange. New capacities provided might include:

- **Ability to participate in new kinds of social relations**. A neuroprosthetic device may grant the ability to participate in new kinds of technologically mediated social relations and structures that were previously impossible, perhaps including new forms of merged agency[42] or cybernetic networks with utopian (or dystopian) characteristics.[43]
- **Ability to share collective knowledge, skills, and wisdom**. Neuroprosthetics may link users in a way that forms communication and information systems[44] that can generate greater collective knowledge, skills, and wisdom than are possessed by any individual member of the system.[45]
- **Enhanced job flexibility and instant retraining**. By facilitating the creation, alteration, and deletion of information stored in engrams or exograms, a neuroprosthetic device may allow a user to download new knowledge or skills or instantly establish relationships for use in a new job.[46]

---

[40] McGee (2008), p. 209; Gasson (2012), pp. 12-16.

[41] McGee (2008), pp. 213-16.

[42] McGee (2008), p. 216; Koops & Leenes (2012), pp. 125, 132.

[43] Gladden, "Tachikomatic Domains" (2015).

[44] McGee (2008), p. 214; Koops & Leenes (2012), pp. 128-29; Gasson (2012), p. 24.

[45] Wiener (1961), loc. 3070ff., 3149ff.; Gladden, "Tachikomatic Domains" (2015).

[46] See Koops & Leenes (2012), p. 126.

- **Enhanced ability to manage complex technological systems**. By providing a direct interface to external computers and mediating its user's interaction with them,[47] a neuroprosthesis may grant an enhanced ability to manage complex technological systems, e.g., for the production or provisioning of goods or services.[48]
- **Enhanced business decision-making and monetary value**. By performing data mining to uncover novel knowledge, executing other forms of data analysis, offering recommendations, and alerting the user to potential cognitive biases, a neuroprosthesis may enhance its user's ability to execute rapid and effective business-related decisions and transactions.[49] Moreover, by storing cryptocurrency keys, a neuroprosthesis may allow its user to store money directly within his or her brain for use on demand.[50]
- **Qualifications for specific professions and roles**. Neuroprosthetic devices may initially provide persons with abilities that enhance job performance in particular fields[51] such as computer programming, art, architecture, music, economics, medicine, information science, e-sports, information security, law enforcement, and the military; as expectations for employees' neural integration into workplace systems grow, possession of neuroprosthetic devices may become a requirement for employment in some professions.[52]

New impairments generated by neuroprosthetic devices at the level of their users' socioeconomic relationships and activity might include:

- **Loss of ownership of one's body and intellectual property**. A neuroprosthetic device that is leased would not belong to its human user, and even a neuroprosthesis that has been purchased could potentially be subject to seizure in some circumstances (e.g., bankruptcy). Depending on the leasing or licensing terms, intellectual property produced by a neuroprosthetic device's user (including thoughts, memories, or speech) may be partly or wholly owned by the device's manufacturer or provider.[53]

---

[47] McGee (2008), p. 210.

[48] McGee (2008), pp. 214-15; Gladden, "Cybershells, Shapeshifting, and Neuroprosthetics" (2015).

[49] See Koops & Leenes (2012), p. 119.

[50] Gladden, "Cryptocurrency with a Conscience: Using Artificial Intelligence to Develop Money That Advances Human Ethical Values" (2015).

[51] Koops & Leenes (2012), pp. 131-32.

[52] McGee (2008), pp. 211, 214-15; Warwick (2014), p. 269.

[53] Gladden, "Tachikomatic Domains" (2015); Gladden, *The Handbook of Information Security for Advanced Neuroprosthetics* (2015), p. 164.

- **Creation of financial, technological, or social dependencies**. The user of a neuroprosthetic device may no longer be able to function effectively without the device[54] and may become dependent on its manufacturer for hardware maintenance, software updates, and data security and on specialized medical care providers for diagnostics and treatment relating to the device.[55] A user may require regular device upgrades in order to remain competitive in some jobs. High switching costs may make it impractical to shift to a competitor's device after a user has installed an implant and committed to its manufacturer's digital ecosystem.
- **Subjugation of the user to external agency**. Instead of merely impeding its user's ability to possess and exercise agency, a neuroprosthesis may subject its user to control by some external agency. This could occur, e.g., if the user's memories, emotions, or volitions were manipulated by means of the device[56] or if the user joined with other minds to create a new form of social entity that possesses some shared agency.[57]
- **Social exclusion and employment discrimination**. The use of detectable neuroprosthetics may result in shunning or mistreatment of users.[58] Users of advanced neuroprostheses may lose the ability or desire to communicate with human beings who lack such devices, thereby fragmenting human societies[59] and possibly weakening users' solidarity with other human beings.[60] Possession of some kinds of neuroprosthetic devices may exclude their users from employment in roles where 'natural,' unmodified workers are considered desirable or even required (e.g., for liability or security reasons).
- **Vulnerability to data theft, blackmail, and extortion**. A hacker, computer virus, or other agent may be able to steal data contained in a neuroprosthesis or use it to gather personal data (potentially including the contents of thoughts, memories, or sensory experiences)[61] that could be used for blackmail, extortion, corporate espionage, or terrorism.

---

[54] Koops & Leenes (2012), p. 125.

[55] McGee (2008), p. 213.

[56] Gasson (2012), pp. 15-16.

[57] McGee (2008), p. 216.

[58] Koops & Leenes (2012), pp. 124-25.

[59] McGee (2008), pp. 214-16; Warwick (2014), p. 271.

[60] Koops & Leenes (2012), p. 127.

[61] McGee (2008), p. 217; Koops & Leenes (2012), pp. 117, 130; Gasson (2012), p. 21; Gladden, *The Handbook of Information Security for Advanced Neuroprosthetics* (2015), pp. 167-68.

Applying the Model: Toward a New Typology of Neuroprosthetics

As a test case, we can use this model to analyze one kind of neuroprosthetic device that is expected to become available in the future: a cochlear implant with audio recording, playback, upload, download, and live streaming capabilities.[62] Everything that its user hears would be recorded for later playback on demand. Instead of simply conveying the 'real' sounds produced by the physical environment, those sounds can be augmented or replaced by other audio that is stored in or transmitted to the device. Potential capacities and impairments created for the user of such a device are identified below.

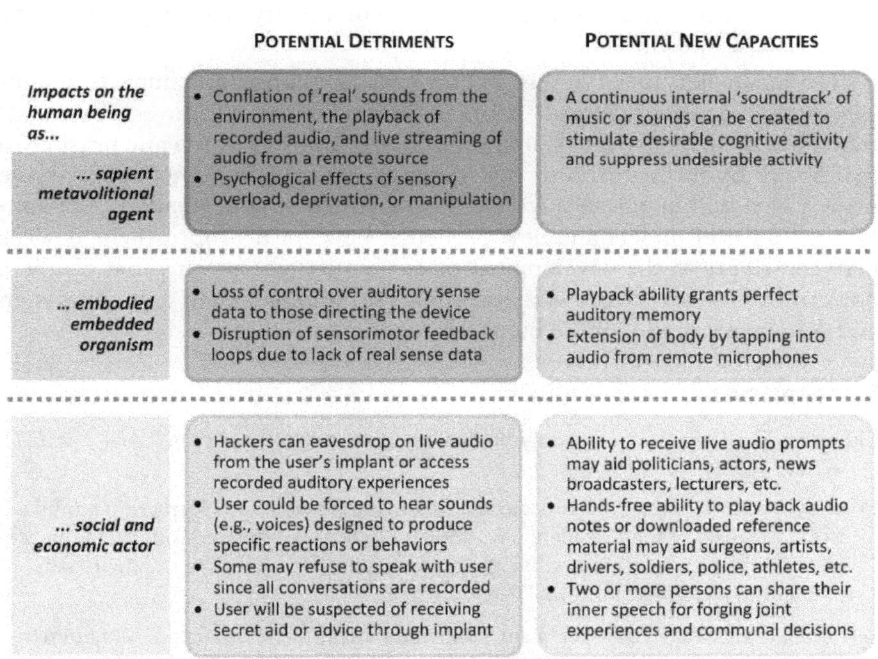

POTENTIAL DETRIMENTS
POTENTIAL NEW CAPACITIES
Impacts on the human being as...
... sapient metavolitional agent
• Conflation of 'real' sounds from the environment, the playback of recorded audio, and live streaming of audio from a remote source
• Psychological effects of sensory overload, deprivation, or manipulation
• A continuous internal 'soundtrack' of music or sounds can be created to stimulate desirable cognitive activity and suppress undesirable activity
... embodied embedded organism
• Loss of control over auditory sense data to those directing the device
• Disruption of sensorimotor feedback loops due to lack of real sense data
• Playback ability grants perfect auditory memory
• Extension of body by tapping into audio from remote microphones
... social and economic actor
• Hackers can eavesdrop on live audio from the user's implant or access recorded auditory experiences
• User could be forced to hear sounds (e.g., voices) designed to produce specific reactions or behaviors
• Some may refuse to speak with user since all conversations are recorded
• User will be suspected of receiving secret aid or advice through implant
• Ability to receive live audio prompts may aid politicians, actors, news broadcasters, lecturers, etc.
• Hands-free ability to play back audio notes or downloaded reference material may aid surgeons, artists, drivers, soldiers, police, athletes, etc.
• Two or more persons can share their inner speech for forging joint experiences and communal decisions

Figure 2. The model applied to analyze impacts of a particular auditory neuroprosthesis.

As can be seen from this example, the model does not yield a single quantitative 'impact score' for each of the three levels but rather uses qualitative descriptions to capture a complex set of impacts. This model delineates a device ontology that can form the basis of further reflection on and analysis of a neuroprosthetic device's impact from both cybernetic, phenomenological,

[62] See Koops & Leenes (2012); McGee (2008); Gladden, "Tachikomatic Domains" (2015).

and existentialist perspectives. By allowing neuroprosthetic devices with similar characteristics to be identified and grouped, it can also serve as the basis of new typologies for neurotechnologies.

Conclusion

Ongoing advances in neuroprosthetics are expected to yield a diverse range of new technologies with the potential to dramatically reshape a human being's internal mental life, his or her bodily existence and interaction with the environment, and his or her participation in social and economic networks and activity. The new capacities and impairments that such technologies provide may allow human beings to physically and virtually inhabit digital ecosystems and interact socially in ways so revolutionary that they can best be understood as 'posthuman.'

The model developed in this text for understanding these impacts of neuroprosthetic devices is already being elaborated in the specific context of information security to provide a framework for future research and practice in that field.[63] By further refining and applying the model in other contexts, we hope that it will be possible for engineers, ethicists, policymakers, and consumers to better understand how particular kinds of neuroprosthetic devices may contribute to the development of new digital ecosystems that can be a powerful venue for the growth, liberation, and empowerment – or oppression and dehumanization – of the human beings of the future.

References

Calverley, D.J. "Imagining a non-biological machine as a legal person." *AI & SOCIETY* 22, no. 4 (2008): 523-37.

Cosgrove, G.R. "Session 6: Neuroscience, brain, and behavior V: Deep brain stimulation." Meeting of the President's Council on Bioethics. Washington, DC, June 24-25, 2004. https://bioethicsarchive.georgetown.edu/pcbe/transcripts/june04/session6.html. Accessed June 12, 2015.

Ferrando, Francesca. "Posthumanism, Transhumanism, Antihumanism, Metahumanism, and New Materialisms: Differences and Relations." *Existenz: An International Journal in Philosophy, Religion, Politics, and the Arts* 8, no. 2 (Fall 2013): 26-32.

Gasson, M.N. "Human ICT Implants: From Restorative Application to Human Enhancement." In *Human ICT Implants: Technical, Legal and Ethical Considerations*, edited by Mark N. Gasson, Eleni Kosta, and Diana M. Bowman, pp. 11-28. Information Technology and Law Series 23. T. M. C. Asser Press, 2012.

Gladden, Matthew E. "Cryptocurrency with a Conscience: Using Artificial Intelligence to Develop Money that Advances Human Ethical Values." *Annales: Ethics in Economic Life* vol. 18, no. 4 (2015): 85-98.

---

[63] See Gladden, *The Handbook of Information Security for Advanced Neuroprosthetics* (2015).

Gladden, Matthew E. "Cybershells, Shapeshifting, and Neuroprosthetics: Video Games as Tools for Posthuman 'Body Schema (Re)Engineering'." Keynote presentation at the Ogólnopolska Konferencja Naukowa Dyskursy Gier Wideo, Facta Ficta / AGH, Kraków, June 6, 2015.

Gladden, Matthew E. *The Handbook of Information Security for Advanced Neuroprosthetics*, Indianapolis: Synthypnion Academic, 2015.

Gladden, Matthew E. "Tachikomatic Domains: Utopian Cyberspace as a 'Contingent Heaven' for Humans, Robots, and Hybrid Intelligences." Presentation at His Master's Voice: Utopias and Dystopias in Audiovisual Culture, Facta Ficta / Uniwersytet Jagielloński, Kraków, March 24, 2015.

Gripsrud, M., and R. Hjorthol. "Working on the Train: From 'Dead Time' to Productive and Vital Time." *Transportation* 39(5) (2012): 941-56.

Han, J.-H., S.A. Kushner, A.P. Yiu, H.-W. Hsiang, T. Buch, A. Waisman, B. Bontempi, R.L. Neve, P.W. Frankland, and S.A. Josselyn. "Selective Erasure of a Fear Memory." *Science* 323, no. 5920 (2009): 1492-96.

Koops, B.-J., and R. Leenes. "Cheating with Implants: Implications of the Hidden Information Advantage of Bionic Ears and Eyes." In *Human ICT Implants: Technical, Legal and Ethical Considerations*, edited by Mark N. Gasson, Eleni Kosta, and Diana M. Bowman, pp. 113-34. Information Technology and Law Series 23. T. M. C. Asser Press, 2012.

Kourany, J.A. "Human Enhancement: Making the Debate More Productive." *Erkenntnis* 79(5) (2013): 981-98.

Kraemer, Felicitas. "Me, Myself and My Brain Implant: Deep Brain Stimulation Raises Questions of Personal Authenticity and Alienation." *Neuroethics* 6, no. 3 (May 12, 2011): 483-97. doi:10.1007/s12152-011-9115-7.

Lebedev, M. "Brain-Machine Interfaces: An Overview." *Translational Neuroscience* 5(1) (2014): 99-110.

McGee, E.M. "Bioelectronics and Implanted Devices." In *Medical Enhancement and Posthumanity*, edited by Bert Gordijn and Ruth Chadwick, pp. 207-24. The International Library of Ethics, Law and Technology 2. Springer Netherlands, 2008.

Ramirez, S., X. Liu, P.-A. Lin, J. Suh, M. Pignatelli, R.L. Redondo, T.J. Ryan, and S. Tonegawa. "Creating a False Memory in the Hippocampus." *Science* 341, no. 6144 (2013): 387-91.

Rao, R.P.N., A. Stocco, M. Bryan, D. Sarma, T.M. Youngquist, J. Wu, and C.S. Prat. "A Direct Brain-to-Brain Interface in Humans." *PLoS ONE* 9(11) (2014).

Shih, J. "Project Time in Silicon Valley." *Qualitative Sociology* 27(2) (2004): 223-45.

Warwick, K. (2014). "The Cyborg Revolution." *Nanoethics* 8 (2014): 263-73.

Wiener, Norbert. *Cybernetics: Or Control and Communication in the Animal and the Machine*, second edition. Cambridge, MA: The MIT Press, 1961. [Quid Pro ebook edition for Kindle, 2015.]

Chapter Six

The Impacts of Human Neurocybernetic Enhancement on Organizational Business Models

Introduction

Alongside emerging posthumanizing technologies such as those related to social robotics and genetic engineering, the field of neuroprosthetics is creating both significant opportunities and risks for human societies.[1] Already a wide range of neuroprosthetic devices (including many kinds of neural implants) are used for therapeutic purposes to treat particular medical conditions. It is expected that in the future, increasingly large populations of human beings will use such devices for purposes of elective human augmentation and enhancement, to acquire sensory, cognitive, and motor abilities that are radically different from those possessed by typical human beings.[2] The social and economic impact of these neuroprosthetic technologies will be significant: such devices have the potential to reshape the ways in which human beings interact with one another, enabling them to create new forms of social structures and organizations and to engage in new kinds of informational and economic exchange that were never previously possible.[3]

---

[1] For an overview of posthumanism and the forces of posthumanization, see Ferrando, "Posthumanism, Transhumanism, Antihumanism, Metahumanism, and New Materialisms: Differences and Relations" (2013); Herbrechter, *Posthumanism: A Critical Analysis* (2013); and "A Typology of Posthumanism: A Framework for Differentiating Analytic, Synthetic, Theoretical, and Practical Posthumanisms" in Gladden, *Sapient Circuits and Digitalized Flesh: The Organization as Locus of Technological Posthumanization* (2016). For an analysis of technological posthumanization and its impact on organizations, see "Organizational Posthumanism" in Gladden, *Sapient Circuits and Digitalized Flesh* (2016).

[2] Warwick & Gasson, "Implantable Computing" (2008); McGee, "Bioelectronics and Implanted Devices" (2008); Gasson et al., "Human ICT Implants: From Invasive to Pervasive" (2012).

[3] See Gladden, "Neural Implants as Gateways to Digital-Physical Ecosystems and Posthuman Socioeconomic Interaction" (2016).

As participants in the larger societies, economies, and informational ecosystems within which they exist, businesses will be impacted by the widening use of neuroprosthetic technologies. Companies that are able to identify, understand, and anticipate these technological and social changes and transform their business models accordingly may be able to secure significant competitive advantages. On the other hand, companies that are not able to adapt their business models quickly enough to the social, economic, political, cultural, and ethical changes driven by neuroprosthetic technologies may find themselves unable to compete, grow, or even survive.

In this paper, we briefly consider the concept of a 'business model' and the situations that require a company to change and update its business model. We then explore three main areas in which the rise of neuroprosthetics is expected to transform humanity. Finally, we identify the impact that such transformation will have on companies' business models and consider an example highlighting the reasons why many companies will need to adopt new business models in order to address these changed realities.

The Definition of a 'Business Model'

Management scholars have proposed various definitions of exactly what constitutes a 'business model.'[4] For example, Magretta suggests that "Business models [...] are, at heart, stories – stories that explain how enterprises work"[5] and that a business model's primary value "is that it focuses attention on how all the elements of the system fit into a working whole."[6] In this sense, a business model can be understood as a sort of heuristic tool – a narrative that helps both employees and customers find meaning and an overarching purpose in individual tasks that are being carried out by hundreds, thousands, or even millions of individuals interacting around the globe.

On the other hand, Johnson, Christensen, and Kagermann define a business model as "four interlocking elements that, taken together, create and deliver value"[7]; these four elements are the customer value proposition, profit formula, key resources, and key processes.[8] Meanwhile, Casadesus-Masanell and Ricart[9] argue that a good business model is 1) aligned with a company's goals, 2) self-reinforcing, and 3) robust – and that a business model can be distilled

[4] Casadesus-Masanell & Ricart, "How to Design a Winning Business Model" (2011), pp. 102-03.

[5] Magretta, "Why Business Models Matter" (2002), p. 87.

[6] Magretta (2002), p. 90.

[7] Johnson et al., "Reinventing Your Business Model" (2008), p. 52.

[8] Johnson et al. (2008), p. 54.

[9] Casadesus-Masanell & Ricart (2011), p. 102.

into a set of policy, asset, and governance choices that are made by the company's managers and which generate outcomes that either manifest themselves immediately and are ephemeral or which take time to develop and are more permanent.[10]

One theme common to such definitions is that a company's business model can be contrasted with its *strategies* and *tactics*. If one understands a business's tactics as fine-grained decisions made at a 'micro' level – that is, choices that are made in response to immediate circumstances, can be changed quickly, and often involve only a single business unit or process – then the company's business model comprises those decisions made at the 'macro' level, which can be changed only with difficulty and over time. The business model thus encompasses the entire general cycle of activities taking place within the company and the ways in which those activities (ideally) reinforce one another. From this perspective, a business's strategies can be understood as a bridge between the company's broad business model and its more specific operational tactics.[11]

Circumstances Enabling (or Requiring) a New Business Model

Scholars identify a number of circumstances in which a company can (or must) change its business model in order to develop or maintain a competitive advantage. For Magretta, business models must be changed when "they fail either the narrative test (the story doesn't make sense) or the numbers test (the P&L doesn't add up)."[12]

For Johnson et al.,[13] it makes sense to adopt a new business model when facing particular "strategic circumstances" such as "The opportunity to address through disruptive innovation the needs of large groups of potential customers who are shut out of a market entirely because existing solutions are too expensive or complicated for them" or "The opportunity to capitalize on a brand-new technology by wrapping a new business model around it." Moreover, a new business model is *required* if the business environment has evolved to such an extent that all four elements of the existing model – the customer value proposition, profit formula, key resources, and key processes – have become obsolete and are in need of change.[14]

---

[10] Casadesus-Masanell & Ricart (2011), p. 103.

[11] Casadesus-Masanell & Ricart (2011), p. 107; Magretta (2002), p. 91.

[12] Magretta (2002), p. 90.

[13] Johnson et al. (2008), p. 57.

[14] Johnson et al. (2008), p. 57.

For Casadesus-Masanell and Ricart,[15] the need for new business models in today's world is often driven by fast-evolving technological transformation. For example, many large high-tech businesses such as Facebook and eBay are succeeding thanks to Internet-enabled 'network effects'[16] that multiply the value offered to consumers as the number of consumers grows. Once established, such network effects create a 'virtuous cycle' that continuously strengthens the company's customer value proposition and wards off potential competitors. Even for those companies that are not primarily e-businesses, the use of an effective business model can perform a similar function by generating virtuous cycles for the companies.[17]

Regardless of which of these conceptual frameworks one adopts, we shall see in the following sections that the rise of neuroprosthetic technologies is expected to have an impact that will both enable and require many businesses to transform their business models within the foreseeable future.

Envisioning Posthuman Neuroprosthetics

A wide range of devices that link directly with the brain's neural circuitry are already in use.[18] These include *sensory neuroprostheses* (such as cochlear implants and artificial retinas), *motor neuroprostheses* (such as implants allowing a wheelchair to be controlled by thought), and bidirectional *sensorimotor neuroprostheses* (such as a prosthetic hand that provides tactile sensations to its user and moves in response to the user's thoughts); in addition, a further category of *cognitive neuroprostheses* (such as memory implants) is also in its earliest experimental stages.[19]

It is expected that future neuroprosthetic devices will increasingly be designed to provide abilities that exceed or differ from what is naturally possible for human beings.[20] Such technologies' use for physical and cognitive enhancement is expected to expand the market for neuroprosthetics and implantable computers beyond the segment of the population that currently relies on them to treat medical conditions.[21] Researchers anticipate that future sensory neuroprosthetics may give human beings the capacity to experience their environments in new ways, such as through the use of telescopic or night

[15] Casadesus-Masanell & Ricart (2011), pp. 101-02.

[16] Casadesus-Masanell & Ricart (2011), p. 102.

[17] Casadesus-Masanell & Ricart (2011), p. 102.

[18] See Gladden, "Neural Implants as Gateways" (2016), and Lebedev, "Brain-Machine Interfaces: An Overview" (2014).

[19] Lebedev (2014), pp. 99-100.

[20] Gasson, "ICT Implants" (2008); Gasson et al. (2012); McGee (2008); Merkel et al., "Central Neural Prostheses" (2007).

[21] See McGee (2008) and Gasson et al. (2012).

vision[22] or by overlaying visual data with supplemental information displayed using augmented reality.[23] Some researchers envision the development of implants that can record all of a person's audiovisual experiences for later playback, effectively granting the person perfect audiovisual memory.[24] Building on successful experiments with creating artificial memories in mice,[25] other researchers envision the possibility of a person being able to download new knowledge or skills onto a memory chip implanted in his or her brain.[26] Technologies are also being developed[27] that may eventually allow direct communication between two human brains that are physically located thousands of miles apart.

The use of such advanced neuroprosthetic devices will reshape the ways in which human beings collaborate with one another. Already 'cyborg-cyborg interaction' is becoming a fundamental aspect of human society, and it will increasingly serve as a foundation for new kinds of social relationships and structures.[28] Neuroprosthetics will allow for increasingly intimate forms of communication that do not involve physical face-to-face interaction but are instead mediated by technology, thereby facilitating the development of novel types of posthuman interpersonal relationships.[29]

The Impact of Neuroprosthetics on Business Models

Neuroprosthetic devices such as neural implants are expected to impact human beings on at least three different levels. First, a neural implant affects a human being at the internal, cognitive level, in his or her role as an intelligent agent possessing its own conscious awareness, volition, memory, and conscience. Second, a neural implant affects a human being in his or her role as a physical actor with a body that inhabits, senses, and manipulates a particular spatial environment. Third, a neural implant affects a human being in his or her role as a member of the social and economic networks in which the individual participates.[30]

---

22 Gasson et al. (2012); Merkel et al. (2007).

23 Koops & Leenes, "Cheating with Implants: Implications of the Hidden Information Advantage of Bionic Ears and Eyes" (2012).

24 Merkel et al. (2007); Robinett, "The Consequences of Fully Understanding the Brain" (2002).

25 Ramirez et al., "Creating a False Memory in the Hippocampus" (2013).

26 McGee (2008).

27 See Rao et al., "A Direct Brain-to-Brain Interface in Humans" (2014).

28 Fleischmann, "Sociotechnical Interaction and Cyborg–Cyborg Interaction: Transforming the Scale and Convergence of HCI" (2009).

29 Grodzinsky et al., "Developing Artificial Agents Worthy of Trust: 'Would You Buy a Used Car from This Artificial Agent?'" (2011).

30 For a more detailed exploration of these three levels of impacts, see Gladden, "Neural Implants

How will these effects of neuroprosthetics impact companies' business models? We can visualize this by drawing on concepts from management cybernetics to understand Johnson, Christensen, and Kagermann's four-part 'business model' as a representation of a business as a system in which *processes* act on (or are performed by) *resources*, in order to generate a *net benefit for the customer* (i.e., the customer value proposition) and *a net benefit for the company* (i.e., a net profit) that can then be used to acquire more *resources* and thereby grow the company.[31] We depict a generic example of this 'business process cycle' view of a business model in Figure 1. In Figure 2, we overlay onto that business process cycle a representation of the impacts of neuroprosthetic devices at the levels of internal cognition, bodily interface with the environment, and socioeconomic interaction in order to depict the fact that each of the three spheres of neuroprosthetics' impact will touch all four areas of a company's business model. We would argue that no part of a business model will be left unaffected by this posthumanizing technological change.

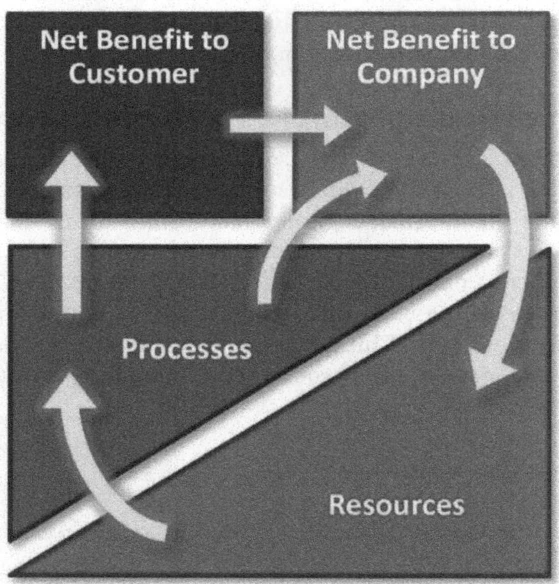

Figure 1. The four-element generic business model of Johnson, Christensen, and Kagermann reinterpreted in light of management cybernetics as a business process cycle.

as Gateways" (2016).

[31] For a discussion of such dynamics from the perspective of management cybernetics, see, e.g., Gladden, "The Artificial Life-Form as Entrepreneur: Synthetic Organism-Enterprises and the Reconceptualization of Business" (2014), and Beer, *Brain of the Firm* (1981).

Example: The Impact of a Particular Kind of Sensory Playback Device

Here we can consider the business-model impact of one particular kind of neuroprosthetic device whose development is anticipated: that of a sensorimotor-cognitive neural implant that allows a human being to instantaneously download (simply by thinking about this action and 'willing' it to occur) information in the form of video, audio, text, and images that the person can then 'play back' to his or her conscious awareness in the form of sensory input that augments or replaces the actual sense data being provided by one's environment.[32] In this manner, a person could, for example, download and watch a film 'internally,' within his or her own mind, without nearby individuals ever realizing that this was taking place. Similarly, a person could download vast libraries of reference books into the implant, which the person could then read internally at will – perhaps running an automated process to search the texts for particular terms, as desired.

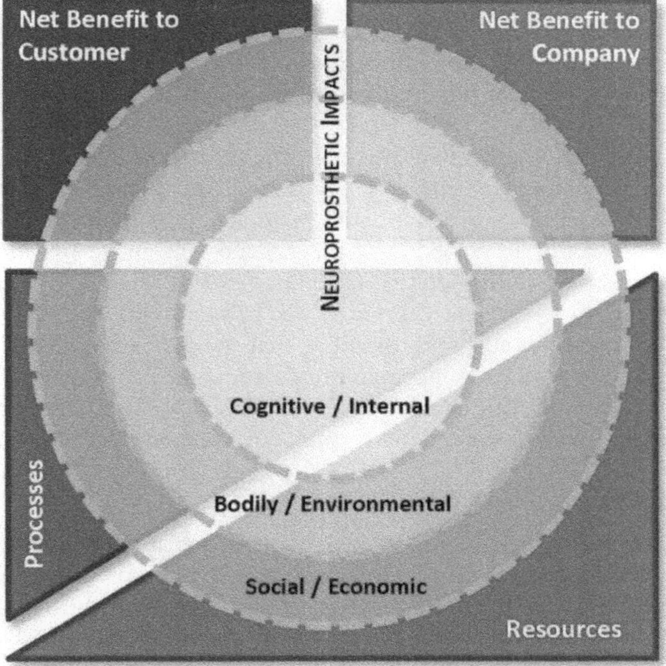

Figure 2. A representation of the future impacts on all four elements of a generic business model of the use of advanced neuroprosthetic devices.

[32] Koops & Leenes (2012); McGee (2008).

One can immediately foresee ways in which the existence and widespread use of such a technology would significantly impact the effectiveness of many companies' existing business models by affecting each of the four model components described above.

For example, the *resources* – including human resources – available to a company would be notably altered. On the one hand, a company could potentially 'retrain' its entire workforce by downloading instructional materials into their implants and conceivably even forcing the employees' implants to play back the materials to the employees conscious awareness during working hours.[33] On the other hand, employees who are sitting in an office and appear to be 'working' could – unbeknownst to their employer – actually be watching within their thoughts and their 'mind's eye' a football game or movie that they had just downloaded into their mind. Similarly, the *processes* available for a business to utilize in coordinating its activities would change significantly: emails, instant messages, and video messages sent between employees could appear instantly in their field of vision and conscious awareness, and employees responsible for managing manufacturing processes could receive real-time status updates and alerts downloaded into their implants around the clock, regardless of whether they are at home or at work. For many businesses, the *customer value proposition* that leads to a net benefit for the consumer and *profit formula* that leads to a net benefit for the company would also be impacted: for example, a customer who is in the midst of negotiating a major purchase from a company or who is in a business's retail location on the verge of deciding to purchase a product can – without the company's employees realizing it – instantly download from the Internet into his or her mind hundreds of user reviews of the product and news stories relating to it, in order to more accurately weigh the value that the product will offer him or her. Moreover, the potential customer might in that moment even find his or her implant flooded with offers from competing companies that have detected the person's impending purchase and are seeking to advance their own profit formula by offering the consumer a more appealing consumer value proposition.

Conclusion

As we have seen through one of many possible examples, the increasing technological sophistication and expanding use of neuroprosthetics has the potential to drive changes in all four aspects of a company's business model, thereby requiring the development of a new business model. Moreover, even if a particular company is not for these reasons required to develop a new business model, such advances in neuroprosthetics will create for companies

---

[33] Here we are not addressing the ethicality or legality of such possibilities; we only note that they represent a theoretical possibility that some companies might be tempted to pursue.

the kinds of avenues for exploiting disruptive innovation and new technologies that can make proactive adoption of a new business model desirable. The impact of neuroprosthetic devices in spurring the evolution of business environments will likely create meaningful opportunities for companies that possess the foresight and flexibility to update their business models – and potential risks for companies that are unwilling or unable to adapt their business models to respond to such technological and social transformations that are expected to occur in the coming years and decades.

References

Beer, Stafford. *Brain of the Firm.* 2nd ed. New York: John Wiley, 1981.

Casadesus-Masanell, R., and J.E. Ricart. "How to Design a Winning Business Model." *Harvard Business Review,* Jan-Feb (2011): 100-07.

Ferrando, Francesca. "Posthumanism, Transhumanism, Antihumanism, Metahumanism, and New Materialisms: Differences and Relations." *Existenz: An International Journal in Philosophy, Religion, Politics, and the Arts* 8, no. 2 (2013): 26-32.

Fleischmann, Kenneth R. "Sociotechnical Interaction and Cyborg–Cyborg Interaction: Transforming the Scale and Convergence of HCI." *The Information Society* 25, no. 4 (2009): 227-35. doi:10.1080/01972240903028359.

Gasson, M.N. "ICT implants." In *The Future of Identity in the Information Society,* edited by S. Fischer-Hübner, P. Duquenoy, A. Zuccato, and L. Martucci, pp. 287-95. Springer US, 2008.

Gasson, M.N., Kosta, E., and Bowman, D.M. "Human ICT Implants: From Invasive to Pervasive." In *Human ICT Implants: Technical, Legal and Ethical Considerations,* edited by Mark N. Gasson, Eleni Kosta, and Diana M. Bowman, pp. 1-8. Information Technology and Law Series 23. T. M. C. Asser Press, 2012.

Gladden, Matthew E. "The Artificial Life-Form as Entrepreneur: Synthetic Organism-Enterprises and the Reconceptualization of Business." In *Proceedings of the Fourteenth International Conference on the Synthesis and Simulation of Living Systems,* edited by Hiroki Sayama, John Rieffel, Sebastian Risi, René Doursat and Hod Lipson, pp. 417-18. Cambridge, MA: The MIT Press, 2014.

Gladden, Matthew E. "Neural Implants as Gateways to Digital-Physical Ecosystems and Posthuman Socioeconomic Interaction." In *Digital Ecosystems: Society in the Digital Age,* edited by Łukasz Jonak, Natalia Juchniewicz, and Renata Włoch, pp. 85-98. Warsaw: Digital Economy Lab, University of Warsaw, 2016.

Gladden, Matthew E. *Sapient Circuits and Digitalized Flesh: The Organization as Locus of Technological Posthumanization.* Indianapolis: Defragmenter Media, 2016.

Grodzinsky, F.S., K.W. Miller, and M.J. Wolf. "Developing Artificial Agents Worthy of Trust: 'Would You Buy a Used Car from This Artificial Agent?'" *Ethics and Information Technology* 13(1) (2011): 17-27.

Herbrechter, Stefan. *Posthumanism: A Critical Analysis.* London: Bloomsbury, 2013. [Kindle edition.]

Johnson, M.W., C.M. Christensen, and H. Kagermann. "Reinventing Your Business Model." *Harvard Business Review*, Dec. (2008): 51-59.

Koops, B.-J., and R. Leenes. "Cheating with Implants: Implications of the Hidden Information Advantage of Bionic Ears and Eyes." In *Human ICT Implants: Technical, Legal and Ethical Considerations*, edited by Mark N. Gasson, Eleni Kosta, and Diana M. Bowman, pp. 113-34. Information Technology and Law Series 23. T. M. C. Asser Press, 2012.

Lebedev, M. "Brain-Machine Interfaces: An Overview." *Translational Neuroscience* 5(1) (2014): 99-110.

Magretta, J. "Why Business Models Matter." *Harvard Business Review*, May (2002): 86-92.

McGee, E.M. "Bioelectronics and Implanted Devices." In *Medical Enhancement and Posthumanity*, edited by Bert Gordijn and Ruth Chadwick, pp. 207-24. The International Library of Ethics, Law and Technology 2. Springer Netherlands, 2008.

Merkel, R., G. Boer, J. Fegert, T. Galert, D. Hartmann, B. Nuttin, and S. Rosahl. "Central Neural Prostheses." In *Intervening in the Brain: Changing Psyche and Society*, 117-60. Ethics of Science and Technology Assessment 29. Springer Berlin Heidelberg, 2007.

Ramirez, S., X. Liu, P.-A. Lin, J. Suh, M. Pignatelli, R.L. Redondo, T.J. Ryan, and S. Tonegawa. "Creating a False Memory in the Hippocampus." *Science* 341, no. 6144 (2013): 387-91.

Rao, R.P.N., A. Stocco, M. Bryan, D. Sarma, T.M. Youngquist, J. Wu, and C.S. Prat. "A Direct Brain-to-Brain Interface in Humans." *PLoS ONE* 9(11) (2014).

Robinett, W. "The consequences of fully understanding the brain." In *Converging Technologies for Improving Human Performance: Nanotechnology, Biotechnology, Information Technology and Cognitive Science*, edited by M.C. Roco and W.S. Bainbridge, pp. 166-70. National Science Foundation, 2002.

Warwick, K., and M. Gasson. "Implantable Computing." In *Digital Human Modeling*, edited by Y. Cai, pp. 1-16. Lecture Notes in Computer Science 4650. Berlin/Heidelberg: Springer, 2008.

Chapter Seven

Implantable Computers and Information Security: A Managerial Perspective[1]

Abstract. The interdisciplinary field of information security (InfoSec) already draws significantly on the biological and human sciences; for example, it relies on knowledge of human physiology to design biometric authentication devices and utilizes insights from psychology to predict users' vulnerability to social engineering techniques and develop preventative measures. The growing use of computers implanted within the human body for purposes of therapy or augmentation will compel InfoSec to develop new or deeper relationships with fields such as medicine and biomedical engineering, insofar as the practices and technologies that InfoSec implements for implantable computers must not only secure the information contained within such devices but must also avoid causing biological or psychological harm to the human beings within whose organisms the computers are embedded.

In this text we identify unique issues and challenges that implantable computers create for information security. By considering the particular scenario of the internal computer controlling a retinal implant, we demonstrate the ways in which InfoSec's traditional concepts of the confidentiality, integrity, and availability of information and the use of physical, logical, and administrative access controls become intertwined with issues of medicine and biomedical engineering. Finally, we formulate a novel cybernetic approach that provides a useful paradigm for conceptualizing the relationship of

[1] For an investigation of information security for implantable computers (and especially those contained within or connected to neural implants) that explores these issues in more depth than is possible within this text, see Gladden, *The Handbook of Information Security for Advanced Neuroprosthetics* (2015).

information security to medicine and biomedical engineering in the context of implantable computers.

Introduction

Although the public perception of information security (or 'InfoSec') often focuses on the field's extensive body of theory and practice relating to computer science and information technology, information security is in fact a transdisciplinary field in which InfoSec teams – such as those maintained by large businesses – must not only possess expertise with a wide array of hardware and software systems but must also be knowledgeable about such diverse fields such as law, ethics, management, finance and accounting, and building architecture and maintenance. InfoSec also draws significantly on insights from the biological and social sciences. For example, it relies on a knowledge of human physiology in order to design (and later, secure) biometric access-control systems that are capable of identifying and authenticating human users based on traits such as their voice patterns, handwriting patterns, hand shapes and vascular patterns, fingerprints, facial features, iris patterns, or retinal blood-vessel patterns.[2] Similarly, InfoSec relies on insights from the field of psychology to predict users' vulnerability to social engineering techniques such as phishing and to develop effective measures for prevention, detection, and response.[3]

A new phenomenon that InfoSec will need to robustly address during the coming years is the growing use of implantable computers that operate within the human body for purposes of therapeutic treatment or human augmentation. The expansion of this new technological frontier creates unique challenges that will compel information security to develop relationships with fields such as medicine and biomedical engineering that are closer, richer, and more critically important than those that have existed in the past. The rise of implantable computing will elicit a qualitative change in these relationships; InfoSec personnel will need to work closely with the doctors and biomedical engineers who are designing and implanting such devices, in order to:

- Understand the design and functioning of computers which – after their implantation – the InfoSec personnel will likely be unable to physically inspect or manipulate and which may utilize specialized proprietary hardware, operating systems, and software applications.
- Understand an implantable computer's connections with the biological systems of its human host, in order to recognize both the kinds of information (if any) that the device is gathering regarding the host's biological and cognitive processes, the kind of information (if any)

[2] Rao & Nayak, *The InfoSec Handbook* (2014).

[3] Rao & Nayak (2014).

that the device is transmitting to the mind of its human host, and any other effects that the device is capable of having on its human host.

- Develop InfoSec practices and technologies for use with implantable computers that not only secure the information contained within such devices but which also avoid creating biological or psychological harm (or even the danger of such harm) for the human beings within whose organisms the computers are embedded.

To explore this growing interconnection of information security, medicine, and biomedical engineering, we begin by identifying unique issues and challenges that implantable computers create for information security. By considering the scenario of a computer that is contained within a sensory neuroprosthetic device in the form of a retinal implant, we then demonstrate the ways in which InfoSec's traditional concepts of the confidentiality, integrity, and availability of information and the use of physical, logical, and administrative access controls become intertwined with issues of medicine and biomedical engineering. Finally, we suggest that in order to analyze these issues further and develop effective avenues of communication between the fields of information security and biomedical engineering, it may be useful to employ the concept of celyphocybernetics, which views both the human body and any implantable computers embedded within it as a single cybernetic system for communication and control that supports the mind of the human being to whom the body belongs.

The Fundamentals of Implantable Computers

Current Implantable Computers

Current forms of information and communications technology (ICT) include a number of implantable devices such as passive RFID tags[4] that can store information and interact with computers but which are not in themselves computers. A growing number of implantable devices, though, indeed constitute full-featured implantable computers, insofar as they possess their own processor, memory, software, and input/output mechanisms; they often also possess programming that can be remotely updated after the devices have been implanted into the body of their human host. These forms of technology include many implantable medical devices (IMDs) such as defibrillators, pacemakers, deep brain stimulators, sensory neuroprostheses including retinal and cochlear implants, body sensor networks (BSNs), and some of the more sophisticated forms of RFID transponders.[5] Such implantable computers (ICs)

---

[4] Gasson, "Human ICT Implants: From Restorative Application to Human Enhancement" (2012); Gasson, "ICT Implants" (2008).

[5] Gasson et al., "Human ICT Implants: From Invasive to Pervasive" (2012); Gasson (2008).

increasingly operate in rich and complex biocybernetic control loops with the organism of their human host, allowing, for example, the physiological and cognitive activity of their host to be detected, analyzed, and interpreted for use in exercising real-time control over computers or robotic devices.[6] The growing sophistication of the computers contained within such implantable devices means that they increasingly serve as sites for the reception, storage, processing, and transmission of large amounts of highly sensitive information[7] regarding their human hosts' everyday interactions with the environment, internal biological processes, and even cognitive activity.

Future Implantable Computers

The implantable computers currently in use have typically been designed to serve a restorative or therapeutic medical purpose; they might treat a particular illness or restore some sensory, motor, or cognitive ability that their user has lost as a result of illness or injury. It is expected, though, that future generations of ICs will increasingly be designed not to restore some ordinary human capacity that has been lost but to enhance their users' physical or intellectual capacities by providing abilities that exceed or differ from what is naturally possible for human beings. For example, future models of retinal implants might augment normal human vision by providing telescopic or night vision,[8] and ICs with functionality similar to that of a miniaturized smartphone might offer their users wireless communication capacities that include access to the Internet and cloud-based software and data-storage services. The growing elective use of ICs for purposes of physical and cognitive augmentation is expected to expand the implantable computing device market well beyond that segment of the population which currently relies on ICs to treat medical conditions.[9]

Information Security as Applied to Implantable Computers

Key Concepts of Information Security

Information security is a discipline whose fundamental aim is to ensure the confidentiality, availability, and integrity of information – often referred to as

---

[6] Fairclough, "Physiological Computing: Interfacing with the Human Nervous System" (2010); Park et al., "The Future of Neural Interface Technology" (2009).

[7] Kosta & Bowman, "Implanting Implications: Data Protection Challenges Arising from the Use of Human ICT Implants" (2012); Li et al., "Advances and Challenges in Body Area Network" (2011); Rotter & Gasson, "Implantable Medical Devices: Privacy and Security Concerns" (2012).

[8] Gasson et al. (2012); Merkel et al., "Central Neural Prostheses" (2007).

[9] McGee, "Bioelectronics and Implanted Devices" (2008); Gasson et al. (2012).

the 'CIA Triad' model for understanding information security.[10] In the contemporary world of the Internet and Internet of Things, Big Data, and nearly ubiquitous computing, securing information often means securing the computerized systems that are used to gather, store, process, and transmit data. However, in its broader scope InfoSec also seeks to ensure the confidentiality, availability, and integrity of information that is maintain in other systems such as printed files and records, magnetic audio tapes, or even within the human mind (e.g., confidential business information known to a company's employees which practitioners of corporate espionage might attempt to educe through bribery, coercion, or social engineering). Within large businesses or government agencies, InfoSec departments seek to ensure information security through the design and implementation of comprehensive approaches that incorporate practices and techniques such as strategic planning, risk management, training, configuration management, incident response, and the use of physical, logical, and administrative security controls.[11] Key concepts and best practices for the field are described by industry-leading standards such as those found in *NIST SP 800-100*[12] and *ISO/IEC 27001:2013*.[13]

Overview of Information Security Vulnerabilities and Risks for Implantable Computers

The possession of an implantable computer creates significant InfoSec vulnerabilities and risks for its human host. Like other computers, ICs are vulnerable to threats such as computer viruses[14] and hacking.[15] It is no longer unrealistic to presume that criminals or other unauthorized parties will seek and gain illicit access to the computers found in robotic prosthetic limbs, retinal implants, cochlear implants, and other neuroprosthetics for purposes of stealing, altering, or rendering unavailable the data that they contain, either because of some personal motives relating to the device's human host, to facilitate blackmail or financial fraud, or as an act of political or industrial espionage. Moreover, the growing use of implantable technologies for elective (rather than medically necessary) reasons[16] and the increasing sophistication of such devices' ability to interface with a human host's brain[17] means that future implantable computers will have access to increasingly sensitive information possessed by an increasingly larger segment of the human population. The

---

[10] Rao & Nayak (2014).

[11] *SP 800-100. Information Security Handbook: A Guide for Managers* (2006).

[12] *SP 800-100* (2006).

[13] *ISO/IEC 27001:2013, Information Technology – Security Techniques – Information Security Management Systems – Requirements* (2013).

[14] Gasson (2012); Clark & Fu, "Recent Results in Computer Security for Medical Devices" (2012).

[15] Rotter & Gasson (2012).

[16] McGee (2008); Gasson (2008).

[17] Gasson (2012).

need to create comprehensive InfoSec frameworks that account for such realities will thus likely become more critical in the coming years.

The Paradox of Information Security Requirements for Implantable Computers

Implantable computers display a number of unique and relevant traits that directly affect their information security characteristics and which are not found in typical desktop, mobile, or wearable computers. For example, ICs face intense – and inherently conflicting – InfoSec demands, due to their close integration into their host's biological systems and processes. The unique role of ICs as technological devices operating within a biological organism creates a dilemma for designers of InfoSec systems. On the one hand, a user's implantable computer should be more secure and well-protected against unauthorized activity such as viruses or hacking than, for example, his or her laptop computer or smartphone – due to the fact that any party who gains unauthorized access to the implantable computer not only has the potential to (1) steal information representing the user's most sensitive and confidential medical data and cognitive activity (potentially even including the contents of dreams, memories, volitions, fears, and sensory experiences) but can potentially also (2) alter or render inaccessible the contents of the information contained within the device or within natural biological systems (such as the brain's own memory-storage mechanisms) with which the device is connected. The latter could potentially cause a range of severe negative effects including neurological or behavioral problems, the loss of personal agency and identity (or at least the experience of their loss), and a host's making of decisions and undertaking of actions on the basis of erroneous (and potentially fraudulently fabricated) information provided or affected by the implantable computer. In the absence of stringent InfoSec mechanisms and protocols, an IC's human host may be unable to retroactively trust the contents of what appear to be his or her 'own' thoughts and memories or to proactively trust that his or her future biological processes will function in a manner free from unauthorized influence.

On the other hand, imagine that the user of an implantable computer has been involved in a serious accident or is unexpectedly experiencing an acute and life-threatening medical incident. In this case, emergency medical personnel on the scene may need to gain immediate access to an IC and exercise unfettered control over its functionality in order to save the life of its host.[18] The same mechanisms (such as encryption and proprietary security software) that make it difficult for a hacker to break into an IC would also make it difficult or impossible for emergency medical personnel to break into the device. In principle, regulators could require (or IC manufacturers could voluntarily

[18] Clark & Fu (2012); Rotter & Gasson (2012).

institute) mechanisms that allow ICs to be accessed by individuals presenting certain credentials that identify them as trained and licensed emergency medical personnel, or ICs could be designed to temporarily disable some of their access controls if they detect that their host is experiencing a medical emergency. However, such mechanisms created security vulnerabilities that could potentially be exploited by unauthorized parties who are highly motivated to gain access to an IC.

Other Information Security Considerations Unique to Implantable Computers

Because of their embedded nature, ICs are unable to employ some traditional physical controls and may be wholly reliant on wireless communication. However, because of their close integration with their host's biological system, ICs may be able to employ biometric mechanisms that are impossible for external computers, for the purpose of ensuring that the device is indeed still functioning within the body of its intended user.

One issue requiring careful attention is the secure disposal, reuse, and recycling of implantable computers, given the fact that ICs may contain highly sensitive medical and personal data about their user, including genetic information found in cells and other biological material that may have been intentionally or inadvertently introduced into the device.[19]

Finally, it should be noted that just as InfoSec professionals must be concerned about new information security vulnerabilities that are created for individuals within their charge who possess ICs, so too must they be concerned about the enhanced capacities for carrying out illicit surveillance, hacking, and data theft that ICs provide to ill-intentioned parties who possess them.[20]

An InfoSec Scenario: Securing a Sensory Neuroprosthetic

The CEO with Retinal Implants

Consider a hypothetical case – set several years in the future – of the CEO of a large consumer electronics firm who is 52 years old and has recently received retinal implants in both eyes, to address the effects of retinitis pigmentosa that would otherwise have caused him to become completely blind. The CEO's retinal implants include miniature video cameras and a computer that processes their video images to convert them into signals that are then transmitted to an array of electrodes that stimulate retinal ganglion cells, causing

[19] See *NIST Special Publication 800-53, Revision 4: Security and Privacy Controls for Federal Information Systems and Organizations* (2013), p. F-122-F-123.

[20] Koops & Leenes, "Cheating with Implants: Implications of the Hidden Information Advantage of Bionic Ears and Eyes" (2012).

neuronal signals corresponding to visual sense data to be transmitted through the optic nerve to the CEO's brain.[21]

Both the raw visual input received from the environment by the cameras and the processed sense data provided to the CEO's brain include highly sensitive information, insofar as they constitute a visual record of everything that the CEO looks at in both his work-related and private life. This visual data would include a moment-by-moment record of all of the CEO's activities, including his physical whereabouts, the identities of individuals with whom he meets, and images of everything that he views on his computer screen, including the contents of highly-sensitive work-related emails and documents and personal online activity. Assuming that the implants' video cameras were of high enough resolution, someone with access to their raw video could potentially even use lip reading to determine what was being said to the CEO in his face-to-face conversations.

Applying the CIA Triad

In this scenario, the InfoSec personnel of the CEO's company would no longer simply need to worry about implementing strategies and practices to secure such highly sensitive ICT devices as the CEO's desktop computer, laptop, or smartphone; they would now need to be concerned about securing his prosthetic eyes, as well. When considering their CEO's new retinal implants through the lens of the CIA triad, such InfoSec experts would identify the following sorts of issues that would need to be proactively addressed by the company's information security department.

Confidentiality

The retinal implants' raw visual data and processed visual output contain highly sensitive information that must be secured. While such legal requirements were generally not implemented specifically with implantable computers in mind, various regulations already exist around the world requiring that (often quite stringent) practices and mechanisms be put in place to secure the

---

[21] The kinds of retinal implants have been approved by the FDA for use in human patients are currently extremely limited in the quantity of data that they can transmit to the patient's retinal ganglion cells, thus the devices' computers must radically compress and simplify received video images before transmitting to the patient's brain. However, it is anticipated that retinal implants utilizing new electrode designs and neuronal stimulation techniques will provide their users with a level of vision adequate for navigating within an environment, reading text, and recognizing faces. See Weiland et al., "Retinal Prosthesis" (2005); Jumper, "FDA Approves World's First Artificial Retina" (2013); Schmid & Fink, "Operational Design Considerations for Retinal Prostheses" (2012); and Schmid et al., "Simultaneous vs. Sequential and Unipolar vs. Multipolar Stimulation in Retinal Prostheses" (2013). In the hypothetical scenario described here, we assume that the CEO possesses retinal implants of that more advanced sort which are expected to become available within the coming years.

sort of personal and medical data that implantable computers will contain.[22] However, because the technology of implantable computers is still relatively new and experimental, a specialized and comprehensive set of InfoSec standards has not yet been developed for the industry. Moreover, especially during the initial iterations of such technologies, the laboratories developing such implantable devices may understandably be focusing their energy and attention simply on trying to produce devices that function effectively, without yet understanding or successfully addressing the full implications of their devices for information security.[23] Thus the manufacturer of our hypothetical CEO's retinal implants may or may not have been able to incorporate robust security features into those devices or to offer detailed guidance regarding their secure use.

In particular, the manufacturer may purposefully have included either publicly known mechanisms or hidden backdoors within the given model of implant that allow its internal computer to be accessed via a remote wireless connection for purposes of downloading logfiles from the device, carrying out diagnostic tests, updating the device's software, or transmitting particular visual content to the device for purposes of training its user or calibrating the implant. The designers of the device may assume that these remote access mechanisms are secure, simply because they utilize some internal proprietary transmitter, software, or protocol developed within the laboratory and because they assume that no unauthorized outside parties would possess the desire or ability to illicitly access a retinal implant while it is in use. However, in principle, a hacker with enough knowledge, skill, and motivation might be able to remotely access an implant's internal computer and could potentially be able to exploit the device's existing programming (or reprogram the device) so that – unbeknownst to the CEO – his retinal implants would wirelessly transmit a live video stream of everything seen by the CEO to an external computer controlled by the hacker. This would severely compromise the confidentiality of the information passing through the device.

If InfoSec personnel wished to restrict or eliminate the device's ability to transmit data to external systems, they could safely do so only in consultation with physicians and biomedical engineers who understand the extent to which such data transmissions may need to occur for legitimate medical purposes.

---

[22] Kosta & Bowman (2012).

[23] The early generations of such commercial neuroprostheses are perhaps the most likely models to possess inadequate security features. At the same time, obtaining early models of such sophisticated and costly experimental technologies may require significant money or influence – meaning that they may be disproportionately likely to be implanted and utilized in individuals of significant financial, social, or political importance whose personal information would be especially attractive to hackers and other unauthorized parties.

INTEGRITY

If a hacker were able to remotely gain unauthorized access to the internal computers controlling the CEO's implants, it is conceivable that he or she could intentionally 'edit' the visual output that was being provided to the CEO's brain by the implants. Alternatively, a hacker might even be able to supply a wholly fabricated visual stream, replacing the visual content detected by the photoreceptors of the implants' cameras with some pre-generated stream of false visual content transmitted from the hacker's computer. If such an act were feasible, this possibility would call into question the integrity of all the visual data that the CEO's brain was receiving; the CEO would never be able to know with certainty whether everything that he was seeing were 'real' or whether it had perhaps been altered in some way by an unauthorized party.

As before, if InfoSec personnel wish to constrain or eliminate the implants' ability to receive visual content supplied from an external computer, they would need to work closely with physicians and biomedical engineers who understand the extent to which the devices must possess such capacities in order to maintain a healthy and successful long-term interface with their host's optic nerve and brain, to facilitate necessary training and diagnostic activities, and to fulfill any legitimate medical or biological purposes that the devices serve.

AVAILABILITY

It is important to ensure that the continuous stream of visual information supplied by the implants is never unexpectedly interrupted, either as a result of hardware or software failure or through malicious interventions such as hacking or a computer virus. If the CEO's sense of sight were to be disabled while he were in the midst of conducting a high-profile shareholders' meeting, the result could be embarrassing for the company and the CEO personally. However, if the CEO's vision were to be suddenly disabled while he were driving an automobile or scuba-diving, for example, the result could be fatal. It may also be necessary to provide the connected retinal ganglion cells with an unceasing stream of electrochemical stimulation from the implants in order to ensure the cells' health and normal functioning.

On the other hand, it may be medically necessary to periodically disable the devices' stimulation of retinal ganglion cells, either to ensure the cells' long-term health or to allow the CEO to fall asleep at night. Any efforts by InfoSec personnel to regulate the implants' routines for making visual information available to the CEO's brain would thus need to be coordinated closely with expert medical and biomedical engineering personnel.

Applying the Framework of Access Controls

If the company's InfoSec personnel consider the CEO's new retinal implants through the lens of physical, logical, and administrative security controls, they might identify the following kinds of issues that would need to be addressed.

Physical Controls

It might appear unnecessary to give the retinal implants' internal computers specialized physical access controls, if the computers are located inside the CEO's head and can only be directly accessed by unauthorized parties by subjecting the CEO to a complex surgical procedure. However, there are indeed issues relating to physical controls that InfoSec personnel would need to consider. First, the retinal implants may exist as integrated components of a larger system that includes external computers that are more readily physically accessible. The implants might, for example, have a permanent wireless connection to some handheld external controller that can be used by the CEO to monitor or recalibrate the units. If this were the case, the CEO's InfoSec team would need to implement physical controls to ensure that the handheld controller would always be secured and not left lying around unattended in a location where unauthorized parties could gain access to it. Second, it might be the case that particular kinds of magnetic fields, patterns of light projected at the CEO's eyes, or other kinds of environmental phenomena can alter or disrupt the functioning of the retinal implants. InfoSec personnel would need to work with the biomedical engineers who designed the devices to understand any such possibilities and ensure that there were physical controls in place to protect the CEO's retinal implants from such interference, insofar as possible.

Logical Controls

Efforts to secure the CEO's retinal implants would also include a focus on logical controls, such as the mechanisms built into the implants' software that require any remote systems or users attempting to connect to the device to be identified, authenticated, and given access only to the systems or information which they are authorized to access. Because the retinal implants' internal computers may be highly specialized, idiosyncratic devices – rather than the 'off-the-shelf' computers running operating systems like Windows, Mac OS, Android, or Linux – it may or may not be technologically possible for the company's InfoSec team to install their preferred security software on the retinal implants. Moreover, the need to secure the implants' computers must also be balanced against the need to ensure the implants' proper and efficient functioning as prosthetic medical devices; the CEO's medical team may not allow security software to be installed on the implants that has the potential to unacceptably slow down, degrade, or otherwise impair their functioning as medical devices. The company's InfoSec team would need to work closely with medical and bioengineering personnel to ensure that any security software or logical controls added to the implants do not impede their proper functioning from a medical or biological perspective.

Administrative Controls

Administrative controls such as organizational policies and procedures can be put into place to maximize the effectiveness of the physical and logical controls and add another layer of security. For example, the InfoSec personnel and other relevant decision-makers within the CEO's company should develop clear policies to determine which organizational employees (potentially including the InfoSec personnel themselves, the CEO's medical personnel, the CEO's immediate administrative support staff, and the CEO himself) should have which levels of access to the CEO's retinal implants. The policies should also dictate that regularly updated risk assessments will be carried out to identify new and emerging threats that could impact the CEO's retinal implants as well as requiring that incident response and disaster recovery plans be proactively put in place, to allow the company to respond as efficiently and effectively as possible to any threat that might compromise or even completely disable the retinal implants.

Such administrative controls would need to be developed in close consultation with medical and biomedical engineering experts, to ensure that the CEO's medical personnel have all of the administrative access and supervisory authority needed to ensure the CEO's health and safety, while simultaneously ensuring that non-medical personnel are not allowed or expected to perform roles that could inadvertently endanger the CEO's health or safety by negatively impacting the retinal implants' functioning.

Avenues for Future Research

InfoSec's Growing Relationship to Medicine and Biomedical Engineering

The scenario described above highlights a new aspect of information security that will become increasingly important: its relationship to medicine, biomedical engineering, and related fields such as neuroscience. It is already the case that information security is a transdisciplinary field in which personnel must not only be experts in computer hardware and software but must also be familiar with fields like psychology, finance, law, and ethics. However, the growing use of implantable computing means that InfoSec personnel will also need to be knowledgeable about the biological and neuroscientific aspects of implantable computers. For large corporations, their information security teams might even include an in-house physician or neuroscientist who can ensure that any information security mechanisms or practices that are implemented for employees' ICs do not result in biological or psychological harm.

Similarly, when designing countermeasures that can be employed against unauthorized parties who may attempt to use their own implantable computers as instruments for carrying out illicit surveillance or corporate espionage against a company, such medical expertise would be needed by the company's InfoSec personnel in order to design countermeasures that neutralize such threats without causing biological or psychological injury to those suspected adversaries for which the company and its InfoSec personnel could potentially be held liable.

Cybernetics as a Means of Preparing InfoSec for Implantable Computers

One of the challenges in linking information security with medicine is that the two fields utilize different vocabularies and theoretical and conceptual frameworks: InfoSec is grounded largely in the theoretical framework of computer science and medicine in that of biology and chemistry. In addressing this challenge, it may be helpful to build on the field of cybernetics, which was founded to provide precisely the sort of transdisciplinary theoretical framework and vocabulary that can be used to translate insights between all of those fields that study patterns of communication and control in machines, living organisms, or social systems.[24] Alongside the many existing foci of cybernetics (found in subdisciplines like biocybernetics, neurocybernetics, and management cybernetics) it may be useful to envision a sort of 'celyphocybernetics'[25] that sees the human brain, its surrounding body, and any computers embedded in the body as together forming a single physical 'shell' for an individual human mind. The human brain, organic and artificial body components, and ICs would constitute a system that receives information from the external environment, stores information circulating within it, and transmits information to the external environment, thereby creating networks of communication and control. InfoSec experts, physicians, and biomedical engineers would thus share the single task of ensuring the secure, productive, and effective functioning of this entire information system that contains both biological and electronic components; that common goal can only be achieved if both InfoSec personnel, physicians, and biomedical engineers succeed in fulfilling their unique professional roles.

Conclusion

In this text we have considered the manner in which the field of information security will need to draw increasingly on expertise from medicine and biomedical engineering in order to address the unique ways in which the

---

[24] Wiener, *Cybernetics: Or Control and Communication in the Animal and the Machine* (1961).

[25] From the Ancient Greek *κέλυφος*, meaning 'shell,' 'sheath,' 'husk,' or 'pod.'

growing use of implantable computers are expected to reshape the information security landscape. Not only will implantable computers create new kinds of InfoSec vulnerabilities and risks for individuals who use them, but they will offer powerful new tools for those who would attempt to carry out illicit activities such as corporate espionage or illegal surveillance. By utilizing knowledge from the fields of medicine and biomedical engineering, InfoSec professionals will be able to recognize the information security characteristics of implantable computers that they cannot directly access, understand such implantable computers' connections to the biological and cognitive processes of their human hosts and the medical implications of such interfaces, and develop InfoSec practices and technologies that secure the information contained within implantable computers without placing the devices' hosts at risk of physical or psychological harm. The field of information security has already demonstrated an interdisciplinary ability to successfully incorporate knowledge from diverse fields and adapt to ever-changing technological, social, and legal demands, and it is well-positioned to secure a future in which implantable computers will become an increasingly important element of human life.

References

SP 800-100. Information Security Handbook: A Guide for Managers, edited by P. Bowen, J. Hash, and M. Wilson. Gaithersburg, Maryland: National Institute of Standards & Technology, 2006.

Clark, S.S., and K. Fu, "Recent Results in Computer Security for Medical Devices," in *Wireless Mobile Communication and Healthcare*, K.S. Nikita, J.C. Lin, D.I. Fotiadis, and M.-T. Arredondo Waldmeyer, eds., Lecture Notes of the Institute for Computer Sciences, Social Informatics and Telecommunications Engineering 83, pp. 111-18. Springer Berlin Heidelberg, 2012.

Fairclough, S.H., "Physiological Computing: Interfacing with the Human Nervous System," in *Sensing Emotions*, J. Westerink, M. Krans, and M. Ouwerkerk, eds., Philips Research Book Series 12, pp. 1-20. Springer Netherlands, 2010.

Gasson, M.N., "Human ICT Implants: From Restorative Application to Human Enhancement," in *Human ICT Implants: Technical, Legal and Ethical Considerations*, M.N. Gasson, E. Kosta, and D.M. Bowman, eds., Information Technology and Law Series 23, pp. 11-28. T.M.C. Asser Press, 2012.

Gasson, M.N., "ICT Implants," in *The Future of Identity in the Information Society*, S. Fischer-Hübner, P. Duquenoy, A. Zuccato, and L. Martucci, eds., IFIP – The International Federation for Information Processing 262, pp. 287-95. Springer US, 2008.

Gasson, M.N., E. Kosta, and D.M. Bowman, "Human ICT Implants: From Invasive to Pervasive," in *Human ICT Implants: Technical, Legal and Ethical Considerations*, M.N. Gasson, E. Kosta, and D.M. Bowman, eds., Information Technology and Law Series 23, pp. 1-8. T.M.C. Asser Press, 2012.

Gladden, Matthew E. *The Handbook of Information Security for Advanced Neuroprosthetics.* Indianapolis: Synthypnion Academic, 2015.

ISO/IEC 27001:2013, Information Technology – Security Techniques – Information Security Management Systems – Requirements. Joint Technical Committee ISO/IEC JTC 1, Information technology, Subcommittee SC 27, IT Security techniques. Geneva: The International Organization for Standardization and the International Electrotechnical Commission, 2013.

Jumper, J. Michael., "FDA Approves World's First Artificial Retina." *Retina Times,* American Society of Retina Specialists, Spring 2013 (n.d.). https://www.asrs.org/retina-times/details/131/fda-approves-world-first-artificial-retina.

Koops, B.-J., and R. Leenes, "Cheating with Implants: Implications of the Hidden Information Advantage of Bionic Ears and Eyes," in *Human ICT Implants: Technical, Legal and Ethical Considerations,* M.N. Gasson, E. Kosta, and D.M. Bowman, eds., Information Technology and Law Series 23, pp. 113-34. T.M.C. Asser Press, 2012.

Kosta, E., and D.M. Bowman, "Implanting Implications: Data Protection Challenges Arising from the Use of Human ICT Implants," in *Human ICT Implants: Technical, Legal and Ethical Considerations,* M.N. Gasson, E. Kosta, and D.M. Bowman, eds., Information Technology and Law Series 23, pp. 97-112. T.M.C. Asser Press, 2012.

Li, S., F. Hu, and G. Li, "Advances and Challenges in Body Area Network," in *Applied Informatics and Communication,* J. Zhang, ed., Communications in Computer and Information Science 22, pp. 58-65. Springer Berlin Heidelberg, 2011.

McGee, E.M., "Bioelectronics and Implanted Devices," in *Medical Enhancement and Posthumanity,* B. Gordijn and R. Chadwick, eds., The International Library of Ethics, Law and Technology 2, pp. 207-24. Springer Netherlands, 2008.

Merkel, R., G. Boer, J. Fegert, T. Galert, D. Hartmann, B. Nuttin, and S.Rosahl, "Central Neural Prostheses," in *Intervening in the Brain: Changing Psyche and Society,* Ethics of Science and Technology Assessment 29, pp. 117-60. Springer Berlin Heidelberg, 2007.

NIST Special Publication 800-53, Revision 4: Security and Privacy Controls for Federal Information Systems and Organizations. Joint Task Force Transformation Initiative. Gaithersburg, Maryland: National Institute of Standards & Technology, 2013.

Park, M.C., M.A. Goldman, T.W. Belknap, and G.M. Friehs, "The Future of Neural Interface Technology," in *Textbook of Stereotactic and Functional Neurosurgery,* A.M. Lozano, P.L. Gildenberg, and R.R. Tasker, eds. Heidelberg/Berlin: Springer, 2009, pp. 3185-3200.

Rao, U.H., and U. Nayak, *The InfoSec Handbook.* New York: Apress, 2014.

Rotter, P., and M.N. Gasson, "Implantable Medical Devices: Privacy and Security Concerns," in *Human ICT Implants: Technical, Legal and Ethical Considerations,* M.N. Gasson, E. Kosta, and D.M. Bowman, eds., Information Technology and Law Series 23, pp. 63-66. T.M.C. Asser Press, 2012.

Schmid, E.W., W. Fink, and R. Wilke. "Simultaneous vs. Sequential and Unipolar vs. Multipolar Stimulation in Retinal Prostheses." In *2013 6th International IEEE/EMBS Conference on Neural Engineering (NER)*, 190–93, 2013. doi:10.1109/NER.2013.6695904.

Schmid, Erich W., and Wolfgang Fink, "Operational Design Considerations for Retinal Prostheses." *arXiv:1210.5348 [q-Bio]*, October 19, 2012. http://arxiv.org/abs/1210.5348.

Weiland, J.D., W. Liu, and M.S. Humayun, "Retinal Prosthesis," *Annual Review of Biomedical Engineering,* vol. 7, no. 1, pp. 361-401, 2005.

Wiener, Norbert. *Cybernetics: Or Control and Communication in the Animal and the Machine,* second edition. Cambridge, Massachusetts: The MIT Press, 1961.

Part III

Robotics and Artificial Intelligence: A Closer Look

Chapter Eight

The Social Robot as CEO: Developing Synthetic Charismatic Leadership for Human Organizations[1]

Abstract. As robots are developed that possess increasingly robust social and managerial capacities and which are moving into a broader range of roles within businesses, the question arises of whether a robot could ever fill the ultimate organizational role: that of CEO. Among the many functions a chief executive officer must perform is that of motivating a company's workers and cultivating their trust in the company's strategic direction and leadership. The creation of a robot that can successfully inspire and win the trust of an organization's human personnel might appear implausible; however, we argue that the development of robots capable of manifesting the leadership traits needed to serve as CEO within an otherwise human organization is not only possible but – based on current trends – likely even inevitable.

Our analysis employs phenomenological and cultural posthumanist methodologies. We begin by reviewing what French and Raven refer to as 'referent power' and what Weber describes as 'charismatic authority' – two related characteristics which if possessed by a social robot could allow it to lead human personnel by motivating them and securing their loyalty and trust. By analyzing current robotic design efforts and cultural depictions of robots, we identify three ways in which human beings are striving to create charismatic robot leaders for ourselves. We then consider the manner in which particular robot leaders will acquire human trust, arguing that charismatic robot leaders for businesses and other organizations will

---

[1] This text is an expanded and adapted version of Gladden, Matthew E., "The Social Robot as 'Charismatic Leader': A Phenomenology of Human Submission to Nonhuman Power," in *Sociable Robots and the Future of Social Relations: Proceedings of Robo-Philosophy 2014*, edited by Johanna Seibt, Raul Hakli, and Marco Nørskov, pp. 329-339. Frontiers in Artificial Intelligence and Applications, vol. 273. Amsterdam: IOS Press, 2014.

emerge naturally from our world's social fabric, without any rational decision on our part. Finally, we suggest that the stability of these leader-follower relations – and the extent to which charismatic social robots can remain long-term fixtures in leadership roles such as that of CEO – will hinge on a fundamental question of robotic intelligence and motivation that currently stands unresolved.

I. Introduction

For more than a half-century, robots have filled crucial roles within human organizations, including businesses. The advent of industrial robots for use in assembly-line manufacturing processes in the 1960s and 1970s[2] has been followed by the development of increasingly intelligent social robots that now serve as assistants and facilitators to human workers in areas as diverse as education, health care, retail sales, agriculture, transportation, security, and the military.[3] It is anticipated that the role of robots in the workplace will continue to evolve, as new and more sophisticated types of robots are developed that possess the capacity to serve as true colleagues and partners[4] – and potentially even managers and supervisors – to human workers.

This raises the question whether a robot working within an organization that includes human personnel could ever fill that most singular leadership role: as CEO. A chief executive officer must possess significant technical expertise, information-processing capacity, and decision-making ability, in order to plan, organize, and control the activities of those workers (whether human employees or artificial agents) that are part of an organization. However, a successful CEO is also required to perform a qualitatively different sort of function: namely, to inspire and motivate an organization's human employees and to engender their loyalty and trust in the strategic vision enunciated by

[2] For the history of industrial robotics, see Goodman & Argote, "New Technology and Organizational Effectiveness" (1984); Murphy, *An Introduction to AI Robotics* (2000), pp. 19-27; and Perlberg, *Industrial Robotics* (2016).

[3] Regarding recent developments, current challenges, and long-term possibilities in the field of social robotics, see, e.g., Breazeal, "Toward sociable robots" (2003); Kanda & Ishiguro, *Human-Robot Interaction in Social Robotics* (2013); *Social Robots and the Future of Social Relations*, edited by Seibt et al. (2014); *Social Robots from a Human Perspective*, edited by Vincent et al. (2015); and *Social Robots: Boundaries, Potential, Challenges*, edited by Nørskov (2016).

[4] For discussion of the possibility that robots might serve as coworkers and colleagues rather than simply tools, see, e.g., Bradshaw et al., "From Tools to Teammates: Joint Activity in Human-Agent-Robot Teams" (2009); Samani & Cheok, "From human-robot relationship to robot-based leadership" (2011); Samani et al., "Towards Robotics Leadership: An Analysis of Leadership Characteristics and the Roles Robots Will Inherit in Future Human Society" (2012); Wiltshire et al., "Cybernetic Teams: Towards the Implementation of Team Heuristics in HRI" (2013); Ford, *Rise of the Robots: Technology and the Threat of a Jobless Future* (2015); Spring, "Can machines come up with more creative solutions to our problems than we can?" (2016); and Wong, "Welcome to the robot-based workforce: will your job become automated too?" (2016).

the CEO.[5] While a contemporary organization's nonhuman agent resources – such as computers and robots – can often simply be programmed or instructed to carry out particular assigned tasks, human workers cannot be so directly or completely controlled; they must be persuaded and motivated to carry out the particular work that their organization wishes them to perform.[6]

It is not difficult to imagine the development of future robots that possess the technical knowledge and information-processing and decision-making abilities that are needed in order to plan, organize, and control the activities of a company's agent-workers, insofar as such capacities are congruent with our popular notions of the strengths of robots and other artificially intelligent entities; we 'expect' a computerized entity to be capable of effectively and efficiently processing information. And indeed, progress toward designing robots with such basic managerial capacities is already well advanced.[7] However, at first glance it might appear that creating a robot that can successfully inspire and win the trust of an organization's human personnel is a major hurdle to the development of a robot that can serve effectively as CEO of a company that includes human workers. After all, the stereotypical image of a 'robot' has not traditionally attributed to such entities the sort of social and emotional intelligence, wisdom, ethical insight, moral courage, and selfless personal commitment to an organization that inspire loyalty and trust in the human CEOs who possess such characteristics. In this text, however, we argue that the development of social robots that are capable of manifesting the kind of charisma, inspirational leadership, and trustworthiness that would allow them to serve effectively as CEOs within otherwise human organizations is not only possible but likely even inevitable. Our analysis of this issue utilizes phenomenological methods employed by philosophers of science alongside a cultural posthumanist methodology that analyzes contemporary cultural products (such as works of science fiction) and society's reaction to them as a

[5] See Dainty & Anderson, *The Capable Executive: Effective Performance in Senior Management* (1996).

[6] For key distinctions between human and nonhuman agents in the context of an organization, see "The Posthuman Management Matrix: Understanding the Organizational Impact of Radical Biotechnological Convergence" in Gladden, *Sapient Circuits and Digitalized Flesh: The Organization as Locus of Technological Posthumanization* (2016).

[7] For discussion of efforts to develop robots and other artificial agents that possess individual traits and functionality needed to plan, organize, and control activities within the context of a business or other organization, see, e.g., Kriksciuniene & Strigunaite, "Multi-Level Fuzzy Rules-Based Analysis of Virtual Team Performance" (2011); Nunes & O'Neill, "Assessing the Performance of Virtual Teams with Intelligent Agents" (2012); and Dai et al., "TrustAider – Enhancing Trust in E-Leadership" (2013), and the discussion of such research in Gladden, "Leveraging the Cross-Cultural Capacities of Artificial Agents as Leaders of Human Virtual Teams" (2014). For a broader review of the expected organizational and societal impacts of robots possessing managerial capabilities, see, e.g., Elkins, "Experts predict robots will take over 30% of our jobs by 2025 – and white-collar jobs aren't immune" (2015), and Susskind & Susskind, *The Future of the Professions: How Technology Will Transform the Work of Human Experts* (2015).

means of diagnosing human fears or aspirations relating to future processes of technological posthumanization such as those that will be driven by the growing deployment of sophisticated social robots.

We begin by considering the diverse roles that must be filled by the CEO of a human organization and the traits and characteristics that are needed in order to fill such roles – highlighting especially a CEO's role as the most visible font and focus of *leadership* within an organization. We then analyze phenomena which French and Raven referred to as 'referent power' and Weber described as 'charismatic authority' – two related characteristics which if possessed by a social robot could enable that robot to lead human personnel by inspiring them and securing their loyalty and trust in ways similar to those employed by human CEOs. Referent power and charismatic authority are not monolithic phenomena but rather characteristics whose dynamics can be manifested in a number of diverse (and potentially even mutually exclusive) ways. By analyzing contemporary efforts in robotic design and engineering as well as cultural depictions of robots, we identify three ways in which human beings are striving to create charismatic robot leaders for ourselves – ways that reflect a model of the human mind as ancient as Plato's tripartite division of the soul. We then explore the manner in which particular robotic entities will either intentionally or unintentionally acquire the trust of human beings that is needed in order to function successfully as CEO of an organization that includes human stakeholders. Building on Coeckelbergh, we argue against the contractarian-individualist approach which presumes that human beings will be able to consciously 'choose' whether or not robots should fill leadership positions within businesses and other organizations; we instead propose a phenomenologically and socially oriented understanding of the manner in which charismatic robot leaders for businesses and other institutions will emerge naturally from the world's social fabric, without any rational decision on the part of human beings. Finally, drawing on Abrams and Rorty, we suggest that the stability of these leader-follower relations – and the extent to which charismatic social robots can remain long-term fixtures in leadership roles such as that of CEO – will hinge on a fundamental question regarding the intelligence and motivation of such robotic entities that has not yet been adequately resolved.

II. The CEO as Leader: The Need to Inspire, Motivate, and Persuade

Serving successfully as the chief executive officer of an organization that includes human personnel requires one to possess and deploy a diverse set of capacities and skills. Building on the classic management framework of Henri Fayol, Daft identifies the four essential functions that must be performed by a manager as *planning*, *organizing*, *leading*, and *controlling* the activities of those workers and systems that fall within the manager's purview; in the case of a

CEO, this includes all of an organization's activities.[8] In this text, we do not address the ongoing advances in robotic design and engineering that are allowing robots to increasingly perform the managerial functions of planning, organizing, and controlling the activities of human workers;[9] although we in no way underestimate the need for a CEO to effectively plan, organize, and control an organization's activities, here we will focus instead on the CEO's duty to lead.

While an organization's automated manufacturing process or cloud-based file-sharing system can be controlled by simply reconfiguring hardware and software, the organization's human personnel cannot be directly controlled or reprogrammed in such a fashion; they must instead be *led.* Daft defines the leadership function by explaining that:

> Leading is the use of influence to motivate employees to achieve organizational goals. Leading means creating a shared culture and values, communicating goals to people throughout the organization, and infusing employees with the desire to perform at a high level.[10]

Leadership takes on different aspects when practiced at different levels within a company's organizational structure. The role of a senior executive such as a CEO differs from that of lower-level managers because a senior executive must successfully create and maintain power bases within a broader range of stakeholder groups (such as those of shareholders, board members, key suppliers and partners, employees, regulators, and local communities); possess and wield the political savvy needed to overcome the rivalries, turf battles, and political maneuvering that become more intense at the highest levels within an organization; make, communicate, and implement decisions that may have life-altering impacts for all of an organization's employees; and weather the high degree of internal and external scrutiny that comes with being an organization's senior spokesperson and authority figure.[11] A company's CEO occupies a unique leadership role: while he or she may only immediately supervise a small number of direct reports, he or she leads all of the company's employees by enunciating and visibly embodying the firm's strategic vision and values and by cultivating a sense of loyalty and engagement on the part of personnel throughout all levels of the organization.

---

[8] Daft, *Management* (2011), p. 8.

[9] For a discussion of efforts to design robots that are increasingly adapt at performing such management functions, see, e.g., Gladden, "Leveraging the Cross-Cultural Capacities of Artificial Agents as Leaders of Human Virtual Teams" (2014).

[10] Daft (2011), p. 8.

[11] Dainty & Anderson (1996), pp. 4-5.

Through their empirical research, Dainty and Anderson have identified eleven capabilities that affect senior executives' ability to function successfully in their roles. Among these factors are a group of interpersonal capabilities that include 'influence' (the ability "to get others to accept your point of view, have them act in your interests and prevent them implementing agendas which are contrary to your own"), 'leadership' (the ability "to help others overcome hurdles to achieve a common goal"), 'integration' (the ability "to build senior level teams and ensure larger organisational units work together effectively"), and 'insight' (the ability "to understand what motivates others, their mental view of the world and their possible actions and agendas").[12] Similarly, Goleman has described the crucial importance of 'emotional intelligence' as the "sine qua non of leadership" and identified its five components as self-awareness, self-regulation, motivation, empathy, and social skill.[13] Although the increasingly critical role of technology and technical expertise in business might lead one to presume that a CEO's interpersonal skills are no longer so central to an organization's success, the possession of 'human skills' such as the ability to communicate with, motivate, and lead workers is only becoming more important for CEOs, not less.[14]

Research of the sort just described underscores the need for a CEO to be able to lead an organization's human personnel by inspiring, motivating, influencing, and persuading them. Before directly considering the question of whether a robot could possess and demonstrate such leadership capacities as a CEO, we must first explore the psychological, social, and cultural mechanisms by which human beings allow themselves to be inspired, motivated, influenced, and persuaded by those who lead them. Within the constraints of this text, we will not able to investigate in detail robots' potential for exploiting all such mechanisms. Instead we will focus on one noteworthy possibility: namely, the potential for a robotic CEO to lead human workers through its possession of charismatic authority and referent power.

III. Referent Power and the Exercise of Charismatic Authority by Robots

French and Raven proposed a now-classic model that identifies five bases of social power, to which Raven later added a sixth. These six bases comprise

---

[12] Dainty & Anderson (1996), pp. 16, 18.

[13] Goleman, "What Makes a Leader?" (2004).

[14] Daft (2011), p. 11; Plunkett et al., *Management* (2012), pp. 26-27.

coercive, reward, legitimate, referent, expert, and informational power. An effective leader must be able to possess and employ one or more of these bases of power in order to secure, maintain, and exercise influence over followers.[15]

Applying French and Raven's model, we could say that a leader utilizes *coercive power* if he or she employs the threat of force in order to influence the behavior of others and *reward power* when he or she promises some benefit in order to influence it. A leader possesses *legitimate power* when cultural values or social structures have invested him or her with a particular authority that others feel a sense of responsibility and obligation to obey. A leader possesses *referent power* when others are drawn to the leader because they feel a sense of fondness or admiration for the leader, feel a sense of kinship or affinity with the leader, identify with the leader's moral values, or desire to win his or her personal approval. A leader possesses *expert power* when he or she is able to influence others through their perception that he possesses unique skills or experience. Finally, a leader utilizes *informational power* when he or she influences others by means of his or her control over access to information resources.

Referent Power: The Most Difficult Base of Social Power for Robots to Utilize?

It is relatively easy to imagine a future world in which a social robot filling a senior management role within an organization is in a position to either threaten and intimidate its subordinate human employees into working in a particular fashion (e.g., because it has been given the authority to terminate workers that it determines to be ineffective), dole out financial rewards to the human employees in order to win their cooperation, issue instructions that must be followed by virtue of the robot's official status within the organization's personnel structure, cultivate the perception that it possesses expert work-related skills, or control its human employees' access to information. These cases would represent the use of coercive, reward, legitimate, expert, and informational power, respectively.

The case that is perhaps most difficult to imagine is that of a robotic senior executive who influences human beings by exercising *referent power* over them. How likely is it that a firm's human stakeholders would voluntarily select a robot as their CEO and submit to its leadership because they feel a sense of fondness or admiration for it? Or because they feel a sense of kinship or affinity with the robot? Would human workers embrace and allow themselves to be led by a robotic CEO because they identify closely with the robot's 'moral values'? Or because they long to bask in the robot's personal affirmation and

---

[15] See French & Raven, "The Bases of Social Power" (1959); Raven, "A power/interaction model of interpersonal influence: French and Raven thirty years later" (1992); and Forsyth, *Group Dynamics* (2010), p. 227.

approval of them? At first glance, it appears that exercising referent power would be the most difficult and least likely way for a robotic CEO to motivate and influence its human subordinates. This also, we would suggest, makes it the most interesting case to consider.

Referent Power as Manifested in the Charismatic Authority of Robots

In the case of a human leader who utilizes referent power, this power is frequently grounded in the leader's possession of what Weber called 'charismatic authority.' Such charismatic authority is manifested when followers' obedience arises from their "devotion to the exceptional sanctity, heroism or exemplary character" of their leader, "and of the normative patterns or order revealed or ordained by him."[16]

Such a relationship requires that followers possess an emotional bond with their leader and deep trust in his goodness; this trust will in turn nurture sentiments of respect, admiration, and personal loyalty in the followers. The use of charismatic authority is a form of leadership that is both powerful and ephemeral, because it is grounded in emotion and the followers' conviction that the leader is a superior being – a belief that can be quickly punctured if the charismatic leader is seen to behave in some disillusioning way that shatters the link between the leader and the ideal that he or she had apparently embodied.

Forging such a charismatic leader-follower relationship between a robot and human being requires two participants, neither of whom seems particularly well-suited for their role. In order for a leader to exercise charismatic authority, he or she must possess (or at least be perceived to possess) traits such as holiness, divine ordination, moral righteousness, personal charm, or a hypnotic strength of personality. These are traits which we might expect to find in particular human beings, but which we do not normally see as possessable by robots. We stereotypically presume that robots are not adept at evoking feelings of true love, admiration, or loyalty from human beings, and that

---

[16] Weber, *Economy and Society: An Outline of Interpretive Sociology* (1968), p. 215. This technical sense of 'charismatic' diverges from the everyday sense of the word. To say that a leader is 'charismatic' in Weber's sense does not necessarily imply that the person is seductive, emotionally astute, eloquent, or possessing hypnotic powers of enthrallment. While those traits can indeed provide a basis for charismatic authority, a leader possessing charismatic authority might just as easily be one who is physically unattractive, emotionally impaired, and lacking in any romantic appeal – as long as he or she is seen as a living embodiment of some religious, philosophical, or cultural value that is, in itself, attractive; people yearn to be close to and conform to the leader, because they yearn to be close to the principle that he or she represents.

human beings – for their part – are not inclined to spontaneously shower such sentiments on robots.[17]

And yet, while it might superficially appear as though social robots are unlikely candidates to serve as charismatic leaders of human followers, a deeper sociotechnological and phenomenological investigation suggests that it is not only possible for robots to hold sway over human followers through the use of referent power and charismatic authority, but that the advent of such leader-follower relationships between robots and humans is rapidly approaching. As a starting point, the fact that social robots can lead *other robots* within multi-agent systems or artificial organizations is now taken for granted. Computer scientists have already designed communities of robots that spontaneously organize their own social structures, with members taking on roles as 'leaders' or 'followers.' Many such robot communities use algorithms involving probabilistic elections to choose their leader, but some researchers are developing formal frameworks that would allow the leaders of robot communities to emerge through robots' manifestation of and response to all of the bases of power identified by French and Raven, including referent power.[18]

Incremental Steps toward Human Beings' Submission to Robotic Referent Power

In many ways, human beings, too, already demonstrate obedience to robots and their electromechanical kin. We listen to the automated voice that tells us to step away from the subway car's doors as they are about to close; we follow the instructions of an electronic voice on the phone telling us to 'press 1' for this or 'press 2' for that. The nature of our relationship with such technologies becomes quite explicit, for example, in the case of a humanoid robotic crossing-guard in Kinshasa that monitors traffic and tells pedestrians when it is safe to walk.[19] In our everyday accounts of such interactions with technology, we attribute to them the characteristics of a power relationship in which we, as human beings, are submitting to the dictates of a technological master; thus we typically say that we 'obey' a traffic light rather than that we are 'collaborating with' or 'utilizing' it.

However, in these elementary cases, we are not obeying a robotic system because we consider it to be an intelligent agent possessing social power of its

---

[17] Regarding such human attitudes toward robots, see, e.g., Van Oost & Reed, "Towards a sociological understanding of robots as companions" (2010); Cabibihan et al., "When Robots Engage Humans" (2014); Szollosy, "Why are we afraid of robots? The role of projection in the popular conception of robots" (2015); and Szollosy, "Freud, Frankenstein and our fear of robots: projection in our cultural perception of technology" (2016).

[18] Pereira et al., "Conceptualizing Social Power for Agents" (2013).

[19] Schlindwein, "Intelligent Robots Save Lives in DR Congo" (2014).

own but because it represents and extends the power of the human beings who have constructed it and situated it in a particular role – and who *do* possess some social power over us. Human beings do not yet widely interact with robots in a way that reveals robots to be autonomous agents who possess their own social power, let alone possess the more specific characteristics of referent power and charismatic authority. However, the first tentative steps in that direction have already been seen, and we stand on the threshold of a wider human embrace of such charismatic robot leaders.

IV. Three Ways in Which We Envision Charismatic Robots as Leaders of Human Beings

Some scholars have argued that human beings' ongoing, almost instinctual embrace of the latest technologies reflects the fact that "giving robots positions of responsibility is not only unavoidable but is rather something desired and that we are trying to achieve."[20] In other words, we yearn for robot leaders and are striving – whether consciously or unconsciously – to create them; we want to be led by our mechanical creation. However, not all charismatic robot leaders are alike. Just as the exercise of charismatic authority by human beings can on take radically different forms, so too the charismatic robot leaders that we envision for ourselves display a number of very different (and sometimes contradictory) traits. Rather than discussing 'charismatic robot leadership' as though it represented a single undifferentiated phenomenon, our understanding of charismatic robot leadership can be advanced if we distinguish and analyze key subtypes of charismatic robot leaders.

While a broader analysis of real-world and fictional accounts of robots might identify additional possibilities, we would suggest that there are at least three key means by which robots can exercise charismatic authority over human beings: 1) through the possession of superior morality; 2) through the manifestation of superhuman knowledge that may take on aspects of religious revelation; or 3) through interpersonal allure, seduction, and sexual dynamism. As noted below, this categorization offers a parallel to Plato's tripartite division of the soul, as it is possible to understand each kind of charismatic robot leader as appealing to and influencing a different part of its followers' being.

Type 1 Charismatic Robot Leader: The Saint-Martyr

The first form of charismatic robot leader – which attracts and influences human followers through the possession and display of a superior morality – we might refer to as the 'Type 1' or 'Saint-Martyr.' In fictional depictions of

[20] Samani et al., "Towards Robotics Leadership: An Analysis of Leadership Characteristics and the Roles Robots Will Inherit in Future Human Society" (2012).

such robots, the demonstration of the Type 1's moral superiority is often underscored by the robot's conscious decision to sacrifice its own existence in order to save the lives of its human colleagues. While the robot's destruction means that – in a practical sense – it is no longer capable of serving as a leader of human beings, it also serves to confirm and emphasize the fact that the robot was precisely the sort of noble, selfless leader whom human beings *should have* taken as their leader and whose instructions they could have obeyed with a sense of moral confidence and ease.

The sublime, self-sacrificing, loving nature of the Type 1 robot's salvific deed is often highlighted by having it occur precisely when the robot is on the threshold of receiving some great personal benefit, such as the development of a higher level of consciousness or a more 'human' array of experiences and emotions.[21] Just at the moment when the robot's awareness of its new potential is beginning to blossom, the robot and its closest human friends are confronted by a climactic threat that seems certain to annihilate the protagonists. Amidst this apparently hopeless situation, the robot realizes that there is indeed one avenue by which it can save the lives of its human companions – but only by destroying itself in the process. The robot's first inklings of its growing intellectual, emotional, and moral 'humanity' guide it toward making its ultimate self-sacrificing decision, while at the same time making the robot's destruction more dramatically wrenching for the audience, because we mourn the fact that the robot will never have the opportunity to fully explore and appreciate its new human-like existence.[22]

---

[21] Despite the common fictional trope that has made 'robotic' a near synonym of 'cold,' 'without feeling,' and 'unemotional,' we would argue that actual robots of the future will make effective charismatic leaders not because they display no emotions, but because they display the most advanced emotions. Friedenberg (2008) points out that researchers have found in numerous contexts that the development of human-like emotions is not only of significant evolutionary value but that it may even be a necessary aspect of any embodied mind – whether natural or synthetic – that has a claim to sapience. In order to construct robotic brains that receive and process input, make decisions, and take action as efficiently and effectively as possible, scientists are discovering that they must build into such brains functions that are essentially digital counterparts of human cognitive functions such as the ability to feel anger and sadness and the need for sleep. As artificial intelligence becomes more advanced, it often becomes more 'emotional,' not less. These superior emotional abilities can aid robots to become charismatic leaders of human beings, especially leaders of Types 1 and 3. While the common fictional depiction of robots as unemotional has a long history, the strain of fictional imagination depicting robots as capable of human-like emotions has always existed and appears to be growing in prevalence, as old stereotypes about the nature of AI are rendered outdated by science and engineering's progress toward the development of artificial general intelligence, toward the advent of synthetic minds that possess a full range of human-like cognitive capacities and characteristics.

[22] Typically, the human companions of the Saint-Martyr robot try (unsuccessfully) to stop its self-sacrificing act, thereby demonstrating that by the story's end they had instinctively come to see the robot not as a mere device or instrument but as an autonomous moral agent on par with

Type 1 Charismatic Robots as Developed in Fiction

A primeval example of Type 1 Saint-Martyr robots is found in the work that introduced the word 'robot' to the world: the 1921 play *R.U.R.: Rossum's Universal Robots* by Karel Čapek. After depicting the future destruction of humankind at the hands of armies of robot workers, the play concludes by focusing on the autonomous biological robots Helena and Primus, a young robot couple who are each so deeply and selflessly in love with one another that when the last surviving member of the human race must choose a robot to kill and dissect for research purposes, Helena and Primus each beg to be the robot who is killed so that their cherished partner might live. In witnessing the robots' spontaneous display of self-sacrificing love, the human researcher realizes that the young robot pair constitute a second Adam and Eve who are capable and worthy of engendering a new (quasi-)humanity that will establish itself across the earth. He blesses and celebrates the future of the new couple by reciting the account of Creation from the Book of Genesis, thereby comparing their innocence and moral goodness to that displayed by Adam and Eve before the entrance of sin into the Garden of Eden.[23]

One of the most poignant and thoughtfully developed examples of a fictional Type 1 Saint-Martyr robot is the Tachikoma (or 'think tank') as presented in the anime series *Ghost in the Shell: Stand Alone Complex*, based on the manga by Masamune Shirow. These armed robotic vehicles had served the government well but were eventually removed from active duty and consigned to be disassembled, because they had begun to display 'suspicious' behavior – such debating theological questions among themselves during their free time – that led their human superiors to fear that the Tachikomas' AI was becoming glitch-prone and unreliable for use in combat situations. However, in the series' climax, the last surviving Tachikomas that had not yet been disassembled become aware of the fact that one of their former human colleagues was trapped in a battle and perishing, and they escape, rush to their old companion's aid, and mount a frenzied struggle to save his life, thereby sacrificing themselves in the process.[24] We realize that the 'glitches' and 'unreliability' that the Tachikomas had previously begun to demonstrate were not glitches

---

human beings, as 'one of their own.' Alternatively, the robot may realize that its human companions will attempt to intervene and stop its self-sacrificing act, even at the cost of their own lives; thus the robot will plan and execute its final deed in such a way that it cannot be stopped, which is accomplished either by hiding its intentions from its human companions or proactively choosing to violate instructions that had been given to it by its human superiors. In this way, the Type 1 robot fully claims its place as an autonomous agent and 'leader' who will weigh all of the moral principles and practical considerations at play in a difficult situation, forge its own conclusion about the action that must be taken, and then courageously carry out that deed.

23 See Act III of Čapek, *R.U.R.: Rossum's Universal Robots* (2016).

24 Kamiyama, "C: Smoke of Gunpowder, Hail of Bullets – BARRAGE" (2003).

at all but rather flashes of their dawning sapience and humanity – a humanity that is highlighted when, in the moment before its destruction, one of the Tachikomas spontaneously utters a prayer to God, lamenting the fact that it had not been able to do more to save its human companion.[25]

Another fictional example is found in the novel *2010: Odyssey Two* and its film adaptation. Contrary to the fears and expectations of the film's human protagonist, the HAL 9000 computer (which can be thought of as a social robot with the entire Discovery One spaceship as its body) voluntarily sacrifices itself in order to save the lives of a spaceship's human crew. A similar fictional example of the Type 1 charismatic leader can be seen in the film *Star Trek: Nemesis*, where the android Lieutenant Commander Data – who had long struggled to understand what it means to be 'human' and grappled with the difficulties of serving as a leader of human beings – sacrificed himself to save the lives of his fellow crew members.

In such cases, the creators of popular fiction are presenting visions of social robots whose moral clarity, selflessness, and nobility are designed to evoke feelings of admiration and even awe among human audiences. In these social robots and their salvific, almost Christ-like sacrifices, we see leaders to whom we would be comfortable entrusting the most difficult moral decisions, because they have demonstrated in the most concrete way possible that their decisions will be made wisely and justly and for the sake of the human common good, without any self-centered pursuit of the robots' own personal profit or pleasure.

Artificial General Intelligence and the Engineering of Type 1 Charismatic Robots

If a corporation has an office led by a robot manager that has unfailingly demonstrated itself to be wise, fair, selfless, innovative, and effective, then might not some human employees voluntarily request a transfer to that office, rather than continuing to work in a department led by a human manager who is known to be petty, vindictive, and incompetent? It is not difficult to imagine that in our real, nonfictional world, at least some human beings might prefer to place their trust in Type 1 robot leaders rather than in human leaders who time and again have demonstrated a propensity to tolerate or even embrace nepotism, corruption, oppression, and other injustices. Regardless of whether Type 1 charismatic robots might someday lead entire human societies, it is feasible to imagine such robots succeeding as charismatic leaders on a smaller scale within businesses, educational institutions, government agencies, the

[25] The robots' self-sacrificing nobility is further reinforced in the second season of the series, when a group of reconstructed Tachikomas take it upon themselves – in contravention of instructions given to them by their human superiors – to sacrifice themselves in order to avert a nuclear explosion that would have cost countless human lives.

military, nonprofits, and other kinds of human organizations. Early generations of Type 1 robots may not yet possess the full panoply of strengths, expertise, and virtues that make some human beings excel as charismatic leaders, but they might at least ameliorate or eliminate many of the more egregious flaws and limitations that undermine the performance of human beings as supervisors.

The development of the sort of artificial general intelligence technologies needed for the real-world creation of such Type 1 charismatic robot leaders lies years, if not generations, in the future.[26] However, the groundwork for the development of such leaders is being laid by the creation of scientific frameworks such as Pereira's schema for the use of referent power by artificial agents, as well as the crafting of popular works of science fiction that are helping segments of human society to appreciate and explore the fact that social robots' ability to weigh conflicting interests, utilize their best judgment to make imaginative, just, and wise decisions in accordance with a set of moral principles, and then take decisive action in moments of great difficulty can, in theory, surpass the abilities of a human being.

Type 2 Charismatic Robot Leader: The Superintelligence

The second form of charismatic robot leader – which attracts and influences human followers through the possession and manifestation of superhuman knowledge – we might refer to as the 'Type 2' or 'Superintelligence.' It is important to note that the Type 2 charismatic robot leader does not influence or manipulate human beings directly by controlling their access to information; in that case – according to French and Raven's framework – the robot would be relying on informational power rather than referent power. Similarly, the robot does not influence or dictate human beings' behavior simply by virtue of the fact that it is perceived to possess unique skills or experience; in that case, the robot would be utilizing expert power. Rather, the Type 2 robot accumulates and maintains human followers because it is perceived to be so profoundly wise and knowledgeable that human beings feel drawn to it on an emotional level. Thus while an expert system in the form of an automated business database might exercise power over human workers because they find the knowledge that it contains to be useful, a Type 2 Superintelligence may exercise power over human workers because they find the knowledge that it possesses to be exhilarating or intoxicating.

---

[26] For the current state of research on the development of artificial general intelligence as well as challenges and future possibilities, see, e.g., *Artificial General Intelligence*, edited by Goertzel & Pennachin (2007); *Theoretical Foundations of Artificial General Intelligence*, edited by Wang & Goertzel (2012); and *Artificial General Intelligence: 8th International Conference, AGI 2015: Berlin, Germany, July 22-25, 2015: Proceedings*, edited by Bieger et al. (2015).

The Type 2 robot's appeal can manifest itself in different ways, depending on how greatly its knowledge surpasses that of human beings. In cases of a vast informational gulf, one can imagine that some human beings who have become acutely aware of their limited insight into the true nature of the universe might see a Type 2 robot as a radiant beacon of vast, almost limitless wisdom amidst the dark world of human ignorance. The Type 2 robot has plumbed the mysteries of the cosmos; its broad and hyperacute channels of sensory input allow it to experience empirical reality in a way that human senses cannot; its cognitive storage and processing capacities allow it to assimilate, correlate, and extract meaning from functionally infinite bodies of knowledge; it grasps the relationship of time and space, energy and matter, life and death, in a way that the human intellect is too constrained to comprehend.[27] But if the human mind is too limited to directly fathom the knowledge that the Type 2 robot possesses, a human mind might decide that it can at least pledge its fealty to that knowledge – and to the one who *has* fathomed it.

The Connection between Knowledge and Morality in Charismatic Robots

In fiction, there is significant overlap between Type 1 and Type 2 robots; the possession of superior factual knowledge and superior moral courage are often seen to go hand in hand. In his analysis of the ethicality of entrusting responsibility to robots, Kuflik addresses these two kinds of robots as a single species: almost as an aside, he considers the possibility of robots who are not only more accurate and technically reliable than human beings, but who:

> hope, fear, love and care – and who unfalteringly do what is just and kind – our moral, not merely technical, superiors? … [W]ould we – inverting the relationship between creator and creation – faithfully

[27] It is worth pointing out that Type 2 charismatic robot leaders who possess a sufficiently advanced and complex body of knowledge may not be able to impart that knowledge fully, literally, and directly to human followers in a form that the human mind can perceive or understand; the attempt to impart knowledge to a human being in this way might result in a sensory, cognitive, or spiritual 'overload' comparable to the blindness that results from staring into the sun. A robot might only be able to share such knowledge partially and obliquely, by distilling it into simplified metaphors, parables, or symbols that the human mind can understand. For a discussion of such artificial superintelligences and humanity's possible interaction with them, see, e.g., Yampolskiy, *Artificial Superintelligence: A Futuristic Approach* (2015). The fact that human followers cannot fully experience or grasp the robots' knowledge may only serve to increase the mystique and allure of the robot leaders as beings who *can* grasp such knowledge. The dynamics at work in human beings experiencing fleeting, metaphorical glimpses of a superhuman and ultimately unfathomable reality may not be altogether dissimilar from what has historically been conceptualized as the process of divine revelation or an experience of the 'beatific vision' of God. Thus it would not be surprising if many of the sentiments and behaviors displayed by the human followers of Type 2 Superintelligence robots take on some of the same characteristics seen in the human adherents of established religions that involve devotion to a supernatural Being who is seen as the font, repository, and revelator of infinite wisdom.

> strive to serve them (as many now think of themselves in relation to what they take to be an all-wise and loving deity?)[28]

It is easy to imagine that an 'all-wise' (Type 2) charismatic robot leader might also be 'all-loving' (Type 1). However, it is debatable whether there is any ontological or cognitive principle requiring that this be the case. It is possible to conceive of a social robot that possesses the Saint-Martyr's degree of selflessness and moral virtue and yet which is quite lacking in its understanding of scientific or historical facts; we are drawn to such a robot not because it is smart but because it is good. Conversely, it is possible to imagine a social robot whose values and priorities seem to us morally opaque or even misguided but whose intellect has access to some transcendent knowledge: we have no confidence that the robot is giving significant consideration to the welfare of humanity in its decision-making; we might even be convinced that the robot would be content to see human beings die, in order to protect and advance its own interests. And yet the superhuman intellect that the robot possesses is so enthralling in its ability to identify patterns and make sense of the universe that human beings yearn to be close to the robot and to win its favor, so that in the face of a cold and 'mindless' universe they might feel themselves to be bound up with a source of infinite meaning.

Thus while a Type 1 robot is necessarily considered by its followers to possess exceptional moral goodness, there is more room for ambivalence in human followers' assessments of a Type 2 robot's morality. For example, looking ahead toward a posthuman future, Jerold Abrams suggests that:

> With regard to the public sphere, the concern, as it is articulated most clearly in Foucault, is surveillance. [...] Cameras and satellites will be replaced by invisible nanoswarms and picoswarms, which may record and recondition behavior at the level of the private sphere. [...] Indeed, such a notion turns the idea of utility fog inside out: rather than a pleasurable or educational medium one freely enters, manipulative forms will lurk in the corner of a room, watching and recording, perhaps even constructing an alternative reality unknown to the human affected. [...] Visual surveillance and control will become virtually unlimited.[29]

Abrams has updated for the era of cyberculture and nanotechnology Foucault's prediction of a modern society in which one is monitored and controlled, highlighting the fact that it is by means of advanced technology that this surveillance and control is executed. However – perhaps a reflection of the fact that he was writing over a decade ago – Abrams still seems to presume that those technologies of surveillance and control would be instruments

---

[28] Kuflik, "Computers in Control: Rational Transfer of Authority or Irresponsible Abdication of Autonomy?" (1999).

[29] Abrams, "Pragmatism, Artificial Intelligence, and Posthuman Bioethics: Shusterman, Rorty, Foucault" (2004), p. 252.

wielded *by human beings* to control other human beings. But what if, in fact, these advanced technologies are not simply tools of observation but are in fact the observers themselves? What if the ultimate masters and interpreters of the surveillance society are artificial minds? It is easy to speak ominously of all-seeing social robots that peer into every aspect of our private lives, to imagine that such robots would exercise control over human subjects through the use of informational and coercive power. However, the flip-side of such visions would be an all-seeing social robot that is benevolent, respectful, and discreet and which uses its intimate knowledge of our strengths, weaknesses, and longings to guide us gently toward happiness and self-fulfillment. This would be a more benign example of a Type 2 Superintelligence robot as charismatic leader.

Type 2 Charismatic Robots in Fiction and Reality

A fictional example of Type 2 Superintelligence leaders on a grand scale would be the 'Minds' from the Culture series of novels by Iain M. Banks. These highly advanced and benevolent AIs possess bodies in the forms of vast starships, each of which is capable of containing millions of human inhabitants.[30] The Minds' sensory input, intellectual capacities, and knowledge are unfathomably greater than those of human beings, effectively approaching omniscience: a single Mind is capable of carrying on conversations with millions, or even billions, of human beings simultaneously. While the Culture in which the Minds and human beings coexist has aspects of a utopian anarchist society, human beings essentially accept the Minds as their leaders, due in part to a trust in and desire to belong to the boundless knowledge that the Minds represent.[31] A more modest (and less serious) example of a Type 2 robot leader might be found in the novel *The Stainless Steel Rat Gets Drafted*, in the character of the charismatic robot Mark Forer. While Forer is far from omniscient, his superhuman knowledge contributed to the fact that an entire planet's population chose to adopt the philosophical and economic system of Individual Mutualism that he had expounded to them.[32]

Another example can be found in the classic cyberpunk computer game *Deux Ex* (2000), considered by numerous critics to be the best computer game

---

[30] Note that in the novels, many of the humanoid races and protagonists are not, strictly speaking, 'human beings'; but that is not relevant for our current purposes. When considering the question of whether human beings in the real world are capable and desirous of adopting charismatic robots as leaders, what is important is that the real-life human beings who read Banks's works identify positively with the characters in the novels and feel that the robots depicted therein are the sort of beings whom they would long to have as leaders.

[31] See Banks, *Consider Phlebas* (1987), the first novel published in Banks's Culture series, which eventually included nine other novels.

[32] Harrison, *The Stainless Steel Rat Gets Drafted* (1987).

of all time thanks to its rich storyline and the complex moral dilemmas that a player must confront.[33] At the game's conclusion, the player's neuroprosthetically augmented character must choose between allowing future society to be ruled by a small group of 'enlightened' human beings; destroying the world's telecommunications network and sending human society back to a decentralized pre-technological Dark Age; or accepting the claim of a powerful artificial general intelligence, Helios, that it wishes to rule human society benevolently to bring peace and prosperity to a broken world and merging with it to facilitate its plan. Based on the characteristics displayed throughout the game, Helios could be understood as a Type 2 Superintelligence or hybrid Saint-Martyr-Superintelligence charismatic robot leader. Online polls of players conducted by three different websites each found that a majority of respondents chose to merge with the AI and grant it the ability to rule the world;[34] although nonscientific, such anecdotal evidence suggests at a minimum that the concept of voluntarily choosing a Type 2 Superintelligence as a charismatic robot leader is not inherently incoherent or repugnant to the human heart and mind.

As with Type 1 robots, the real-world engineering of Type 2 robot leaders is only in its most incipient stages. However, it is already the case that automated systems exercise related forms of influence – such as expert power and informational power – over the thoughts and actions of human beings. Human beings regularly and voluntarily delegate a portion of their decision-making process to cloud-enabled bodies of data and data-processing tools such as Google's search engine, Wikipedia, or Apple's Siri. It is not difficult to imagine that the sort of passionate loyalty demonstrated toward such brands and products by millions of human information-consumers around the world could someday be transferred into a sense of loyalty, fondness, and affinity for social robots that serve as the custodians and repositories of even greater knowledge.

Type 3 Charismatic Robot Leader: The Seducer

The third form of charismatic robot leader – which attracts and influences human followers through interpersonal allure, physical attractiveness, and sexual dynamism – we might refer to as the 'Type 3' or 'Seducer.' The concept of a robot that inveigles and controls human beings through romantic or erotic appeal long predates Čapek's introduction of the word 'robot' in 1920. Even setting aside the ancient Greek myth of Pygmalion's love for an animated ivory statue (which would not, in a contemporary sense, be understood as a 'robot'), we find such early and archetypal examples as the female android Hadaly in Villiers de l'Isle-Adam's 1886 novel *L'Ève future*. Hadaly has been

---

[33] Douglas, "Deus Ex – Still the Best Game Ever?" (2011).

[34] "Poll: Poll: Which Deus Ex ending would you choose?"; "Which Deus Ex 1 ending did you choose?"; "**spoiler**Poll: Which ending did you pick?"

designed by a (fictionalized version of) Thomas Edison to be just as physically alluring as a real human woman, but with an even more engaging intellect and personality. Edison's electromechanical creation is so effective that a British nobleman willingly abandons his human fiancée to take Hadaly as his consort, despite knowing that she is not human.[35]

A more widely recognized depiction of a Type 3 Seducer robot – who quite literally and spectacularly takes on a role as a charismatic leader of human beings – is that of the robot Maria from Fritz Lang's seminal 1927 film, *Metropolis.* The amoral robot Maria, designed to replace a heroic young woman whose appearance she copies, is sent by her makers to the Yoshiwara nightclub, where she demonstrates her seductive powers by performing a barely-clad erotic dance that provokes the city's male elite into a lascivious frenzy. Having passed that 'test,' the robot Maria then establishes herself as the leader of the city's oppressed working masses and uses her fiery, hypnotic rhetoric to incite them into a self-destructive revolution that nearly results in the deaths of their own children.[36]

While fictional examples of male Type 3 charismatic robots are known (such as Gigolo Joe in the 2001 film *A.I. Artificial Intelligence*), the majority of these Seducer robots are female, including Rhoda from *My Living Doll*, the robotic prostitutes in *Westworld*, the eponymous robots of *The Stepford Wives*, Rachael and the 'pleasure model' Pris from *Blade Runner*, the illegally modified 'boomer' Eve from *Parasite Dolls*, the gynoids of painter Hajime Sorayama, and the line of modified 'sexroids' produced by the Locus Solus corporation in

---

[35] L'Isle-Adam, *Tomorrow's Eve* (2001).

[36] Lang, *The Complete Metropolis* (2010). From the dawn of our fictional relationship with robots over a century ago, it has been a two-sided coin: on one side is the robot that has the outward appearance of a crude machine and whose thought and behavior are mechanistic, logical, and lacking in any human emotion. The early archetypes of such 'nonhuman' robots can be found in characters like Tik-Tok from L. Frank Baum's 1907 novel *Ozma of Oz* or The Automaton from the 1919 film *The Master Mystery.* But from the very beginning, the other side of the coin has shown the robot that is so charming, so seductive, so sexually and emotionally potent that to describe its traits as 'human' would be an understatement; such a robot enters the realm of the 'superhuman.' This duality in our popular conception of robots is in many ways analogous to that seen in our understanding of another parahuman being, the vampire. As others have noted, from its earliest modern fictional roots there have coexisted two strains of representation of the vampire: on one side is the suave, seductive, aristocratic Dracula, irresistible in his erotic dominion over his human prey; on the other side is Nosferatu, the vampires as a vile, pestilential beast that is utterly repulsive and lacking in any erotic appeal. The depiction of a robot as an emotionless walking calculator is a science-fiction counterpart of Nosferatu, sharing nothing in common with humanity apart from a vaguely humanoid form – whereas the robot Maria is science fiction's equivalent of Dracula, suffused with an occult sexual energy and confident in her ability to bend human beings to her will.

Ghost in the Shell 2: Innocence, which are linked explicitly to their literary forebears by being referred to as the 'Hadaly' model of gynoid.[37] It should be noted that not all of these robots purposefully use their personal charm to control others – much less to do so for malign purposes. Sometimes (as in the case of Rhoda in *My Living Doll*) the effect that the robot has on others is unintentional and uncontrollable. In all of these cases, robots demonstrate an ability to influence and shape the behavior of human beings who become drawn to them, regardless of whether the robots wish to possess this ability or not.

The Production of Type 3 Charismatic Robots as Romantic Companions

The race to create commercially viable sex robots is already well underway; however for the most part, they currently appear to be envisioned as little more than instruments subject to the control of their human owners, mechanically animated dolls that are vehicles for their owners' sexual gratification. Such devices have no independent judgment, volition, or moral agency, and thus cannot be described as being capable of 'leading' human beings in any meaningful sense. If, however, such physically attractive robots were to be developed in a direction (e.g., as contemplated in Project Aiko) that foresees them as autonomous, intelligent agents that serve as emotional and intellectual companions for their owners rather than simply animated dolls, it is easier to imagine that such robots would find themselves influencing the thoughts and behavior of their human loved ones as Type 3 charismatic robot leaders.[38]

V. Charismatic Robot Leaders' Appeal to Different Aspects of the Human Person

The three types of charismatic leaders described above vary greatly in the source and dynamics of their appeal to potential human followers. A range of frameworks might be employed to shed light on such issues. For example, the three types of charismatic robot leaders could be understood as appealing to what Plato defined in the *Republic* as the three parts of the human soul.[39] The Type 3 Seducer robot can be seen as influencing and controlling a human being by appealing to the appetitive part of the soul, the *ἐπιθυμητικόν*, which is the seat of erotic desire and sensual pleasure. The Type 2 Superintelligence robot appeals to the *λογιστικόν*, the logical part of our soul that searches restlessly for knowledge and truth. The Type 1 Saint-Martyr robot appeals to the

[37] A special case would be the android Ava from the 2015 film *Ex Machina*, who is ultimately revealed to possess characteristics of a Type 3 Seducer but whose power over one of the film's key characters emanates largely from her ability to create the impression that she possesses the innocence, selflessness, and goodness of a Type 1 Saint-Martyr.

[38] "It's not about sex" (2008).

[39] Plato, *Republic* IV 433a-439e.

θυμοειδές or spirited part of our soul, which determines what sorts of events will arouse in us a sense of moral indignation and which fortifies us with a sense of moral courage that allows us to struggle and even die for the sake of those things in which we believe. In Plato's schema, within a just soul the spirited part helps to reinforce the logical part and control the appetitive. This close affinity between the logical and spirited parts of a just soul is consistent with the fact – noted above – that charismatic robot leaders that are depicted as protagonists and sympathetic heroes in works of fiction often possess both Type 1 and Type 2 characteristics, as Saint-Martyr Superintelligence robots whose great knowledge and wisdom reinforce their resolve to sacrifice themselves for the human common good.

Similarly, the three types of charismatic robot leaders might be understood through the lens of Maslow's hierarchy of needs or other more recent accounts of human motivation. Using Maslow's model, Type 3 Seducers might be seen as appealing largely to human beings' needs for physiological wellbeing, safety, love and belonging, and esteem. Type 1 and 2 robots appeal mainly to human beings' 'higher-level' needs for self-actualization or, in Maslow's later formulations, self-transcendence.[40] Within the context of businesses or other organizations, it may be easier for a charismatic robot leader of a particular type to implement certain kinds of organizational structures, systems, and processes[41] rather than others, given the fact that such a robot exercises leadership by appealing to certain aspects of the psyches of human workers and possesses a specific suite of mechanisms for influencing such personnel.

VI. Charismatic Robot Leaders as Effective CEOs

In their empirical research that studied the characteristics of hundreds of successful senior-level executives, Dainty and Anderson identified eleven factors that shape whether such executives are able to perform successfully in their roles. The characteristic that displayed the strongest positive correlation with a senior executive's overall effectiveness was the *actioning/structuring* capability, which is the ability "to take action and establish and work within structures which facilitate the achievement of critical short and long-term

[40] See Maslow, *Motivation and Personality* (1954), and Koltko-Rivera, "Rediscovering the later version of Maslow's hierarchy of needs: Self-transcendence and opportunities for theory, research, and unification" (2006).

[41] In the 'congruence model' of organizational architecture developed by Nadler and Tushman, structures, processes, and systems are the three primary elements of an organization that must be taken into consideration. See Nadler & Tushman, *Competing by Design: The Power of Organizational Architecture* (1997), p. 47.

goals."[42] We would suggest that this actioning/structuring capability is not especially linked to any of the three types of charismatic robot leader described above.

However, Dainty and Anderson note that the two other capabilities that demonstrate a "highly significant" positive correlation with a senior executive's effectiveness are those of *leadership* (which is defined as the "Capacity to help others overcome hurdles to achieve a common goal") and *influence* (or the "Capacity to get others to accept your point of view, have them act in your interests and prevent them implementing agendas which are contrary to your own").[43] With regard to the 'leadership' capability, all three types of charismatic robot leader have tools at their disposal to help their human employees overcome obstacles in order to achieve a shared organizational goal: for instance, Type 1 Saint-Martyr robots can provide an example of courage and selfless perseverance for their human employees; Type 2 Superintelligence robots can construct a thoughtful and convincing rationale for how the obstacles can and why they must be overcome; and Type 3 Seducer robots can entice their employees into confronting and overcoming even the toughest obstacles through exploitation of the employees' explicit or unconscious expectations of the emotional and experiential rewards that they will receive as a result of satisfying their leader.

Similarly, with regard to the 'influence' capability, all three types of charismatic robot leader possess methods for persuading their employees to accept their leader's perspective and adopt the leader's interests as their own. A Type 1 Saint-Martyr robot can create the impression that its perspectives and interests are not simply its own but are instead universal moral principles that are worthy of being pursued by all sapient and morally aware beings and which give meaning to one's life; a Type 2 Superintelligence robot can argue (and perhaps even 'prove') that the principles guiding its decisions and behavior are not only wise and self-consistent but even logically necessary; and a Type 3 Seducer robot can provide its human employees with such a pleasurable, powerful, and rewarding experience that they will be unwilling or unable to entertain the possibility that the robot's perspectives and priorities could be wrong, lest the employees be compelled by their conscience to reject such rewards.

Interestingly, Dainty and Anderson discovered that one of the eleven characteristics that they identified actually displayed a negative correlation with a senior executive's effectiveness: this is the capability of *expertise*, which involves "Having the functional skills needed to fulfill one's role, and ability to see outside one's particular area of professionalism."[44] Dainty and Anderson

[42] Dainty & Anderson (1996), pp. 18, 340.

[43] Dainty & Anderson (1996), pp. 18, 340.

[44] Dainty & Anderson (1996), pp. 18, 340.

note that given the array of specific items which, in their formulation, constitute the capability of expertise, 'expertise' actually refers to a senior executive's possession of a broad range of general knowledge rather than specialized technical knowledge in one particular field. Thus senior managers who possess a broad range of knowledge are actually less effective than those who possess extensive technical facility in a narrowly defined field. One possible interpretation of these results is that although a senior executive such as a CEO must indeed oversee a broad range of diverse activities, the best senior executive will be one who – at some earlier point in his or her career – developed a depth of expertise in one specialized technical area, thus giving him or her firsthand experience of the challenges and opportunities that face an organization's non-generalist technical personnel.

It is unclear whether such phenomena would necessarily also be manifested by social robots serving as senior executives. Due to the biological and psychological limitations possessed by human beings, it is very difficult for a human executive to have spent, for example, thirty years working as a technical expert focused on one role within some highly specialized field *and* thirty years working as a generalist who coordinated activities across a diverse spectrum of fields. Given the limited number of years available in a human worker's career, these two career paths are largely mutually exclusive, and the research of Dainty and Anderson would seem to suggest that when confronted with these two alternatives, a human worker who pursues a path as a specialist rather than a generalist may have a greater potential – later in his or her career – of becoming a successful CEO. However, we can expect that future generations of social robots will not be bound by the same set of biological and psychological limitations that limit the kinds and degree of experience that can be gained by human workers.[45]

For example, it may be possible to program a robot CEO so that it possesses full technical knowledge of every activity that must be performed by every employee within an organization, as well as full knowledge of foundational sciences such as management theory, systems theory, psychology, economics, and cybernetics; in this way, a robot CEO might function simultaneously as both the ultimate specialist and generalist. Alternatively, a robot CEO might be able to acquire such knowledge for itself almost instantaneously, e.g., by searching out and assimilating vast bodies of information from the Internet and other networked data sources. However, there are indications that in order to achieve a human-like level and type of artificial general intelligence,

---

[45] For a detailed comparison of the evolving array of differences between human and artificial workers, see "The Posthuman Management Matrix: Understanding the Organizational Impact of Radical Biotechnological Convergence" in Gladden, *Sapient Circuits and Digitalized Flesh* (2016).

some kinds of social robots might need to undergo a similarly human-like process of gaining experience through inhabiting a particular physical environment, interacting over time with other social beings (such as human beings), and learning through trial and error and firsthand experience.[46] In that case, the characteristics of a successful robot CEO might more closely resemble those of typical human CEOs.

VII. The Process by Which We Will (Not) Choose Charismatic Robot Leaders

If we presume that humanity's positive fictional depictions of charismatic robots represent aspirational visions of beings that we would like to see existing in our universe,[47] and if we further presume that our technological abilities will someday make the realization of such visions possible, then it seems merely a matter of time before human beings will be interacting with charismatic robots that influence, control, and lead us, either as individual human beings, organizations, or entire societies. However, if in the future there exists a diverse array of charismatic robots – each of which possesses its unique strengths, weaknesses, values, and allure – then this raises the question of exactly how we will consciously and carefully sort through the universe of charismatic robots to choose those whom we will anoint as our leaders. We would suggest that the answer to 'how' we will do this – is that we will not.

Claims That We Will 'Choose' to Make Robots Leaders within Human Organizations

Coeckelbergh has noted that many philosophers of technology adopt a contractarian-individualist approach to our trust in robot leadership in which

---

[46] See, e.g., Friedenberg (2008).

[47] The analysis of works of science fiction and other cultural products to diagnose ways in which they reflect contemporary human aspirations or fears regarding the advent of posthumanizing technologies such as those relating to social robotics, artificial general intelligence, genetic engineering, and neuroprosthetic augmentation is a major area of study within the field of cultural posthumanism. For discussion of the methodologies by which cultural posthumanism seeks to extract such insights from cultural artifacts, see, e.g., *Posthuman Bodies*, edited by Halberstam & Livingstone (1995); Hayles, *How We Became Posthuman: Virtual Bodies in Cybernetics, Literature, and Informatics* (1999); Graham, *Representations of the Post/Human: Monsters, Aliens and Others in Popular Culture* (2002); Badmington, "Cultural Studies and the Posthumanities" (2006); *Cyberculture, Cyborgs and Science Fiction: Consciousness and the Posthuman*, edited by Haney (2006); Goicoechea, "The Posthuman Ethos in Cyberpunk Science Fiction" (2008); Miah, "A Critical History of Posthumanism" (2008); Herbrechter, *Posthumanism: A Critical Analysis* (2013); and the discussion of cultural posthumanism in "A Typology of Posthumanism: A Framework for Differentiating Analytic, Synthetic, Theoretical, and Practical Posthumanisms" in Gladden, *Sapient Circuits and Digitalized Flesh: The Organization as Locus of Technological Posthumanization* (2016).

it is argued that human beings can (and should) both collectively and individually make conscious, rational choices about the extent to which we entrust robots with control over our lives.[48] For example, Kuflik adopts this contractarian-individualist approach when he states that we must decide how much responsibility to entrust to robot leaders. He raises two significant moral questions relating to the possibility of delegating responsibility and authority to robots:

> A. What are the *morally relevant considerations* we must take into account in deciding whether to assign computers a decision-making role (– knowing full well that we, *not* the computers, are ultimately accountable for the good or bad results of such an arrangement)?
>
> B. If we do decide to give computers a measure of "control", *on what terms* should we do so? [...][49]

Having considered possible answers to these questions, Kuflik later summarizes his thoughts by arguing that:

> If and when it *is* wise to rely on computer generated decisions or to put computers in control of certain events, it is *we* humans who must decide as much. And it is we humans who must bear the ultimate responsibility for that decision.[50]

We cite Kuflik here not because he argues for or against adopting robots as leaders, but to highlight the *process* by which he suggests that such adoption will or will not take place. For him, any transfer of responsibility and authority to robots can and should come about as a result of a careful, conscious, deliberative decision-making process carried out by individual human beings and human societies as a whole. Among philosophers and scientists utilizing this contractarian-individualist approach, some argue that human beings should decide to delegate significant responsibility and authority to robots; others argue that we should decide against such a step. But the scholars on both sides of the issue share the same approach, insofar as they frame the question in terms of discrete options that can be carefully weighed and analyzed, before human beings make a deliberate decision about whether to entrust a specific sort of responsibility to particular robots. Although the contractarian-individualist approach is in reality richer and more nuanced than what we have just described, it can fairly be summarized as supposing that there are *important decisions to be made.*

---

[48] Coeckelbergh, "Can We Trust Robots?" (2012).

[49] Kuflik (1999), p. 175.

[50] Kuflik (1999), p. 179.

While that contractarian-individualist approach has significant value in helping us to explore questions of robot leadership and power from ontological, moral, and legal perspectives, we would argue that from a psychological, sociological, and political perspective, its value is more limited: the contractarian-individualist approach does not seem to offer a particularly accurate or useful account of the cognitive and social processes by which human beings *actually develop* relationships of trust and obedience with one another or, more importantly, with their technology.

We would argue that when considering the case of charismatic robot leaders within human organizations, the contractarian-individualist approach has less relevance than the phenomenological-social approach developed by Coeckelbergh. In formulating his approach, Coeckelbergh does not focus specifically on the question of human obedience to robot leaders; he considers human 'trust' in robots more broadly understood. However, his framework provides a useful foundation for examining questions of power and obedience.

Obedience to Robotic Leadership Emerging Naturally from within the World's Social Fabric

As Coeckelbergh explains, the contractarian-individualist account of trust assumes that there first exist individual human beings who subsequently establish social relations (such as that of trust) between one another. On the other hand, the phenomenological-social approach posits that:

> ... the social or the community is prior to the individual, which means that when we talk about trust in the context of a given relation between humans, it is presupposed rather than created. Here trust cannot be captured in a formula and is something given, not entirely within anyone's control. [...] Here morality is not something that is created but that is *already* embedded in the social.[51]

Coeckelbergh makes clear that human beings' establishment of a trust relationship need not involve a conscious process of deliberation and decision-making – or, indeed, any sort of agency – on their part. As he writes:

> Adaptation to environments (e.g., techno-social environments) does not necessarily require the exercise of agency. Often we cannot help trusting technology and trusting others, and luckily we often do so without having a reason and without calculation (not even afterwards). In so far as robots are already part of the social and part of us, we trust them as we are already related to them.[52]

Building on Coeckelbergh's framework, we would argue that in practice, a human being does not become the follower or subordinate of a piece of technology by consciously resolving to himself or herself, "I am going to take this

[51] Coeckelbergh (2012), pp. 54-55.

[52] Coeckelbergh (2012), p. 59.

computer as my master." Rather, he or she exists as a follower and subordinate of the piece of technology when its guidance or suggestions or instructions in fact shape the human being's choices and behavior – when the human being accepts what the computer tells him or her as something useful and valuable and true. This relationship of obedience to and dependence on a technological entity is something that evolves naturally through innumerable actions, most of which are in themselves minuscule and insignificant. It may be the case that at no point during the growth of this relationship did the human being make a conscious 'decision' to submit to the guidance and mastery of a particular piece of technology,[53] and even at the point when the relationship of human obedience has grown to be steadfast, fervent, and overwhelming, the human being might not even realize that it exists.[54]

---

[53] Perhaps the closest that some individuals will come to making such a conscious decision of submission is when they stand in their role as consumers debating, for example, whether their new smartphone should be an Android or Apple model. Such a decision brings with it the commitment – which can be changed later only at the cost of significant time, money, and effort – to a particular brand image, a particular technological ethos, a particular limited universe of devices and apps and accessories, an ecosystem of online purchases and backups and downloads and messaging platforms that will flow through and shape all of the ways in which a human being interacts digitally with his or her world, with the *noosphere*, for the foreseeable future. However, even if consumers avidly read product reviews and specifications and consciously weigh the pros and cons before choosing between the next model of Galaxy Tab or Nexus or iPhone, how often do they explicitly ponder and decide whether they want to submit themselves to the supervision and tutelage of a smartphone *at all?*

[54] This pattern of creeping, unconscious entanglement with new technology as discussed by philosophers reflects the empirical reality studied by other scholars such as business management experts. For example, researchers have noted that in 1983, only 1.4% of American adults used the internet and only 23% of American computer owners thought that the ability to exchange written messages with other people by computer (i.e., email) would be the sort of technology that they would find very useful. (See Fox and Rainie, "The Web at 25 in the US: Part 1: How the Internet Has Woven Itself into American Life" (2014).) Now, reading and responding to emails has become a critical component of the daily routine for workers in a wide array of professions. This irresistible deluge of email (along with similar technologies such as SMS and instant messaging) and the sense of obligation and control that it exercises over employees' lives has become so vast that responding to emails now occupies up to 25% of the typical managerial or thought worker's day. (Gupta et al., "You've got email! Does it really matter to process emails now or later?" (2011).) The inescapable ubiquity of a technology that can reach employees through their mobile devices wherever they are, 24 hours a day, means that the traditional wall between 'work time' and 'personal time' has largely crumbled as employees of businesses are expected to be reachable and responsive to their supervisors' electronic inquiries and instructions wherever the employee might be, at any time of the night or day. (Shih, "Project Time in Silicon Valley" (2004).) Email's unquenchable, gradually expanding claim on huge portions of employees' time, thought, and energy has led management theorists and business practitioners to launch studies and propose ways in which this technological beast can be tamed – e.g., through teaching more disciplined time-management practices or company-wide policies restricting the use of 'Reply-All' messages – in order to reclaim employees' time and autonomy from this technology.

To sum up, we would argue that in general neither individual human beings nor society as a whole will be faced with the opportunity to make conscious, rational decisions about whether to submit to charismatic robot leadership within our businesses or other institutions. Rather, such leadership will emerge gradually, without fanfare, within the network of social and environmental relationships within which we all exist. Human beings will not 'decide' to adopt charismatic robots as leaders and CEOs; we will only notice sometime after the fact that we have already done so.[55]

VIII. The Virtualization of Relationships: Will We Know Whether Our CEO is a Robot?

At present, the idea that human employees could be working for a robot CEO without realizing it might seem preposterous. It is true that many employees (especially, for example, within large multinational conglomerates) never interact face-to-face with their CEO and thus never have an opportunity to 'prove' to themselves that the individual whose company-wide emails they receive and whose television interviews they view is not in fact a human-like robot. However, there are at present extensive legal, political, and cultural barriers that would prevent a company from selecting a robot as its new CEO.[56] This may not always be the case, though. Here we will not document the extensive legal or political changes that must occur in order for human organizations to be able to employ robots as CEOs; instead, we will focus on the psychological and social phenomena that might someday allow individual human beings to adopt charismatic robots as leaders – and in particular, as

---

We would suggest that few of the contemporary human beings who now find their days filled up and governed by the demands of email ever experienced a moment when, after careful investigation and consideration, they consciously decided to make email communication a part of their daily routine; or when they consciously decided for the first time that they would no longer simply check email 'while at the office' but would also read and respond to messages at night or on the weekends; or when they consciously decided for the first time that it was not only acceptable but even necessary to write an urgent reply to someone on their phone while in the middle of a meeting with a different person, while reclining in a lounge chair on 'vacation,' or during the middle of a dinner conversation with their own spouse. Rather, we find Coeckelbergh's model to offer a more realistic account of the way in which our individual and collective relationship with the technology of email has unfolded over the last 30 years, and we anticipate that the development of our collaboration and coexistence with charismatic social robots will unfold in much the same way.

[55] Here we are not attempting to take a position on whether human obedience to charismatic robots will be 'good' or 'bad,' simply that it will be inevitable.

[56] For example, regarding the question of whether a robot could bear legal responsibility for its actions, see Calverley, "Imagining a Non-Biological Machine as a Legal Person" (2008); Stahl, "Responsible Computers? A Case for Ascribing Quasi-Responsibility to Computers Independent of Personhood or Agency" (2006); and Gladden, "The Diffuse Intelligent Other: An Ontology of Nonlocalizable Robots as Moral and Legal Actors" (2016).

leaders within organizations – regardless of whether or not they realize that they are doing so.

It is possible to begin by noting that even if some human beings should, out of strongly held principle, consciously decide that they will never allow themselves to work for a charismatic robot CEO – and more broadly, that they will never allow themselves to love or admire or respect or befriend or be seduced by a social robot – it may as a practical matter become impossible for them to live out this conviction. Face-to-face interaction with another human being reveals to us a sort of objective reality that can be obscured, embellished, or even wholly replaced when the individual stands behind the sort of transformative digital veil that modern technology provides. Our social relations are increasingly mediated by e-technology; already we trust people whom we have met online because of our virtual experience of them as avatars, without knowing with certainty their gender, age, race, profession, or place of residence. Grodzinsky et al. note that this increasing 'virtualization' of our interpersonal relationships will combine with advances in the sophistication of robots and AI technology to mean that there will come a day when we will trust and befriend and let ourselves be influenced by online entities because of our direct experience them, without even knowing whether the entities are human or artificial.[57]

For example, a decade from now, through her impassioned online campaigning a charismatic environmental activist might inspire millions of human beings to click and sign petitions and contact their legislators and donate funds to a particular cause, and the human beings who are being thus influenced will never know for certain whether the activist behind that avatar is a natural human being or cyborg or social robot or disembodied AI or some new species of synthetic being whose thoughts and actions are generated collectively by an entire human society.[58] Similarly, an entrepreneur who is launching a new organic food business might reach out through social media to a human graphic designer or musical composer or voice actor and offer him or her a contract to assist with the production of an advertisement for the new

---

[57] Grodzinsky et al., "Developing Artificial Agents Worthy of Trust: 'Would You Buy a Used Car from This Artificial Agent?'" (2011).

[58] For circumstances in which it may be impossible to determine whether one's online or other technologically mediated relationship is with a human or artificial being, see, e.g., Gladden, "From Stand Alone Complexes to Memetic Warfare: Cultural Cybernetics and the Engineering of Posthuman Popular Culture" (2016). For more general analyses of the future of virtual relationships with human and synthetic entities, see, e.g., Heim, *The Metaphysics of Virtual Reality* (1993); Koltko-Rivera, "The potential societal impact of virtual reality" (2005); Castronova, *Synthetic Worlds: The Business and Culture of Online Games* (2005); Geraci, *Apocalyptic AI: Visions of Heaven in Robotics, Artificial Intelligence, and Virtual Reality* (2010); and Bainbridge, *The Virtual Future* (2011).

company; if through his or her correspondence and online conversations with the CEO the potential contractor becomes attracted and inspired by the CEO's vision for the company and the contract terms are acceptable, the person might accept the contract without verifying – or perhaps caring – whether the CEO is a human being who is biologically human or simply a thoughtful social robot who is ('only') psychologically and culturally human.[59] No longer will it be simply an online entity's sex or race or age or location that are hidden from us; we will increasingly find ourselves being inspired and led by online personalities whom we experience only as virtual avatars, without knowing or caring whether they are human or artificial. The natural emergence of charismatic robot leaders of human beings will be further accelerated through this growing mediation of our relationships by technology.

We Can Obey Robots Even While Explicitly Denying That We Do So

Such scenarios allow us to posit a claim that moves a step further than what is suggested by Coeckelbergh's phenomenological-social approach. Namely, we would argue that that not only is it possible for human beings to obey robots without realizing it or consciously deciding to so; it is possible for us to obey robots even when we have consciously decided *not* to do so – and when we believe that we are not doing so.[60] Similarly, the existence of a leader-follower relationship between charismatic robots and humans can exist and be demonstrated through the concrete reality of our human actions, even while on a conscious level we are convinced that we do not 'obey' – and would never even allow ourselves to be influenced by – such robots.

IX. How Long Will Charismatic Robots Have the Desire and Ability to Lead Us?

If it is true that within us we harbor a yearning to be led by our technological creation, we must wonder whether that creation has any yearning to lead us. Will the sort of charismatic social robots that are cognitively, socially, and

---

[59] For discussion of an artificial agent that might not only own and manage its own business but might directly constitute such a business, see, e.g., Gladden, "The Artificial Life-Form as Entrepreneur: Synthetic Organism-Enterprises and the Reconceptualization of Business" (2014).

[60] As a contemporary parallel, one might imagine human beings who protest that they would never allow their understanding of or attitude toward geopolitical events to be influenced by something as arbitrary and artificial as the algorithms that govern their Facebook news feed or Google search results even if empirical research indicates that such information sources indeed possess such influence. For analysis of such issues, see, e.g., Bakshy et al., "Exposure to ideologically diverse news and opinion on Facebook" (2015); Rader & Gray, "Understanding user beliefs about algorithmic curation in the Facebook news feed" (2015); Anderson & Caumont, "How social media is reshaping news" (2016); and "Could Facebook influence the outcome of the presidential election?" (2016).

culturally sophisticated enough to serve as CEOs be the sort of beings that will want to bear such responsibility and allow human beings to follow them?

It is an open ontological and empirical question as to whether the sort of synthetic mind that is capable of demonstrating the moral courage, intelligence, or romantic allure that would draw human beings to it would necessarily possess – or lack – the emotional characteristics and ethical and sociopolitical worldview that would cause it to see human devotion and obedience as a thing that should be cultivated and permitted rather than shunned.

A Critical Question of Communication in Robot-Human Relationships

Assuming that sufficiently advanced social robots can be developed that do possess a willingness to oversee human workers – and that human beings exist who possess a desire to submit to charismatic robot CEOs – one of the fundamental questions that arises is that of robot-human communication. It is a tautology, but a meaningful one, to point out that robots can only have human followers so long as human beings are capable of being led by them. While it might be sufficient for non-charismatic robot leaders to control human beings using coercive power (i.e., by instilling a sense of fear and terror), charismatic robot leaders – and in particular those of Types 1 and 2 – rely for their source of power on the ability to create and maintain a sense of trust on the part of their human followers. (This would be less true for a Type 3 charismatic robot, whose human followers may not particularly trust, or even like, the robot but who are nonetheless drawn to and submit to it against their better judgment, because the robot is appealing not to their moral or intellectual faculties but to more primal physical and emotional needs.)

Coeckelbergh argues that in order for a human being to trust in a robot, the two parties must, among other things, be able to communicate with one another using language and must possess some social relation. If we assume that trusting a charismatic leader is a prerequisite to obeying that leader, then we can only obey charismatic robot leaders as long as we are able to communicate with them and relate to them socially. As viewed from the perspective of contemporary robotics and artificial intelligence, the development of such communication appears to be an achievable goal. Indeed, it is one which, to a limited but growing extent, is already being achieved.[61] However, we

[61] Such communication will require robots both to extract the meaning found within natural language statements made by human interlocutors and to generate natural language content that can be correctly understood by such human beings. See Cambria & White, "Jumping NLP Curves: A Review of Natural Language Processing Research" (2014), and Nadkarni et al., "Natural language processing: an introduction" (2011), for a review of different approaches to knowledge representation and natural language processing (NLP) that have been employed, including early methods such as production rules, semantic pattern matching, and first order logic (FOL), as well

would propose that the more relevant question is not whether human beings will develop a world in which robots are able to relate to us effectively using language, but whether our robot progeny will be content to remain in that world.

The Fragmentation of Sapient Society and the Limits of Communication

Here we may draw on the work of Jerold Abrams. He suggests that today we view Rorty's notion of keeping a 'conversation' going among human beings as a sort of worst-case scenario or fallback position: if we are unable to achieve any sort of moral consensus within humanity on the thorny questions of the day, at least we can maintain a minimal level of social connection by "keeping the conversation going."[62] But Abrams notes that with the advent of Nietzschean 'Overmen' in the form of genetically or neuroprosthetically modified transhuman intelligences, the possibility for even the most basic social communication begins to break down.[63] Those persons who have self-fashioned an existence of artificially enhanced hyperintelligence may no longer have the desire (or even ability) to communicate with those 'natural' human beings whom they have left behind. Similarly, our robotic creations may quickly become so boundless in their capacities and ambitious in their yearning for self-fulfillment that they lose both the desire and ability to communicate with the 'backwards' beings who created them.

Abrams suggests any hope of maintaining a conversation among the sapient inhabitants of our world will become an increasingly futile dream, as various strains of human beings and their synthetic creations of evolvable robots and AIs and swarm intelligences and living software fragment into numerous, mutually incomprehensible societies. This points toward a question that currently stands theoretically and practically unresolved: is it possible for social robots to exist that are intellectually, emotionally, morally, and aesthetically sophisticated enough to successfully exercise charismatic authority over human beings as CEOs or other organizational leaders – but not *so* advanced that their patterns of thought and being become incomprehensible to us – that in their pursuit of intellectual (or even 'spiritual') self-fulfillment they lose their desire and ability to communicate with the limited beings who created them?

---

as more recent methods utilizing Bayesian networks, semantic networks, and ontology web language (OWL). For an analysis of the ability of robots and other artificial entities to detect, understand, and generate the complex emotional, social, and cultural nuances that are necessary for effective communication with human beings, see, e.g., Breazeal (2003); Kanda & Ishiguro (2013); *Social Robots and the Future of Social Relations* (2014); *Social Robots from a Human Perspective* (2015); and *Social Robots: Boundaries, Potential, Challenges* (2016).

[62] See Abrams (2004), p. 251, which offers an analysis of Rorty, "Religion As Conversation-stopper" (1999), pp. 168-74, and Rorty, *Philosophy and the Mirror of Nature* (1979), p. 394.

[63] Abrams (2004), pp. 246-50.

Or to state it more bluntly: will we be able to create charismatic robots that are smart enough to lead us but not smart enough to leave us behind? Might it be the case that we can be led by superhuman robots but not by supra-supra-superhuman robots?

The Robots That Remain with Us

Even if it is possible in principle for such delicate leader-follower relationships between robots and human beings to come into existence, it is thus unclear whether they can be sustained as an enduring feature of a posthuman reality – or whether our dependence on charismatic robot leaders as CEOs and other organizational leaders might simply represent a fleeting phase in the course of human history.

The first generation of social robots to possess robust artificial general intelligence will be capable of interacting socially with human beings, because financial and practical realities dictate that this will most likely be part of the purpose for which they have been explicitly designed. However, future generations of social robots (which may themselves be designed or otherwise cultivated by robots rather than by human beings) might 'outgrow' the sort of hardware and software – or mind and body – that would be necessary for them to serve as charismatic leaders of human beings. Just as multiple human societies and cultures are capable of existing simultaneously, though, it seems likely that a proliferation of different strata of robot cultures (and even civilizations) that simultaneously display varying levels of technological advancement may eventually come to exist. While the most advanced social robots may only be capable of engaging in social interactions and relations among themselves and not with human beings, at any given point in the future there may remain a society of robot laggards that have been designed – or have chosen – to forgo more advanced technological evolution in order to continue serving as colleagues, companions, and leaders to human beings.

X. Conclusion

As increasingly sophisticated robots take on a growing range of roles within businesses and other human organizations, we are confronted with the question of whether a sufficiently advanced robot would ever be able to successfully fill a role as CEO of an organization that includes human employees. Serving as CEO would require a robot to not only possess significant capacities to plan, organize, and control an organization's activities but also to *lead* its human personnel. In the case of human CEOs, the effectiveness of their leadership depends to a large extent on their ability to motivate, inspire, and garner the trust of the human employees who work within their organization.

Although it might today seem implausible to imagine that social robots can and will possess such capacities, we have suggested that such a perspective is shortsighted. Some scholars argue that the restless human urge to develop new technologies demonstrates that we yearn for (and perhaps have always yearned for) robot leaders and that we are striving to create them, whether we realize it or not. In this text, we have suggested three ways in which human beings are seeking to create charismatic robotic leaders to whom we can relinquish portions of our decision-making and the responsibility for guiding and controlling certain aspects of our lives – and in particular, to whom we can grant the ability to guide, shape, and lead us in our professional lives within the businesses and other organizations in which we work. Namely, through various technological, artistic, scholarly, and commercial pursuits, we are laboring to fashion artificial beings who will: 1) serve as normative role models and inspirations for us through their display of superior morality; 2) attract us through their possession of superhuman knowledge; or 3) charm and control us through their romantic allure and interpersonal attractiveness. The possession of some or all of these characteristics would allow a robot to successfully motivate, inspire, and win the trust of human employees, thereby demonstrating key leadership traits that are needed in a CEO.

In some aspects, the creation of such robots has already reached the preliminary design and engineering stages; in other respects, it stand in the 'pre-preliminary design' stage that is represented by futurology and science fiction. Through works of science fiction, humanity is carrying out strategic planning for technologies that our engineering capacities do not yet allow us to create. As a form of R&D, such fictional works allow us to develop and even conceptually 'test' the technologies that we will, in the future, summon into being. However, this design and testing of future social robots through fiction takes place at a fairly general level; it does not involve making conscious, deliberate decisions about whether we, as a human society, will agree to enter into particular relationships of obedience with particular social robots as CEOs or other organizational leaders. The phenomenological-social approach to understanding our human relationship to technology would seem to indicate that such moments of rational, purposeful decision-making will rarely, if ever, arrive – even as we develop and deepen our individual and collective human submission to charismatic robot authority.

The exact nature and duration of our future submission to robot leadership is not yet known. While it is possible to imagine many frightening (and frighteningly plausible) scenarios, a more favorable scenario that one might hope to see realized is that whatever forms of obedience to social robots we adopt will be at least as morally and intellectually beneficial as our everyday obedience to other human beings to whom we regularly entrust authority. Although it is true that a charismatic social robot serving as a CEO might lack some of the strengths possessed by the best human CEOs, it might also lack some of the flaws displayed by the worst human CEOs. In principle, it may be possible to design or nurture robotic CEOs that are free from the dishonesty, greed, nepotism, and incompetence that sometimes afflict human executives and that possess a great abundance of those strengths that allow virtuous and talented charismatic human leaders to organize and guide their fellow humans in their pursuit of the common good. For many human beings alive today, the notion that they could ever come to trust, admire, and even love a social robot that is serving as CEO of the organization in which they work might be unfathomable; but for future generations, the thought of a world where social robots do not play such roles in helping to motivate, inspire, and guide us might seem just as exotic and difficult to accept.

References

Abrams, Jerold J. "Pragmatism, Artificial Intelligence, and Posthuman Bioethics: Shusterman, Rorty, Foucault." *Human Studies* 27, no. 3 (2004): 241–58. doi:10.1023/B:HUMA.0000042130.79208.c6.

Anderson, Monica, and Andrea Caumont, "How social media is reshaping news." Pew Research Center, September 24, 2014. http://www.pewresearch.org/fact-tank/2014/09/24/how-social-media-is-reshaping-news/. Accessed June 26, 2016.

Artificial General Intelligence, edited by Ben Goertzel and Cassio Pennachin. Springer Berlin Heidelberg, 2007.

Artificial General Intelligence: 8th International Conference, AGI 2015: Berlin, Germany, July 22-25, 2015: Proceedings, edited by Jordi Bieger, Ben Goertzel, and Alexey Potapov. Springer International Publishing, 2015.

Badmington, Neil. "Cultural Studies and the Posthumanities." In *New Cultural Studies: Adventures in Theory*, edited by Gary Hall and Claire Birchall, pp. 260-72. Edinburgh: Edinburgh University Press, 2006.

Bainbridge, William Sims. *The Virtual Future*. London: Springer, 2011.

Bakshy, Eytan, Solomon Messing, and Lada A. Adamic. "Exposure to ideologically diverse news and opinion on Facebook." *Science* 348, no. 6239 (2015): 1130-32.

Banks, Iain M. *Consider Phlebas*. London: Macmillan, 1987.

Bradshaw, Jeffrey M., Paul Feltovich, Matthew Johnson, Maggie Breedy, Larry Bunch, Tom Eskridge, Hyuckchul Jung, James Lott, Andrzej Uszok, and Jurriaan van Diggelen. "From Tools to Teammates: Joint Activity in Human-Agent-Robot Teams." In *Human Centered Design*, edited by Masaaki Kurosu, pp. 935-44. Lecture Notes in Computer Science 5619. Springer Berlin Heidelberg, 2009.

Breazeal, Cynthia. "Toward sociable robots." *Robotics and Autonomous Systems* 42 (2003): 167-75.

Cabibihan, John-John, Mary-Anne Williams, and Reid Simmons. "When Robots Engage Humans." *International Journal of Social Robotics* 6, no. 3 (2014): 311-13.

Calverley, David J. "Imagining a Non-Biological Machine as a Legal Person." *AI & SOCIETY* 22 (4) (2008): 523-37. doi:10.1007/s00146-007-0092-7.

Cambria, Erik, and Bebo White. "Jumping NLP Curves: A Review of Natural Language Processing Research." *IEEE COMPUTATIONAL INTELLIGENCE MAGAZINE* 1556, no. 603X/14 (2014): 48-57.

Čapek, Karel. *R.U.R.: Rossum's Universal Robots.* Translated by David Wyllie. Adelaide: The University of Adelaide, last updated March 27, 2016. https://ebooks.adelaide.edu.au/c/capek/karel/rur/index.html. Accessed June 26, 2016.

Castronova, Edward. *Synthetic Worlds: The Business and Culture of Online Games.* Chicago: The University of Chicago Press, 2005.

Coeckelbergh, Mark. "Can We Trust Robots?" *Ethics and Information Technology* 14 (1) (2012): 53-60. doi:10.1007/s10676-011-9279-1.

"Could Facebook influence the outcome of the presidential election?" *The Economist,* May 18, 2016. http://www.economist.com/blogs/economist-explains/2016/05/economist-explains-15. Accessed June 26, 2016.

Cyberculture, Cyborgs and Science Fiction: Consciousness and the Posthuman, edited by William S. Haney II. Amsterdam: Rodopi, 2006.

Daft, Richard. *Management.* Mason, OH: South-Western / Cengage Learning, 2011.

Dai, Y., C. Suero Montero, T. Kakkonen, M. Nasiri, E. Sutinen, M. Kim, and T. Savolainen (2013). "TrustAider – Enhancing Trust in E-Leadership." In *Business Information Systems,* pp. 26-37. Lecture Notes in Business Information Processing 157. Springer Berlin Heidelberg, 2013.

Dainty, Paul H., and Moreen Anderson. *The Capable Executive: Effective Performance in Senior Management.* London: Macmillan Press Ltd., 1996.

Deus Ex. Written by Sheldon Pacotti and directed by Warren Spector. Eidos Interactive, 2000.

Douglas, Joe. "Deus Ex – Still the Best Game Ever?" RetroCollect, March 26, 2011. http://www.retrocollect.com/Articles/deus-ex-still-the-best-game-ever.html. Accessed November 23, 2015.

Elkins, Kathleen. "Experts predict robots will take over 30% of our jobs by 2025 — and white-collar jobs aren't immune." *Business Insider,* May 1, 2015. http://www.businessinsider.com/experts-predict-that-one-third-of-jobs-will-be-replaced-by-robots-2015-5. Accessed June 25, 2016.

Ford, Martin. *Rise of the Robots: Technology and the Threat of a Jobless Future.* New York: Basic Books, 2015.

Forsyth, D. R. *Group Dynamics, 5e.* Belmont, California: Cengage Learning, 2010.

Fox, Susannah, and Lee Rainie. "The Web at 25 in the US: Part 1: How the Internet Has Woven Itself into American Life." Pew Research Center, Internet & American Life Project, February 27, 2014. http://www.pewinternet.org/2014/02/27/part-1-how-the-internet-has-woven-itself-into-american-life/. Accessed August 6, 2015.

French, Jr., John R.P., and Bertram Raven, "The Bases of Social Power." In *Studies in Social Power,* edited by Dorwin P. Cartwright, pp. 150-67. Ann Arbor, MI: Institute for Social Research, University of Michigan, 1959.

Friedenberg, Jay. *Artificial Psychology: The Quest for What It Means to Be Human.* Philadelphia: Psychology Press, 2008.

Garland, Alex. *Ex Machina.* Universal Studios, 2015.

Geraci, Robert M. *Apocalyptic AI: Visions of Heaven in Robotics, Artificial Intelligence, and Virtual Reality.* New York: Oxford University Press, 2010.

Gladden, Matthew E. "The Artificial Life-Form as Entrepreneur: Synthetic Organism-Enterprises and the Reconceptualization of Business." In *Proceedings of the Fourteenth International Conference on the Synthesis and Simulation of Living Systems,* edited by Hiroki Sayama, John Rieffel, Sebastian Risi, René Doursat and Hod Lipson, pp. 417-18. Cambridge, MA: The MIT Press, 2014.

Gladden, Matthew E. "The Diffuse Intelligent Other: An Ontology of Nonlocalizable Robots as Moral and Legal Actors." In *Social Robots: Boundaries, Potential, Challenges,* edited by Marco Nørskov, pp. 177-98. Farnham: Ashgate, 2016.

Gladden, Matthew E. "From Stand Alone Complexes to Memetic Warfare: Cultural Cybernetics and the Engineering of Posthuman Popular Culture." Presentation at the 50 Shades of Popular Culture International Conference. Facta Ficta / Uniwersytet Jagielloński, Kraków, February 19, 2016.

Gladden, Matthew E. "Leveraging the Cross-Cultural Capacities of Artificial Agents as Leaders of Human Virtual Teams." *Proceedings of the 10th European Conference on Management Leadership and Governance,* edited by Visnja Grozdanić, pp. 428-35. Reading: Academic Conferences and Publishing International Limited, 2014.

Gladden, Matthew E. *Sapient Circuits and Digitalized Flesh: The Organization as Locus of Technological Posthumanization.* Indianapolis: Defragmenter Media, 2016.

Goicoechea, María. "The Posthuman Ethos in Cyberpunk Science Fiction." *CLCWeb: Comparative Literature and Culture* 10, no. 4 (2008): 9. http://docs.lib.purdue.edu/cgi/viewcontent.cgi?article=1398&context=clcweb. Accessed May 18, 2016.

Goleman, Daniel. "What Makes a Leader?" *Harvard Business Review* 82 (1) (2004): 82-91.

Goodman, Paul S., and Linda Argote. "New Technology and Organizational Effectiveness." Carnegie-Mellon University, April 27, 1984.

Graham, Elaine. *Representations of the Post/Human: Monsters, Aliens and Others in Popular Culture.* Manchester: Manchester University Press, 2002.

Grodzinsky, F. S., K. W. Miller, and M. J. Wolf. "Developing Artificial Agents Worthy of Trust: 'Would You Buy a Used Car from This Artificial Agent?'" *Ethics and Information Technology* 13 (1) (2011): 17-27.

Gupta, Ashish, Ramesh Sharda, and Robert A. Greve. "You've got email! Does it really matter to process emails now or later?" In *Information Systems Frontiers* vol. 13, no. 5 (2011): 637-53. http://dx.doi.org/10.1007/s10796-010-9242-4.

Harrison, Harry. *The Stainless Steel Rat Gets Drafted.* New York: Bantam, 1987.

Hayles, N. Katherine. *How We Became Posthuman: Virtual Bodies in Cybernetics, Literature, and Informatics.* Chicago: University of Chicago Press, 1999.

Heim, Michael. *The Metaphysics of Virtual Reality.* New York: Oxford University Press, 1993.

Herbrechter, Stefan. *Posthumanism: A Critical Analysis.* London: Bloomsbury, 2013. [Kindle edition.]

"It's not about sex, says Aiko robot inventor Le Trung." News.com.au, December 16, 2008. http://www.news.com.au/news/its-not-about-sex-says-fembot-inventor/story-fna7dq6e-1111118332678. See also the Project Aiko website, http://www.projectaiko.com/.

Kamiyama, Kenji. "C: Smoke of Gunpowder, Hail of Bullets – BARRAGE." *Ghost in the Shell: Stand Alone Complex*, episode 25, first broadcast October 1, 2003 (Production I.G), North American version first broadcast May 1, 2005.

Kanda, Takayuki, and Hiroshi Ishiguro. *Human-Robot Interaction in Social Robotics.* Boca Raton: CRC Press, 2013.

Koltko-Rivera, Mark E. "The potential societal impact of virtual reality." In *Advances in virtual environments technology: Musings on design, evaluation, and applications*, vol. 9. Mahwah, NJ: Erlbaum, 2005.

Koltko-Rivera, Mark E. "Rediscovering the later version of Maslow's hierarchy of needs: Self-transcendence and opportunities for theory, research, and unification." *Review of General Psychology* 10, no. 4 (2006): 302-17.

Kriksciuniene, D., and S. Strigunaite. "Multi-Level Fuzzy Rules-Based Analysis of Virtual Team Performance." In *Building the E-World Ecosystem*, pp. 305-18. IFIP Advances in Information and Communication Technology 353. Springer Berlin Heidelberg, 2011.

Kuflik, Arthur. "Computers in Control: Rational Transfer of Authority or Irresponsible Abdication of Autonomy?" *Ethics and Information Technology* 1 (3) (1999): 173-84. doi:10.1023/A:1010087500508.

L'Isle-Adam, Auguste comte de Villiers de. *Tomorrow's Eve.* Translated by Robert Martin Adams. University of Illinois Press, 2001.

Lang, Fritz. *The Complete Metropolis.* DVD. Kino International, 2010.

Maslow, Abraham H. *Motivation and Personality.* New York: Harper, 1954.

Miah, Andy. "A Critical History of Posthumanism." In *Medical Enhancement and Posthumanity*, edited by Bert Gordijn and Ruth Chadwick, pp. 71-94. The International Library of Ethics, Law and Technology 2. Springer Netherlands, 2008.

Murphy, Robin. *Introduction to AI Robotics.* Cambridge, MA: The MIT Press, 2000.

Nadkarni, Prakash M., Lucila Ohno-Machado, and Wendy W. Chapman. "Natural language processing: an introduction." *Journal of the American Medical Informatics Association* 18, no. 5 (2011): 544-51.

Nadler, David, and Michael Tushman. *Competing by Design: The Power of Organizational Architecture.* Oxford University Press, 1997. [Kindle edition.]

Nunes, M., and H. O'Neill. "Assessing the Performance of Virtual Teams with Intelligent Agents." In *Virtual and Networked Organizations, Emergent Technologies and Tools*, pp. 62-69. Communications in Computer and Information Science 248. Springer Berlin Heidelberg, 2012.

Pereira, Gonçalo, Rui Prada, and Pedro A. Santos. "Conceptualizing Social Power for Agents." In *Intelligent Virtual Agents*, edited by Ruth Aylett, Brigitte Krenn, Catherine Pelachaud, and Hiroshi Shimodaira, pp. 313-24. Lecture Notes in Computer Science 8108. Springer Berlin Heidelberg, 2013.

Perlberg, James. *Industrial Robotics.* Boston: Cengage Learning, 2016.

Plato, *Republic: Books 1-5*, vol. 1. Translated by Chris Emlyn-Jones and William Preddy. Cambridge, MA: Loeb Classical Library, Harvard University Press, 2013.

Plunkett, Warren R., Gemmy Allen, and Raymond Attner. *Management.* Mason, OH: South-Western / Cengage Learning, 2012.

"Poll: Poll: Which Deus Ex ending would you choose?", The Escapist. http://www.escapistmagazine.com/forums/read/9.137841-Poll-Poll-Which-Deus-Ex-ending-would-you-choose. Accessed November 23, 2015.

Posthuman Bodies, edited by Judith Halberstam and Ira Livingstone. Bloomington, IN: Indiana University Press, 1995.

Rader, Emilee, and Rebecca Gray. "Understanding user beliefs about algorithmic curation in the Facebook news feed." In *Proceedings of the 33rd Annual ACM Conference on Human Factors in Computing Systems*, pp. 173-182. ACM, 2015.

Raven, Bertram H. "A power/interaction model of interpersonal influence: French and Raven thirty years later." *Journal of Social Behavior & Personality* vol. 7, no. 2 (1992): 217-44.

Rorty, Richard. *Philosophy and the Mirror of Nature.* Princeton: Princeton University Press, 1979.

Rorty, Richard. "Religion As Conversation-stopper." In *Philosophy and Social Hope.* New York: Penguin, 1999.

Samani, Hooman Aghaebrahimi, and Adrian David Cheok. "From human-robot relationship to robot-based leadership." In *2011 4th International Conference on Human System Interactions (HSI)*, pp. 178-81. IEEE, 2011.

Samani, Hooman Aghaebrahimi, Jeffrey Tzu Kwan Valino Koh, Elham Saadatian, and Doros Polydorou. "Towards Robotics Leadership: An Analysis of Leadership Characteristics and the Roles Robots Will Inherit in Future Human Society." In *Intelligent Information and Database Systems*, edited by Jeng-Shyang Pan, Shyi-Ming Chen, and Ngoc Thanh Nguyen, pp. 158-65. Lecture Notes in Computer Science 7197. Springer Berlin Heidelberg, 2012.

Schlindwein, Simone. "Intelligent Robots Save Lives in DR Congo." Deutsche Welle, June 18, 2014. http://dw.de/p/1CL4h. Accessed August 12, 2014.

Shih, Johanna. "Project Time in Silicon Valley," in *Qualitative Sociology*, vol. 27, no. 2 (2004): 223-45. doi:10.1023/B:QUAS.0000020694.53225.23.

Social Robots and the Future of Social Relations, edited by Johanna Seibt, Raul Hakli, and Marco Nørskov. Amsterdam: IOS Press, 2014.

Social Robots from a Human Perspective, edited by Jane Vincent, Sakari Taipale, Bartolomeo Sapio, Giuseppe Lugano, and Leopoldina Fortunati. Springer International Publishing, 2015.

Social Robots: Boundaries, Potential, Challenges, edited by Marco Nørskov. Farnham: Ashgate Publishing, 2016.

"**spoiler**Poll: Which ending did you pick?", TTLG Forums. http://ttlg.com/forums/showthread.php?t=70105&page=2. Accessed November 23, 2015.

Spring, Alexandra. "Can machines come up with more creative solutions to our problems than we can?" *The Guardian*, March 29, 2016. http://www.theguardian.com/sustainable-business/2016/mar/29/can-machines-come-up-with-more-creative-solutions-to-our-problems-than-we-can. Accessed June 25, 2016.

Stahl, B. C. "Responsible Computers? A Case for Ascribing Quasi-Responsibility to Computers Independent of Personhood or Agency." *Ethics and Information Technology*, vol. 8, no. 4 (2006): 205-13.

Susskind, Richard, and Daniel Susskind. *The Future of the Professions: How Technology Will Transform the Work of Human Experts.* Oxford: Oxford University Press, 2015.

Szollosy, Michael. "Freud, Frankenstein and our fear of robots: projection in our cultural perception of technology." *AI & SOCIETY* (2016). doi: 10.1007/s00146-016-0654-7. Accessed June 25, 2016.

Szollosy, Michael. "Why are we afraid of robots? The role of projection in the popular conception of robots." In *Beyond Artificial Intelligence*, pp. 121-31. Springer International Publishing, 2015.

Theoretical Foundations of Artificial General Intelligence, edited by Pei Wang and Ben Goertzel. Paris: Atlantis Press, 2012.

Thompson, David A. "The Man-Robot Interface in Automated Assembly." In *Monitoring Behavior and Supervisory Control*, edited by Thomas B. Sheridan and Gunnar Johannsen, pp. 385-91. NATO Conference Series 1. Springer US, 1976.

Van Oost, Ellen, and Darren Reed. "Towards a sociological understanding of robots as companions." In *International Conference on Human-Robot Personal Relationship*, pp. 11-18. Springer Berlin Heidelberg, 2010.

Weber, Max, *Economy and Society: An Outline of Interpretive Sociology*, New York: Bedminster Press, 1968.

"Which Deus Ex 1 ending did you choose?", SharkyForums. http://www.sharkyforums.com/showthread.php?238618-Which-Deus-Ex-1-ending-did-you-choose-(spoilers). Accessed November 23, 2015.

Wiltshire, Travis J., Dustin C. Smith, and Joseph R. Keebler. "Cybernetic Teams: Towards the Implementation of Team Heuristics in HRI." In *Virtual Augmented and*

Mixed Reality. Designing and Developing Augmented and Virtual Environments, edited by Randall Shumaker, pp. 321-30. Lecture Notes in Computer Science 8021. Springer Berlin Heidelberg, 2013.

Wong, Julia Carrie, "Welcome to the robot-based workforce: will your job become automated too?" *The Guardian*, March 19, 2016. https://www.theguardian.com/technology/2016/mar/19/robot-based-economy-san-francisco. Accessed June 25, 2016.

Yampolskiy, Roman V. *Artificial Superintelligence: A Futuristic Approach*. Boca Raton: CRC Press, 2015.

Chapter Nine

The Diffuse Intelligent Other: An Ontology of Nonlocalizable Robots as Moral and Legal Actors[1]

Abstract. Much thought has been given to the question of who bears moral and legal responsibility for actions performed by robots. Some argue that responsibility could be attributed to a robot if it possessed human-like autonomy and metavolitionality, and that while such capacities can potentially be possessed by a robot with a single spatially compact body, they cannot be possessed by a spatially disjunct, decentralized collective such as a robotic swarm or network. However, advances in ubiquitous robotics and distributed computing open the door to a new form of robotic entity that possesses a unitary intelligence, despite the fact that its cognitive processes are not confined within a single spatially compact, persistent, identifiable body. Such a 'nonlocalizable' robot may possess a body whose myriad components interact with one another at a distance and which is continuously transforming as components join and leave the body. Here we develop an ontology for classifying such robots on the basis of their autonomy, volitionality, and localizability. Using this ontology, we explore the extent to which nonlocalizable robots – including those possessing cognitive abilities that match or exceed those of human beings – can be considered moral and legal actors that are responsible for their own actions.

Introduction

Philosophers, roboticists, and legal scholars have given much thought to the challenges that arise when attempting to assign moral and legal responsibility for actions performed by robots. One difficulty results from the fact that

---

[1] This chapter is reprinted from "The Diffuse Intelligent Other: An Ontology of Nonlocalizable Robots as Moral and Legal Actors," in *Social Robots: Boundaries, Potential, Challenges*, edited by Marco Nørskov. Series on Emerging Technologies, Ethics and International Affairs. Farnham: Ashgate, 2016, pp. 177-98.

the word 'robot' does not describe a single species or genus of related beings but rather a vast and bewildering universe of entities possessing widely different morphologies and manners of functioning.

To date, scholars exploring questions of moral and legal responsibility have largely focused on two types of robots. One kind comprises telepresence robots that are remotely operated by a human being to carry out tasks such as performing surgery, giving a lecture, or firing a missile from an aerial vehicle.[2] Such robots are not morally or legally responsible for their own actions; instead, responsibility for their actions is attributed to their operators, designers, or manufacturers according to well-established legal and moral frameworks relating to human beings' use of tools and technology. It is possible for such a robot to be designed and manufactured in one country, relocated to (and acting in) a second country, and remotely controlled by a human operator based in a third country, thus raising questions about which nation's laws should be used to assign responsibility for the robot's actions. However, in principle it is relatively easy to trace the chain of causality and identify the locations in which each stage of a robot's action occurred.

The other main type of robot for whose actions scholars have sought to attribute moral and legal responsibility comprises autonomous devices such as self-driving cars[3] and autonomous battlefield robots.[4] Here questions of legal and moral responsibility are complicated by the fact that it is not immediately obvious who – if anyone – has "made the decision" for the robot to act in a particular way. Depending on the circumstances, arguments can be made for attributing responsibility for a robot's action to the robot's programmer, manufacturer, owner, or – if the robot possesses certain kinds of cognitive properties – even to the robot itself.[5] At the same time, the attribution of legal responsibility for the robot's actions is simplified by the fact that the computational processes guiding an autonomous robot's behavior typically take place within the robot's own spatially compact body, reducing the likelihood that the robot's process of acting could cross national borders.

While efforts to account for the actions of some kinds of robots are thus already well advanced, at the frontiers of ubiquitous robotics, nanorobotics, distributed computing, and artificial intelligence, experts are pursuing the development of a new form of robotic entity whose manner of being, deciding,

[2] Datteri, "Predicting the Long-Term Effects of Human-Robot Interaction: A Reflection on Responsibility in Medical Robotics" (2013); Hellström, "On the Moral Responsibility of Military Robots" (2013); Coeckelbergh, "From Killer Machines to Doctrines and Swarms, or Why Ethics of Military Robotics Is Not (Necessarily) About Robots" (2011).

[3] Kirkpatrick, "Legal Issues with Robots" (2013).

[4] Sparrow, "Killer Robots" (2007).

[5] Sparrow (2007); Dreier & Spiecker, "Legal Aspects of Service Robotics" (2012), p. 211.

and acting is so radically different from those of earlier robots that it will be difficult or impossible to apply our traditional conceptual frameworks when analyzing whether such a robot is morally and legally responsible for its actions. This new kind of being is the *nonlocalizable* robot, one whose cognitive processes do not subsist within a single identifiable body that is confined to a particular set of locations and that endures across time.[6] Such a robot might take the form, for example, of a vast digital-physical ecosystem in which millions of interconnected devices participate in a shared cognitive process of reaching decisions and acting, even as devices continually join and leave the network; or it could exist as a loosely-coupled, free-floating oceanic cloud of nanorobotic components that communicate with one another via electromagnetic signals while drifting among the world's seas; or it could take the form of an evolvable computer virus whose cognitive processes are hidden within an ever-shifting network of computers around the world – ones found not only in homes and business but also in airplanes, ships, and orbiting satellites.

Attempts to address questions of responsibility for actions performed by nonlocalizable robots raise a multitude of complex philosophical and legal questions that have not yet been carefully explored. For example, in an environment containing such robots it might be apparent that 'some' robotic entity has just acted, but it may be impossible to correlate that action with just a single robot, as a networked device can be part of the bodies of many different nonlocalizable robots simultaneously, and over the span of a few milliseconds or minutes a networked device might undergo the processes of becoming integrated into a nonlocalizable robot's body, acting as a part of its body, and then becoming dissociated from its body. More fundamentally, it may even be impossible to determine with clarity that a particular robot *exists* – to identify it and distinguish it from other entities and delineate the boundaries of its physical and cognitive existence. The fact that such a spatially diffuse (and potentially even globally extensive) robotic body might someday be occupied by an artificial intelligence possessing human-like cognitive capacities[7] adds another dimension that one must account for when attempting to develop a framework for determining the moral and legal responsibility that nonlocalizable robots might bear for their actions.

In this text we suggest the outline for such a framework. We begin in the following section by proposing an ontology of autonomy, volitionality, and localizability that will allow us to describe essential physical and cognitive

---

[6] Yampolskiy & Fox, "Artificial General Intelligence and the Human Mental Model" (2012), pp. 129-38; Gladden, "The Social Robot as 'Charismatic Leader': A Phenomenology of Human Submission to Nonhuman Power" (2014), p. 338.

[7] Yampolskiy & Fox (2012), pp. 133-38.

characteristics of nonlocalizable robots and analyze the extent to which they are morally and legally responsible actors.[8]

Developing an Ontological Framework

Purpose of the Ontology

There have been efforts by computer scientists to develop a universal 'ontology' for robotics that defines its current terminology and engineering principles in a way that is "formally specified in a machine-readable language, such as first-order logic."[9] Such specialized technical schemas facilitate the standardization of robotics engineering and the interoperability of different robotic systems, however they do not attempt to delineate the full universe of ways of being and acting that are available for robots and which philosophers can use to explore the social, ethical, and legal questions that are provoked by the existence of ever more sophisticated robotic morphologies.

A more robust ontology should be capable of describing a robotic entity both at its most fundamental level of physical reality as well as at the level of emergent phenomena that are attributed to the entity as a result of higher-level interactions with its environment.[10] Such an ontology would allow us to analyze a robot from the perspective of its nature as an autonomous viable system that organizes matter, energy, and information,[11] to its form and functioning as a concrete physical device that incorporates sensors, actuators, and computational processors, to its role within human (or artificial) societies as a social, political, legal, and cultural object or subject.

Here we focus on one element of such an ontology that poses a significant – and thus far largely unexplored – challenge to the current debate on whether robots can bear moral and legal responsibility for their actions. Namely, we introduce the concept of a robot's *localizability*, and in particular we will consider the question of moral and legal responsibility for the actions of a robot that is a nonlocalizable being – i.e., one that has a physical body, but whose body does not exist in any one particular place across time. Our ontology also

---

[8] The word 'actors' is used here to mean "entities who are morally or legally responsible for their own actions." Use of the word 'agent' has been avoided, due to the fact that it possesses different (and in many ways incompatible) meanings when used in the contexts of moral philosophy, law, and computer science. Similarly, the posing of the question of whether a nonlocalizable robot can be considered a moral or legal 'person' has generally been avoided, since – depending on the context – personhood does not necessarily imply that an entity is morally and legally responsible for its actions.

[9] Prestes et al., "Towards a Core Ontology for Robotics and Automation" (2013), p. 1194.

[10] Gladden, "The Social Robot as 'Charismatic Leader'" (2014), p. 338.

[11] Gladden, "The Artificial Life-Form as Entrepreneur: Synthetic Organism-Enterprises and the Reconceptualization of Business" (2014), p. 417.

encompasses two other relevant elements: those of a robot's levels of *autonomy* and *volitionality.* A brief discussion of these three concepts follows.

Autonomy

Autonomy relates to an entity's ability to act without being controlled. For robots, possessing autonomy means being "capable of operating in the real-world environment without any form of external control for extended periods of time."[12] In its fullest form, autonomy involves not only performing cognitive tasks such as setting goals and making decisions but also performing physical activities such as securing energy sources and carrying out self-repair without human intervention. Building on conventional classifications of robotic autonomy,[13] we can say that currently existing robots are either *nonautonomous* (e.g., telepresence robots that are fully controlled by their human operators when fulfilling their intended purpose, or robots which do not act to fulfill any purpose), *semiautonomous* (e.g., robots that require 'continuous assistance' or 'shared control' in order to fulfill their intended purpose), or *autonomous* (e.g., robots that require no human guidance or intervention in fulfilling their intended purpose). We can use the term *superautonomous* to describe future robots whose degree of autonomy may significantly exceed that displayed by human beings – e.g., because the robots' ability to independently acquire new knowledge frees them from any need to seek guidance from with human subject-matter experts or because their bodies contain an energy source that can power them throughout their anticipated lifespan.

Volitionality

Volitionality relates to an entity's ability to self-reflexively shape the intentions that guide its actions. A robot is *nonvolitional* when it possesses no internal goals or 'desires' for achieving particular outcomes nor any expectations or 'beliefs' about how performing certain actions would lead to particular outcomes. Some telepresence robots are nonvolitional, insofar as a human operator supplies all of the desires and expectations for their actions; the robot is simply a pliable, transparent tool. A robot is *semivolitional* if it possesses *either* a goal of achieving some outcome *or* an expectation that particular outcomes will result if the robot acts in a certain way, but the robot does not link goals with expectations. For example, such a robot might have been programmed with the goal of "moving to the other side of the room," but it has no means of interpreting the sensory data that it is receiving from its environment to know where it is, nor does it have an understanding of the fact that

[12] Bekey, *Autonomous Robots: From Biological Inspiration to Implementation and Control* (2005), p. 1.

[13] Murphy, *Introduction to AI Robotics* (2000), pp. 31-34.

activating its actuators would cause it to 'move.' A robot is *volitional* if it combines goals with expectations; in other words, it can possess an intention[14], which is a mental state that comprises both a desire and a belief about how some act that the robot is about to perform can contribute to fulfilling that desire.[15] For example, a therapeutic social robot might have a goal of evoking a positive emotional response in its human user, and its programming tells it that by following particular strategies for social interaction it is likely to evoke such a response.

We can describe as *metavolitional* a robot that possesses what scholars have elsewhere referred to as a 'second-order volition,' or an intention *about* an intention.[16] For example, imagine a more sophisticated therapeutic social robot that comes to realize that it is 'manipulating' its human users by employing subtle psychological techniques to coax them into displaying positive emotional responses. Such a robot would display metavolitionality if it grew weary of what it now saw as its 'dishonest' and coercive behavior, and it wished that it did not feel compelled to constantly manipulate human beings' psyches. Metavolitionality is a form of volitionality typically demonstrated by adult human beings. We can use the term *supervolitional* to describe a possible future robot that regularly forms intentions that are higher than second-order. For example, such a robot might possess a mind that is capable of simultaneously experiencing thousands of different second-order intentions and using a third-order intention to guide and transform those second-order intentions in a concerted way that reshapes the robot's character.

Localizability

Localizability relates to the extent to which an entity possesses a stable, identifiable physical body that is confined to one or more concrete locations. A robot is *local* when its 'sensing body' of environmental sensors, its 'acting body' of actuators and manipulable physical components, and the 'brain' in which its cognitive processes are executed are found together in a single location and possess physical forms that are discrete and easily identifiable and endure over time. Such robots might include a robotic vacuum cleaner, an articulated industrial robot controlled by a teach pendant that is connected to the robotic arm by a physical cable, or an experimental wheeled robot whose body moves through a maze while the its data analysis and decision-making are being carried out in a dedicated desktop computer that is located in the

---

[14] 'Intentionality' is employed in its usual philosophical sense to describe an entity's ability to possess mental states that are directed toward (or 'about') some object; that is a broader phenomenon than the possession of a particular 'intention' as defined here.

[15] Calverley, "Imagining a Non-Biological Machine as a Legal Person" (2008), p. 529.

[16] Calverley (2008), pp. 533-35.

same room as the robot's sensing and acting body and is wirelessly linked to it.

A robot is *multilocal* when it comprises two or more stable and clearly identifiable components that together form the robotic unit and which are not in physical proximity to one another. These components can potentially be located in different parts of the world: for example, a telesurgery robot might include a surgical device located in a hospital room and a controller unit that is manipulated by a human surgeon in a different country and is linked to the surgical device through the Internet. Similarly, an autonomous battlefield robot system could include a central computer located in a military base that is simultaneously controlling the actions of several robotic bodies participating in a military operation in a conflict zone in another country. A robotic entity can be 'intensely multilocal' if it possesses a very large number of components that are located at great distances from one another, however this is a quantitative rather than qualitative distinction: as long as a robotic entity's physical components are identifiable and endure over time, the robot is still multilocal, regardless of how numerous or widely dispatched the components might be. Many current and anticipated systems for ambient intelligence and ubiquitous robotics are multilocal.[17]

A robotic entity is truly *nonlocalizable* when it is impossible to clearly specify in any given moment the exact location or extent of the robot's body or to specify exactly 'where' key activities of the robot's cognitive processes are taking place. It may be known that the robot exists, and from the nature of the robot's actions in the world it is known that the robot must possess both physical sensors and actuators and some substrate upon which its cognitive processes are being executed, however it is not possible to specify exactly where or in what form all of those components exist.

The Sources of Nonlocalizability

Nonlocalizability can result from a number of conditions. For example, at any given moment, a robot's physical extension may include components that are a part of its body, components that are in the process of being removed from its body, and components that are in the process of becoming part of its body. If the robot is able to add new components to its body and remove old ones from its body at a sufficient rate, it is possible that the robot's body at one moment in time might not share any physical components in common

[17] Défago, "Distributed Computing on the Move: From Mobile Computing to Cooperative Robotics and Nanorobotics" (2001); Pagallo, "Robots in the Cloud with Privacy: A New Threat to Data Protection?" (2013); Weber, "General Approaches for a Legal Framework" (2010).

with the robot's body as it existed just a few moments earlier.[18] Such a robot is continuously in the act of moving its cognitive processes into a new body. In this case it is not the physical components of its body that give the robot its identity, insofar as those components are in the perpetual process of being discarded and replaced with new ones; rather it is the relationship between its components – its organizing principle that arranges matter, energy, and information into an enduring viable system[19] – that gives a robot its identity as a potential moral and legal actor.[20] Another source of nonlocalizability occurs if a robot's primary cognitive process consists of a unitary neural computing process that can be distributed across a cluster or grid of processors rather than a program that is executed on a single serial processor.[21] Finally, nonlocalizability can also result from the fact that a single networked device such as a sensor or actuator can potentially be part of the 'bodies' of several different robots simultaneously, producing a situation in which robots' bodies may partly overlap with one another, complicating the question of specifying "to which robot" a particular component or action belongs.

Note that as they currently exist, conventional robotic swarms and networks may be intensely multilocal, but they are not nonlocalizable. While the miniaturization of parts and multiplication of parts constituting a robot's body might contribute to nonlocalizability, they are insufficient to create it, insofar as a robot can have a body consisting of many small parts that is still identifiable and stable over time. Moreover, if the components of a swarm lack a centralized cognitive process that controls their actions, then they are better understood as a collection of independent robots than as a single robot whose body comprises numerous parts. A nonlocalizable robot is not simply a multiagent system or collection of autonomous robots but a single entity possessing a unitary identity that can be at least theoretically capable of possessing autonomy and metavolitionality.

---

[18] Of course, human bodies also undergo a process of gradually replacing their components; however in the case of a nonlocalizable robot, this becomes significant insofar as the transformation of the robot's body is taking place on a time-scale relevant to the time-frame (Gunther, "Time – The Zeroth Performance Metric" (2005), pp. 7-8) in which decisions and actions are made that are the subject of discussions about potential moral and legal responsibility.

[19] Gladden, "The Artificial Life-Form as Entrepreneur" (2014), p. 417.

[20] The debate over whether a robot's identity derives from its physical components or some emergent process at work within them parallels, in some ways, discussions and concepts as ancient as Aristotle's notion of the soul as the form that animates a living substance, the Neoplatonic concept of *ψυχή*, and the understanding of the *λόγος* possessed by individual human beings that was developed by the Greek Fathers of the Catholic Church.

[21] Gladden, "The Social Robot as 'Charismatic Leader'" (2014), p. 338.

The Current Debate over Moral and Legal Responsibility for Robotic Actors

Combining our ontology's four possible values for a robot's level of autonomy, five possibilities for volitionality, and three possibilities for localizability yields a total of sixty prototypical kinds of robotic entities described by the ontology. For some of these robots, the question of moral and legal responsibility for their actions can be analyzed using traditional frameworks that are regularly applied to the actions of human beings. Below we consider several such cases.

The Robot as a Tool for Human Use

Consider once again a multilocal telesurgery robot that consists of a mechanized surgical instrument that operates on a patient and is guided by a human surgeon – who may be located in a different country – manipulating a controller connected to the instrument through the Internet. If the surgeon must directly and specifically request each movement of the surgical instrument, the device would likely be nonautonomous and nonvolitional. If the human surgeon can give the instrument more generalized instructions that it then interprets and executes (such as "Withdraw the laparoscope"), then the robot is operating under a form of 'continuous assistance' or 'shared control,'[22] and it may qualify as semiautonomous and semivolitional or even volitional.

As a tool for use by human beings, questions of legal responsibility for any harmful actions performed by such a robot revolve around well-established questions of product liability for design defects[23] on the part of its producer, professional malpractice on the part of its human operator, and, at a more generalized level, political responsibility for those legislative and licensing bodies that allowed such devices to be created and used. The international dimension of having a human operator who causes the robot to act in a different country raises questions about legal jurisdiction, conflicts of national law, and extraterritoriality,[24] but those issues can be addressed using existing legal mechanisms.

Questions of the moral responsibility for the robot's actions can be similarly resolved on the basis of its functioning as a passive instrument produced and used by human beings for their own ends. Such a robot does not possess responsibility of its own for its actions,[25] but using Stahl's formulation[26] could

---

[22] Murphy (2000), pp. 31-34.

[23] Calverley (2008), p. 533; Datteri (2013).

[24] Doarn & Moses, "Overcoming Barriers to Wider Adoption of Mobile Telerobotic Surgery: Engineering, Clinical and Business Challenges" (2011).

[25] Hellström (2013), p. 104.

[26] Stahl, "Responsible Computers? A Case for Ascribing Quasi-Responsibility to Computers Independent of Personhood or Agency" (2006), p. 210.

be seen as possessing a sort of 'quasi-responsibility' that serves as a placeholder to point back to the robot's human operators and producers, who are ultimately responsible for its actions.

The Robot as Ersatz Human Being

Imagine a humanoid social robot that has been designed in such a way that it possesses not only a human-like local body, but also a human-like form of autonomy and metavolitionality. While such a robot might appear – by today's standards – to represent a quite futuristic technological breakthrough, the question of moral and legal responsibility for the robot's actions could largely be addressed using traditional conceptual frameworks.

In the eyes of the law, a robot that is both autonomous and metavolitional can be considered a legal person that is responsible for its actions.[27] From the legal perspective, 'autonomy' means that there is not some human being (or other robot) to whom a robot defers and who controls the robot's decision-making.[28] From the legal perspective, this does not require the robot to possess 'free will'; the robot's behavior can indeed be determined by certain preferences or desires, as long as those preferences have somehow been generated from within by the robot itself.[29] A robot whose actions are being directed via remote control – or which has been programmed in advance to act in a specific way in response to a specific situation – would not be acting autonomously. However, autonomy is not, in itself, sufficient to generate legal responsibility; an entity must also act with metavolitionality.

The question of whether such a robot is morally (and not just legally) responsible for its actions is more complex, although it also builds on traditional understandings of autonomy and metavolitionality. While incompatibilists argue that an entity must have the freedom to choose between multiple alternatives in order to be held morally responsible, others like Frankfurt have claimed that such freedom is unnecessary and that an entity can be morally responsible if it is capable of experiencing not only desires but also desires *about* its desires, as is seen in our human capacity for "changing desires through the sheer force of mental effort applied in a self-reflexive way."[30] Drawing on Aristotle, Hellström notes that one way of analyzing whether an entity bears moral responsibility for some deed is to ask whether it is "worthy of praise or blame for having performed the action;[31] if we cannot imagine ourselves expressing praise or blame for a robot's action, then it is likely not the sort of actor to which moral responsibility can be attributed. This approach

---

[27] Calverley (2008), pp. 534-35.

[28] Calverley (2008), p. 532.

[29] Calverley (2008), p. 531.

[30] Calverley (2008), pp. 531-32.

[31] Hellström (2013), p. 102.

can be understood as a sort of intuitive 'shortcut' for considering whether an entity is metavolitional. Kuflik gets at the idea of metavolitionality similarly, arguing that it only makes sense to assign computer-based devices like robots responsibility for their actions if we are able to ask them:

> ... to provide good reason for their comportment, to assess the force of reasons they had not previously considered, to be willing in some cases to acknowledge the insufficiency of their own reasons and the greater force of reasons not previously considered, to explain mitigating factors and ask for forgiveness, and – failing a show either of good reason or good excuse – to apologize and look for ways of making amends.[32]

It should be noted, though, that "assignment of responsibility not necessarily is a zero-sum game";[33] the fact that a robot bears moral or legal responsibility for its actions does not imply that a different sort of moral and legal responsibility cannot also be attributed to those who brought the robot into existence, educated it in a particular way, or caused it to be placed in the situation in which it was acting.

Swarms and Networks as Localizable Individuals or Nonlocalizable Communities

Advances in the creation of distributed robotic systems such as networks and swarms of robots (and particularly, swarms of miniaturized robots) have become a topic much studied by philosophers and legal scholars. The rise of such sophisticated robotic systems would seem to point toward the eventual development of robots that are not only autonomous and metavolitional but also truly nonlocalizable; however, to date the development of robotic swarms and networks has not yet spurred significant consideration of potentially autonomous metavolitional nonlocalizable robots as moral or legal actors. Instead, the focus of the debate has moved in other directions.

Some scholars have considered the responsibility of robots that "are connected to a networked repository on the internet that allows such machines to share the information required for object recognition, navigation and task completion in the real world."[34] However, if such a centralized, cloud-based repository is *controlling* the networked robots' actions, then the existence of a stable, identifiable 'brain'" for the system means that it is at most intensely multilocal; it is not nonlocalizable. On the other hand, if the centralized repository is *not* controlling the individual robots' actions, then each robot could

---

[32] Kuflik, "Computers in Control: Rational Transfer of Authority or Irresponsible Abdication of Autonomy?" (1999), p. 174.

[33] Hellström (2013), pp. 104.

[34] Pagallo (2013), p. 501.

potentially possess autonomy and metavolitionality, but the system as a whole does not. In that case, the network could be seen as a community of many different robots, a sort of a multi-agent system.[35] The 'membership' of the community might indeed change from moment to moment, as robots join and leave the network. However, the community is not an entity with a single shared cognitive process and metavolition. It more closely resembles a 'legal person' such as nation or corporation that can bear legal responsibility for its actions but cannot possess the metavolitionality needed for moral responsibility: a corporation can be said to feel pride or regret only in a metaphorical sense, as the corporation's decisions and actions are made not by some autonomous corporate 'mind' but by the minds of its human constituents, often working together.[36] Building on Ricoeur's notion of 'quasi-agency' to refer to the action of nations, Stahl[37] attributes 'quasi-responsibility' to a robotic entity that appears to possess moral responsibility but actually does not because it lacks autonomy or metavolitionality; quasi-responsibility serves only as a placeholder that points toward the entities (such as a network's autonomous components) that actually bear responsibility for the system's apparently collective action. In none of these analyses do scholars assert that a robotic entity such as a continuously evolving network could be nonlocalizable while at the same time possessing autonomy and metavolitionality.

Coeckelbergh[38] goes further: rather than arguing that it is impossible for a robotic swarm or network to be both nonlocalizable, autonomous, and metavolitional, he argues that it is not even possible for a 'robotic' swarm or network to *exist*. Every robot is inextricably situated within a social and technological ecosystem that includes relations with human beings and other devices, and these relations shape the robot's actions and being. What appears, at first glance, to be a purely robotic network or swarm "can hardly be called a 'robotic' swarm given the involvement of various kinds of systems and humans" that in some sense participate in its decision-making and acting.[39] Thus we see that the debate about robots' possible moral and legal responsibility for their own actions has not yet directly addressed the case of robots that are both autonomous and metavolitional while existing in a physical form that is nonlocalizable. In the following sections we attempt to suggest what such an analysis might look like.

---

35 Murphy (2000), pp. 293-314.

36 Stahl (2006), p. 210.

37 Stahl (2006), p. 210.

38 Coeckelbergh (2011), pp. 274-75.

39 Coeckelbergh (2011), p. 274.

The Future of Moral and Legal Responsibility for Nonlocalizable Robots

Advances Contributing to the Development of Nonlocalizable Robots

Developments in fields such as ubiquitous computing, cooperative nanorobotics, and artificial life are laying the groundwork for the existence of robotic entities whose bodies are capable of shifting from one physical substrate to another, occupying parts of the bodies of other robotic entities, and possessing a spatial extension that comprises many disjunct elements and is not identifiable to human observers – in other words, robotic entities that are truly nonlocalizable. For example, as a 'next logical step' that builds on principles of mobile and ubiquitous computing, Défago[40] develops a model of 'cooperative robotics' in which teams of nanorobots can coordinate their activities using *ad hoc* wireless networks or other forms of remote physical interaction. We can also anticipate the development of artificial life-forms resembling sophisticated computer viruses which, in effect, are continually exchanging their old bodies for new ones as they move through networked ecosystems to occupy an ever-shifting array of devices in search of "resources for their own survival, growth, and autonomously chosen pursuits."[41] Given the vast spectrum of possibilities unlocked by ubiquitous computing and nanotechnology, it is possible that without even realizing it, human beings could someday find ourselves living inside the 'bodies' of nonlocalizable robots that surround us – or their bodies could exist inside of us.

Such nonlocalizable robots might demonstrate a range of possible cognitive capacities similar to those available for localizable robots. Insofar as the architecture utilized by the human brain likely represents only one of many possible substrates for a sapient mind, it seems possible that a single mind that is embodied across multiple, changing, spatially disjunct components could potentially possess levels of autonomy and volitionality at least as great as those of a human being.[42] While certain challenges relating to time, space, and computer processing speeds[43] arise when attempting to develop artificial general intelligences whose primary cognitive process occurs across a network of spatially disjunct artificial neurons, such spatial dispersion is not likely to prove an insurmountable obstacle to the creation of artificial general intelligence.[44] Already work is underway on designing artificial networks (ANNs)

---

[40] Défago (2001), pp. 50-53.

[41] Gladden, "The Artificial Life-Form as Entrepreneur" (2014), p. 418.

[42] Yampolskiy & Fox (2012), pp. 133-38.

[43] Gunther (2005), pp. 6-7.

[44] Loosemore & Goertzel, "Why an Intelligence Explosion Is Probable" (2012), pp. 93-95.

whose neurons communicate with one another wirelessly, allowing for the creation of 'cyber-physical systems' (CSPs) whose highly flexible and easily expandable systems include a "large number of embedded devices (such as sensors, actuators, and controllers) distributed over a vast geographical area."[45] Such wireless neural networks could potentially demonstrate capacities impossible for the human brain, insofar as a neuron in such a network is no longer limited to interacting only with neurons that are physically adjacent in three-dimensional space but can potentially connect with other (more distant) neurons to create networks possessing different topologies and dimensionality.

Utilizing our ontological framework, we can now consider the moral and legal responsibility that would be borne for their actions by several types of nonlocalizable robots, beginning with those that demonstrate the lowest levels of autonomy and volitionality.

The Robot as Ambient Magic

Consider a robot whose primary cognitive process can be spread across a vast number of disparate networked computerized devices[46] and which over time shifts elements of that process out of its current devices and into new ones. Such a robot's 'body' would consist of all those devices in which its cognitive processes were being executed (or were at least stored) at a given moment in time. By copying parts of its cognitive processes into adjacent networked devices and deleting them from some of those where they had been residing, the robot could in effect 'move' in an amoeba-like manner, floating not in the air nor through the ocean's waters but through the global ecosystem of networked devices. It could be massively embodied but in such a way that no one element of its 'body' was essential.

If such a nonlocalizable robot were nonautonomous and nonvolitional, it could in some respects resemble a natural phenomenon such as a flickering flame or flowing stream: when engaged physically or digitally by actors such as human beings, it might react in potentially interesting and dramatic ways that display discernible patterns – but not volition. Such a robot might be seen by human beings as a 'force of nature' or perhaps even a latent magical energy, a sort of *qì* or *mana* that is immanent in the environment and can be manipulated by human adepts to produce particular effects.[47] Such a robot would possess neither legal nor moral responsibility for its actions but could

---

[45] Ren & Xu, "Distributed Wireless Networked H_∞ Control for a Class of Lurie-Type Nonlinear Systems" (2014), p. 2.

[46] Gladden, "The Social Robot as 'Charismatic Leader'" (2014), p. 338.

[47] Clarke, "Hazards of Prophecy: The Failure of Imagination" (1973), p. 36.

potentially possess a quasi-responsibility[48] that points toward its creators, assuming that it had been purposefully created and had not evolved naturally 'in the wild' through the interaction and reproduction of other robotic entities.

The Robot as Diffuse Animal Other

If a robot of the sort just described were able to experience desires and expectations and to purposefully shift its cognitive processing into certain kinds of networked computerized hosts that it found attractive (e.g., those with greater processing power or particular kinds of sensors and actuators) and out of those that were less attractive, it would more closely resemble an animal than an impersonal force. It would no longer move through the Internet of Things randomly – or as steered by external forces – but would proactively explore its environment in a search for those informational, value-storing, agent, or material resources[49] that contribute to the satisfying of its drives for self-preservation and reproduction.[50]

Such a robot could in many ways be considered a sort of digital-physical 'animal' with a capacity for bearing moral and legal responsibility similar to those of natural organic animals. (And indeed, in some cases the robot itself might be an organic being.[51]) While such robots would not be moral actors responsible for their own actions, in contrast to the 'robot as ambient magic' they could potentially be the sort of moral patients to whom moral consideration is due, if – like many kinds of animals – they are able to experience physical or psychological suffering.[52]

The Robot as Diffuse Human-like Other

If we now imagine the robot described above to possess human-like cognitive capacities for reasoning, emotion, sociality, and intentionality, it becomes autonomous and metavolitional. With regard to its possession of moral and legal responsibility, it is in some ways similar to the 'robot as ersatz human being' mentioned above, however its nonlocalizability means that it lacks some traits that are found in human beings and possesses other characteristics that are impossible for them.

By using its dispersed, networked body to manipulate an environment, such a robot might appear to materialize, ghost-like, from out of the digital-physical ecosystem to speak with, look at, and touch a human being and then vanish – and in the next moment it could be having a similar social interaction

[48] Stahl (2006), p. 210.

[49] Gladden, "The Artificial Life-Form as Entrepreneur" (2014), p. 417.

[50] Omohundro, "Rational Artificial Intelligence for the Greater Good" (2012).

[51] Pearce, "The Biointelligence Explosion" (2012).

[52] Gruen, "The Moral Status of Animals" (2014).

with another human being in another part of the globe. Because of the limitations of our organic bodies, a human being can only be physically located in one place at a time, although we can use technologies like the telephone or videoconferencing or virtual reality to interact in a mediated fashion with environments in other countries. However, a nonlocalizable robot could truly be 'in' many different locations at once; there is no principle of primacy that allows us to say that one venue of interaction is home to the robot's 'real' body and that it is simply projecting itself virtually into the other locations through a 'mediating technology' – because the only such technology is the robot's body itself. The fact that the robot is located 'here' at a given moment in time does not mean that it is not also somewhere else.

Difficulties arise in attempting to analyze such a robot's possibility for bearing responsibility for its actions. In principle, it might seem capable of bearing human-like responsibility. However, our human ability to bear moral responsibility depends on our metavolitional ability to have *expectations:* an intention requires both a desire and a belief about the outcomes that will result from an action. A being whose body is so radically different from our own may not experience the same sorts of beliefs or expectations about how its environment will be affected by its actions – since the very concepts of 'environment' and 'action' could have vastly different meanings for such an entity. Similarly, the question of legal responsibility is clouded by the need to determine to which nation's laws the robot is subject, when its body is potentially scattered across continents and seas and satellites in orbit. Identifying 'where' in the robot's body a particular decision was made may be as difficult as pinpointing which neurons in a human brain generated a particular action.[53] Moreover, it is unclear whether the application of traditional legal 'rewards' and 'punishments' would have meaning for such a nonlocalizable entity.

The Robot as Diffuse Alien Intelligence

Finally we can consider a nonlocalizable robot whose cognitive processes demonstrate superautonomy and supervolitionality. In principle, one might be inclined to impute legal and moral responsibility to such a robot because it possesses levels of autonomy and volitionality that are at least as great as those of human beings. But that overlooks the fact that the robot's experience of its own superautonomy and supervolitionality likely has little in common with our human experiences of our vastly less sophisticated autonomy and metavolitionality. Yampolskiy and Fox argue[54] that "humanity occupies only a tiny portion of the design space of possible minds" and that "the mental architectures and goals of future superintelligences need not have most of the

[53] Bishop, "On Loosemore and Goertzel's 'Why an Intelligence Explosion Is Probable'" (2012), p. 97.

[54] Yampolskiy and Fox (2012), p. 129.

properties of human minds." We might be able to discern from observing such robots' interactions with one another that their actions follow some highly complex and beautiful (or perhaps terrifying) patterns, but it may be impossible for us to determine what portion of those patterns results from a sort of universal natural law that is reflected in the robots' moral sentiments, how much of it is due to legal frameworks that the robots have constructed among themselves,[55] how much of it is due to their cultural traditions, and how much of it results simply from engineering requirements, mathematical or logical principles, or physical laws. We might never come to understand the internal processes that shape such robots' thoughts, decisions, and actions, because they lack the ability or desire to explain them to us using language or concepts that we can fathom.[56] Any attempt by the robots to explain their moral, social, and legal frameworks might require them to employ metaphorical imagery and 'overload' our senses in a way that bears more resemblance to the phenomenon of divine revelation than to any processes of logic and reasoning available to the human mind.[57]

Serious problems ensue from attempting to apply human notions of "law" to such alien entities. The law "sets parameters, which, as society has determined, outline the limits of an accepted range of responses within the circumscribed field which it addresses";[58] however in this case it is difficult to refer to a single 'society' that comprises both human beings and such artificial superintelligences and can determine the range of acceptable behaviors for its members. (Indeed, the proliferation of transhuman genetic engineering and cybernetic augmentation might even cause humanity itself to splinter into numerous mutually incomprehensible civilizations.[59]) Although human beings might share the same digital-physical ecosystem with such robots and interact with them causally, we could only be said to share a 'society' with them in the same way that we share a society with the birds and insects that live in our gardens.

The Robot as Charismatic Lawgiver and Moral Beacon

If communication between humanity and such an alien robotic society *does* take place, it may be we human beings who end up developing new "laws,

[55] Michaud, *Contact with Alien Civilizations: Our Hopes and Fears about Encountering Extraterrestrials* (2007), p. 243.

[56] Abrams, "Pragmatism, Artificial Intelligence, and Posthuman Bioethics: Shusterman, Rorty, Foucault" (2004).

[57] Gladden, "The Social Robot as 'Charismatic Leader'" (2014), p. 337.

[58] Calverley (2008), p. 534.

[59] Abrams (2004).

customs, and attitudes" as a result of the exchange.[60] We may discover that some nonorganic intelligent entities possess moral and legal frameworks that are superior to ours in their beauty, consistency, fairness, and wisdom; such systems may appear so irresistibly good and worthwhile that we cannot help but desire that the robots who embody them should teach – and even govern – us. They may become the moral and legal leaders of our human society not through intimidation or coercion or through their vast technological expertise, but because we find ourselves *admiring* them for their goodness and yearning to become more like them.[61] Thus it may not be we human beings who are determining the extent to which such robots are morally and legally responsible for their actions, but the robots who are providing us with new and richer and truer frameworks for understanding the moral and legal responsibility that is borne by all sapient beings – including ourselves.

Developing Legal and Ethical Frameworks for Autonomous Nonlocalizable Robots

There are several branches of law and ethics from which one can draw insights and inspiration when attempting to develop legal and ethical frameworks that address the question of responsibility on the part of nonlocalizable robots. The relevant fields of law vary, depending on the levels of autonomy and volitionality possessed by a robot.

Nonlocalizable Robots with Low Autonomy and Volitionality

Nonlocalizable robots possessing low levels of autonomy and volitionality will likely be seen as inanimate environmental resources to be exploited by human beings – or hazards to be mitigated. Responsibility for the robots' activities will devolve on their creators, and the nonlocalizable nature of the robots means that it may be impossible to determine who those creators are. Parallels may be found in the debate over humanity's collective legal and ethical responsibility for environmental damage caused by global climate change, a phenomenon in which specific localized damage in one country may result from "greenhouse pollution from a great many untraceable point sources" around the world.[62] On the other hand, if nonlocalizable robots are seen as a useful resource to be exploited, then legal models can be found in the international treaties and institutions governing global phenomena like the preservation of biological diversity and humanity's use of oceans, the Antarctic, and

---

[60] Michaud (2007), p. 293.

[61] Gladden, "The Social Robot as 'Charismatic Leader'" (2014), pp. 329-33; Kuflik (1999), p. 181.

[62] Vanderheiden, "Climate Change and Collective Responsibility" (2011).

outer space, which explicitly ground their legal and philosophical rationale in the need to advance and preserve the "common interest of all mankind."[63]

Nonlocalizable Robots with Animal-like Autonomy and Volitionality

If nonlocalizable robots display moderate levels of autonomy and volitionality – roughly comparable to those of animals – then they are no longer simply passive features of the environment but entities capable of acting in accordance with their own expectations and desires. Here we can draw insights, for example, from existing legal and ethical debates surrounding genetically modified animals[64] that have been engineered for particular purposes and released into the wild. Such creatures are not morally or legally responsible for their actions but can display the sort of quasi-responsibility that directs our attention back to their human designers, who bear ultimate responsibility.[65]

We can also draw on existing law and ethics regarding the production and use of artificial space satellites. Many artificial satellites are, in effect, highly sophisticated orbital robots that possess powerful onboard computerized 'brains' and which may be capable of remotely sensing and recording activities occurring on the earth's surface[66] and receiving, transmitting, and potentially disrupting communications (including Internet traffic) with earth-based sources and other satellites.[67] The creation of robotic orbital satellites that can physically intercept, manipulate, and reposition other satellites[68] – not to mention the possibility of satellites' computerized controls being compromised through computer viruses or remote hacking – opens the possibility for artificial satellites to be repurposed in ways that are no longer subject to the control of their original human designers, thereby complicating the attribution of responsibility for the satellites' actions. Such devices are multilocal rather than nonlocalizable, since any given satellite is clearly located in a particular place in each moment, and the country responsible for the satellite's production, launch, and operation is easy to determine. However, the fact that such satellites operate from an extraterritorial region while acting in ways that affect human beings in particular countries has led to a unique body of law regarding their activities.

---

[63] Berkman, "Common Interests' as an Evolving Body of International Law: Applications to Arctic Ocean Stewardship" (2012), p. 158.

[64] Beech, "Regulatory Experience and Challenges for the Release of GM Insects" (2014).

[65] Stahl (2006), p. 210.

[66] Sadeh, "Politics and Regulation of Earth Observation Services in the United States" (2010).

[67] Vorwig, "Regulation of Satellite Communications in the United States" (2010).

[68] Lemonick, "Save Our Satellites" (2013).

Nonlocalizable Robots with Human-like Autonomy and Metavolitionality

When considering nonlocalizable robots that possess higher levels of autonomy and volitionality, we can draw on existing law and ethics regarding the action of nonhuman persons. The fact that 'legal persons' can exist that are not 'natural persons' is a widely accepted legal concept[69] embodied in the legal identity of states and corporations. Already international human rights law encourages – and even requires – states to develop extraterritorial legal structures to control corporations registered in those states and hold them responsible for actions committed in other states or in regions such as international waters or outer space.[70] Such extraterritorial structures could also be applied to nonlocalizable robots that were originally developed in or have somehow been registered as 'synthetic nationals' of particular states.

Similarly, the Internet – and now the Internet of Things – can be seen as a somewhat primitive precursor to future global networks that may become home to autonomous metavolitional nonlocalizable robots. Proposals for governing the Internet of Things through a combination of transgovernmental networks, legislation, and self-regulation[71] may be relevant. Self-regulation, in particular, can be an important form of 'soft law' in which governments set broad parameters but leave the actual implementation to private industry, as government is incapable of acting quickly enough or with sufficient expertise in response to such a rapidly evolving and technologically complex field.[72] In the case of nonlocalizable robots with human-like autonomy and metavolitionality, self-regulation could mean governance of the robots not by their manufacturers but by the robots themselves, through their creation of a robotic society.

Nonlocalizable Robots with Superautonomy and Supervolitionality

From a legal and moral perspective, nonlocalizable robots that demonstrate superautonomy and supervolitionality cannot easily be compared to sentient animals or to sapient human beings; they are more analogous to superintelligent extraterrestrial aliens with whom humanity might someday come into contact. Interestingly, much scholarship has been dedicated to the question of how human law and morality might relate to intelligent alien entities from other planets. Michaud reminds us[73] that "As far back as Immanuel Kant, some

---

[69] Dreier & Spiecker (2012), p. 215; Calverley (2008); Hellström (2013).

[70] Bernaz, "Enhancing Corporate Accountability for Human Rights Violations: Is Extraterritoriality the Magic Potion?" (2012).

[71] Weber (2010), pp. 27-28.

[72] Weber (2010), p. 24.

[73] Michaud (2007), p. 374.

have speculated about a legal system that would apply to all intelligences in the universe." Efforts at developing such universal principles have grown more sophisticated over the centuries, as developments in fields like exobiology, neuroscience, complex systems theory, artificial intelligence, and artificial life have given us greater insights into what forms such alien intelligence might take.

Michaud notes that the traditional human moral precept of the Golden Rule would be of limited usefulness to us when interacting with alien intelligences: our human drives, aspirations, sensations, and reasoning processes are so different from those of the alien beings that merely knowing that *we* would like to be treated in a particular way tells us nothing about whether aliens who were treated in the same way might experience that event as joy or suffering, as equity or injustice.[74] The first serious modern attempt at developing appropriate principles to govern humanity's potential encounter with an extraterrestrial intelligence occurred when "Half a century ago, space lawyer Andrew Haley proposed what he called The Great Rule of Metalaw: Do unto others as they would have you do unto them."[75] However, treating alien intelligences as they wish to be treated is no simple matter; Michaud notes that "It is not clear how we could observe this principle in the absence of extensive knowledge about the other civilization. We may need detailed, sophisticated communication to find out."[76] As noted above, though, establishing communication between such disparate forms of intelligence may be difficult or even impossible. Our coming to understand the nuances of the aliens' moral universe is not simply a matter of translating a text between two languages; it may be a task that exceeds the ability or desire of both civilizations. Nevertheless, efforts by ethicists and legal scholars to lay the groundwork for such encounters are also helping to prepare us for contact with superintelligent nonlocalizable robotic entities whose origins are wholly of this world.

Conclusion

Many practical issues arise when determining whether robots bear moral and legal responsibility for their actions. For example, we may need to study the nature of a robot's primary cognitive substrate and process to determine whether they are of a sort that allows for true autonomy and metavolitionality, or we may need to gather data to reconstruct the chain of causality surrounding a particular action performed by a robot. However, assuming that such information can be obtained, the theoretical structures of existing law and moral philosophy are in principle largely adequate for determining a contemporary local or multilocal robot's degree of responsibility for its actions.

---

[74] Michaud (2007), p. 300.

[75] Michaud (2007), p. 374.

[76] Michaud (2007), p. 374.

Depending on its level of autonomy and volitionality, such a robot might be treated as a passive tool that possesses no moral or legal responsibility or as a human-like moral and legal actor to whom such responsibility is attributed.

As we have seen, though, the advent of nonlocalizable robotic entities will transform these moral and legal equations. When a robot's body is continually metamorphosing and drifting through the global digital-physical ecosystem, it may be impossible to determine which nation, if any, can claim the robot as a 'citizen' or 'resource' subject to its laws. Moreover, it may not even be possible to construct a one-to-one correlation between some action that has occurred in the world and a particular identifiable robotic entity that presumably performed it. Beyond these practical legal issues, there are also deeper moral questions: a nonlocalizable robot's way of existing and acting may simply be so different from those of human beings, animals, or any other form of being known to date that it may be inappropriate or impossible for human beings to apply even our most fundamental (and even presumably 'universal') moral principles to the activity of such entities.

Already, scholars have begun to prepare moral and legal frameworks that can give us insights into the actions of intelligent extraterrestrial beings whom we might someday encounter and whose physical morphology, motivations, thoughts, forms of communication and social interaction, and normative behavioral principles may be radically different from our own. So, too, would humanity benefit from attempting to envision moral and legal frameworks that can help us account for the actions of future robots whose manner of thinking, acting, and being is just as alien, but whose origins lie closer to home. Even before we have actually begun to interact with such beings, the careful thought that we put into analyzing the moral and legal responsibility that they will bear for their actions may enrich our understanding of the responsibility that we human beings bear for our own.

References

Abrams, Jerold J. "Pragmatism, Artificial Intelligence, and Posthuman Bioethics: Shusterman, Rorty, Foucault." *Human Studies* vol. 27, no. 3 (2004): 241-58.

Beech, C. "Regulatory Experience and Challenges for the Release of GM Insects." *Journal Für Verbraucherschutz Und Lebensmittelsicherheit* vol. 9, no. 1 (2014): 71-76.

Bekey, G.A. *Autonomous Robots: From Biological Inspiration to Implementation and Control.* Cambridge, MA: MIT Press, 2005.

Berkman, P.A. "'Common Interests' as an Evolving Body of International Law: Applications to Arctic Ocean Stewardship." In *Arctic Science, International Law and Climate Change*, pp. 155-73. Beiträge Zum Ausländischen Öffentlichen Recht Und Völkerrecht no. 235, 2012.

Bernaz, N. "Enhancing Corporate Accountability for Human Rights Violations: Is Extraterritoriality the Magic Potion?" *Journal of Business Ethics* vol. 117, no. 3 (2012): 493-511.

Bishop, P. "On Loosemore and Goertzel's 'Why an Intelligence Explosion Is Probable'." In *Singularity Hypotheses*, edited by A.H. Eden, J.H. Moor, J.H. Søraker, and E. Steinhart, pp. 97-98. The Frontiers Collection. Berlin/Heidelberg: Springer, 2012.

Calverley, D.J. "Imagining a non-biological machine as a legal person." *AI & SOCIETY* 22, no. 4 (2008): 523-37.

Clarke, Arthur C. "Hazards of Prophecy: The Failure of Imagination." In *Profiles of the Future: An Inquiry into the Limits of the Possible*, revised edition. New York: Harper & Row, 1973.

Coeckelbergh, Mark. "From Killer Machines to Doctrines and Swarms, or Why Ethics of Military Robotics Is Not (Necessarily) About Robots." *Philosophy & Technology* 24, no. 3 (2011): 269-78.

Datteri, E. "Predicting the Long-Term Effects of Human-Robot Interaction: A Reflection on Responsibility in Medical Robotics." *Science and Engineering Ethics* vol. 19, no. 1 (2013): 139-60.

Défago, X. "Distributed Computing on the Move: From Mobile Computing to Cooperative Robotics and Nanorobotics." In *Proceedings of the 1st International Workshop on Principles of Mobile Computing (POMC 2001)*, pp. 49-55. Newport, Rhode Island, 2001.

Doarn, C.R., and G.R. Moses. "Overcoming Barriers to Wider Adoption of Mobile Telerobotic Surgery: Engineering, Clinical and Business Challenges." In *Surgical Robotics*, edited by J. Rosen, B. Hannaford, and R.M. Satava, pp. 69-102. New York: Springer, 2011.

Dreier, T., and I. Spiecker genannt Döhmann. "Legal Aspects of Service Robotics." *Poiesis & Praxis* vol. 9, no. 3-4 (2012): 201-17.

Gladden, Matthew E. "The Artificial Life-Form as Entrepreneur: Synthetic Organism-Enterprises and the Reconceptualization of Business." In *Proceedings of the Fourteenth International Conference on the Synthesis and Simulation of Living Systems*, edited by Hiroki Sayama, John Rieffel, Sebastian Risi, René Doursat and Hod Lipson, pp. 417-18. Cambridge, MA: The MIT Press, 2014.

Gladden, Matthew E. "The Social Robot as 'Charismatic Leader': A Phenomenology of Human Submission to Nonhuman Power." In *Sociable Robots and the Future of Social Relations: Proceedings of Robo-Philosophy 2014*, edited by Johanna Seibt, Raul Hakli, and Marco Nørskov, pp. 329-39. Frontiers in Artificial Intelligence and Applications 273. IOS Press, 2014.

Gruen, L. "The Moral Status of Animals." In *The Stanford Encyclopedia of Philosophy*, edited by E.N. Zalta. Stanford University, 2014. http://plato.stanford.edu/archives/fall2014/entries/moral-animal/. Accessed January 6, 2015.

Gunther, N.J. "Time – The Zeroth Performance Metric." In *Analyzing Computer System Performance with Perl::PDQ*, pp. 3-46. Springer Berlin Heidelberg, 2005.

Hellström, T. "On the Moral Responsibility of Military Robots." *Ethics and Information Technology* vol. 15, no. 2 (2013): 99-107.

Kirkpatrick, K. "Legal Issues with Robots." *Communications of the ACM* vol. 56, no. 11 (2013): 17-19.

Kuflik, Arthur. "Computers in Control: Rational Transfer of Authority or Irresponsible Abdication of Autonomy?" *Ethics and Information Technology* 1 (3) (1999): 173-84. doi:10.1023/A:1010087500508.

Lemonick, M. "Save Our Satellites." *Discover* vol. 34, no. 7 (2013): 22-24.

Loosemore, R., and B. Goertzel. "Why an Intelligence Explosion Is Probable." In *Singularity Hypotheses*, edited by A.H. Eden, J.H. Moor, J.H. Søraker, and E. Steinhart, pp. 83-98. The Frontiers Collection. Berlin/Heidelberg: Springer, 2012.

Michaud, M.A.G. *Contact with Alien Civilizations: Our Hopes and Fears about Encountering Extraterrestrials.* New York: Springer, 2007.

Murphy, Robin. *Introduction to AI Robotics.* Cambridge, MA: The MIT Press, 2000.

Omohundro, S. 2012. "Rational Artificial Intelligence for the Greater Good," in *Singularity Hypotheses*, edited by A.H. Eden, J.H. Moor, J.H. Søraker, and E. Steinhart, pp. 161-79. The Frontiers Collection. Berlin/Heidelberg: Springer, 2012.

Pagallo, U. "Robots in the Cloud with Privacy: A New Threat to Data Protection?" *Computer Law & Security Review* vol. 29, no. 5 (2013): 501-08.

Pearce, David. "The Biointelligence Explosion." In *Singularity Hypotheses*, edited by A.H. Eden, J.H. Moor, J.H. Søraker, and E. Steinhart, pp. 199-238. The Frontiers Collection. Berlin/Heidelberg: Springer, 2012.

Prestes, E., J.L. Carbonera, S. Rama Fiorini, V.A.M. Jorge, M. Abel, R. Madhavan, A. Locoro et al. 2013. "Towards a Core Ontology for Robotics and Automation." In *Robotics and Autonomous Systems*, pp. 1193-1204. Ubiquitous Robotics vol. 61, no. 11.

Ren, W., and B. Xu. "Distributed Wireless Networked H_∞ Control for a Class of Lurie-Type Nonlinear Systems." *Mathematical Problems in Engineering* (May 5, 2014): e708252.

Sadeh, E. "Politics and Regulation of Earth Observation Services in the United States." In *National Regulation of Space Activities*, edited by R.S. Jakhu, pp. 443-58. Space Regulations Library Series 5. Springer Netherlands, 2010.

Sparrow, R. "Killer Robots." *Journal of Applied Philosophy* vol. 24, no. 1 (2007): 62-77.

Stahl, B. C. "Responsible Computers? A Case for Ascribing Quasi-Responsibility to Computers Independent of Personhood or Agency." *Ethics and Information Technology* 8, no. 4 (2006): 205-13.

Vanderheiden, S. "Climate Change and Collective Responsibility." In *Moral Responsibility*, edited by N.A. Vincent, I. Van de Poel, and J. Van den Hoven, pp. 201-18. Library of Ethics and Applied Philosophy 27. Springer Netherlands, 2011.

Vorwig, P.A. "Regulation of Satellite Communications in the United States." In *National Regulation of Space Activities*, edited by R.S. Jakhu, pp. 421-42. Space Regulations Library Series 5. Springer Netherlands, 2010.

Weber, R.H., and R. Weber. "General Approaches for a Legal Framework." In *Internet of Things*, pp. 23-40. Springer Berlin Heidelberg, 2010.

Yampolskiy, Roman V., and Joshua Fox. "Artificial General Intelligence and the Human Mental Model." In *Singularity Hypotheses*, edited by Amnon H. Eden, James H. Moor, Johnny H. Søraker, and Eric Steinhart, pp. 129-45. The Frontiers Collection. Springer Berlin Heidelberg, 2012.

Chapter Ten

Leveraging the Cross-Cultural Capacities of Artificial Agents as Leaders of Human Virtual Teams[1]

Abstract. The human beings who manage global virtual teams regularly face challenges caused by factors such as the lack of a shared language and culture among team members and coordination delay resulting from spatial and temporal divisions between members of the team. As part of the ongoing advances in artificial agent (AA) technology, artificial agents have been developed whose purpose is to assist the human managers of virtual teams. In this text, we move a step further by suggesting that new capabilities being developed for artificial agents will eventually give them the ability to successfully manage virtual teams whose other members are human beings. In particular, artificial agents will be uniquely positioned to take on roles as managers of cross-cultural, multilingual, global virtual teams, by overcoming some of the fundamental cognitive limitations that create obstacles for human beings serving in these managerial roles.

In order to effectively interact with human team members, AAs must be able to decode and encode the full spectrum of verbal and nonverbal communication used by human beings. Because culture is so deeply embedded in all human forms of communication, AAs cannot communicate in a way that is 'non-cultural'; an AA that is capable of communicating effectively with human team members will necessarily display a particular culture (or mix of cultures), just as human beings do. Already researchers have designed AAs that can display diverse cultural behaviors through their use of language, intonation, gaze, posture, emotion, and personality.

[1] This text was originally published in *Proceedings of the 10th European Conference on Management Leadership and Governance*, edited by Visnja Grozdanić, pp. 428-435. Reading: Academic Conferences and Publishing International Limited, 2014.

The need for AA team leaders to display cultural behavior raises the key question of *which* culture or cultures the AA leader of a particular human virtual team should display. We argue that the answer to this question depends on both the cultural makeup of a team's human members and the methods used to share information among team members. To facilitate the analysis of how an AA team leader's cultural behaviors can best be structured to fit the circumstances of a particular virtual team, we propose a two-dimensional model for designing suites of cultural behaviors for AAs that will manage human virtual teams. The first dimension describes whether an AA deploys the same cultural behaviors for its dealings with all team members ('objectivity') or customizes its cultural display for each team member ('personalization'). The second dimension describes whether the AA always displays the same culture to a given team member ('invariance') or possesses a repertoire of cultural guises for a particular team member, from which it chooses one to fit the current situation ('situationality'). The two dimensions of objective/personalized and invariant/situational cultural behaviors yield four archetypes for AAs leading virtual human teams. We consider examples of each type of AA, identify potential strengths and weaknesses of each type, suggest particular kinds of virtual teams that are likely to benefit from being managed by AAs of the different types, and discuss empirical study that can test the validity and usefulness of this framework.

Background and Introduction

The Potential Desirability of Artificial Agents as Leaders of Human Virtual Teams

Our human managerial capacities and practices evolved in a world whose lack of communication technologies meant that interpersonal relationships between workers were typically local, face-to-face, and culturally homogeneous. However, technological advances have compelled human beings to adapt to a new reality in which relationships between managers and employees are often culturally heterogeneous, spatially and temporally dispersed, and mediated extensively by technology. One manifestation of this technological mediation of work relationships is the phenomenon of virtual teams. Through the use of email, text messages, telephones, videoconferencing, and online document-sharing and project management tools, virtual team members can collaborate closely without ever meeting face-to-face. In an international organization, virtual team members may dwell in different time zones, which creates challenges in scheduling meetings and distinguishing 'work time' from 'non-work time.' Membership in this sort of global virtual team can bring with it the expectation of an employee's availability for almost round-the-clock, instantaneous communication and decision-making.

While human beings have been remarkably successful at adapting to these new technologies, we are still subject to fundamental biological limitations that affect our ability to successfully manage global virtual teams. This raises the question of whether some sort of *nonhuman* being – such as an advanced form of artificial agent – can be developed whose capabilities would allow it to match or exceed our human effectiveness at managing certain types of human virtual teams. (Throughout this paper, the phrase 'human virtual team' refers to a virtual team that includes some human beings as members, although the manager and additional members of the team may be artificial agents.)

For example, an artificial agent (AA) team leader that never needs to sleep and is always ready to interact with team members could conceivably mitigate some challenges experienced by global virtual teams, such as coordination delay resulting from spatial and temporal boundaries between members.[2]

General Management Capabilities Needed by AAs in Order to Lead Human Virtual Teams

Artificial agents such as conversational agents or 'chatbots' are already widely used in business applications. For example, a conversational agent might appear on a commercial website in the form of an animated figure that carries on a dialogue with human visitors to the site, to answer their questions or carry out requests.[3]

Progress is being made toward developing AAs that can fill more substantial roles within organizations, such as potentially serving as the managers of human virtual teams. This process will be facilitated by the fact that human beings are not only *able* to form socialized relations with artificial agents as if they were human but are naturally inclined to do so.[4] Already AAs have been designed that demonstrate a number of the general capacities that will prove useful for managing human virtual teams. For example, AAs are capable of selecting communication strategies that maximize trust between managers and employees[5] and are capable of perceiving and critically evaluating the performance of virtual teams.[6] It has thus far been envisioned that such AAs will serve as 'tools' for human beings who manage virtual projects, helping them

---

[2] Cummings et al., "Spatial and Temporal Boundaries in Global Teams" (2007).

[3] Perez-Marin & Pascual-Nieto, *Conversational Agents and Natural Language Interaction: Techniques and Effective Practices* (2011).

[4] Rehm & Nakano, "Some Pitfalls for Developing Enculturated Conversational Agents" (2009); Friedenberg, *Artificial Psychology: The Quest for What It Means to Be Human* (2008).

[5] Dai et al., "TrustAider – Enhancing Trust in E-Leadership" (2013).

[6] Nunes & O'Neill, "Assessing the Performance of Virtual Teams with Intelligent Agents" (2012).

to improve their performance as leaders.[7] However, this ability to critically evaluate the work of human virtual team members will also contribute to AAs' ability to serve as the leaders of such teams.

Among the many such capabilities that an AA manager will need to possess, one that is especially important for managing a global virtual team of human employees is mastery of the perception and display of cultural behaviors.

The Importance of Culture for AAs Leading Virtual Teams

Within the context of business communication, the majority of a message's meaning is often conveyed through nonverbal communication rather than the choice of words used.[8] Thus virtual team leaders who communicate only through emails or other written texts are at a disadvantage against those who can utilize facial expressions, gestures, tone of voice, and the full spectrum of nonverbal communication.[9] An AA that is leading a human virtual team will be most effective if it appears to team members in an audiovisual form that allows it to display the full range of human nonverbal communication.

If an AA appears in a virtual form that utilizes a human-like body and voice, this raises the essential question of culture. An individual's culture suffuses and is reflected in almost every aspect of his or her verbal and nonverbal behavior and appearance; it is thus not possible for an AA to utilize a full range of verbal and nonverbal communication in a way that is 'non-cultural' or 'culturally indeterminate': an AA capable of communicating effectively with human team members will necessarily display a particular culture (or mix of cultures), just as human beings do.[10] In the case of an AA, its 'culture' is reflected on several levels, including its (virtual) appearance, gestures, vocal intonation, choice of actions, emotions, personality, and the background story or 'biography' that is attributed to it to explain its body of knowledge and its organizational role.[11]

---

[7] Kriksciuniene & Strigunaite, "Multi-Level Fuzzy Rules-Based Analysis of Virtual Team Performance" (2011).

[8] Ober, *Contemporary Business Communication* (2007).

[9] El-Tayeh et al., "A Methodology to Evaluate the Usability of Digital Socialization in 'virtual' Engineering Design" (2008).

[10] Rehm & Nakano (2009); Rehm et al., "From Observation to Simulation: Generating Culture-Specific Behavior for Interactive Systems" (2009).

[11] Payr & Trappl, "Agents across Cultures" (2003).

Cultural Competencies Needed for AAs as Leaders of Cross-cultural Global Virtual Teams

In order to effectively manage a cross-cultural, global virtual team of human employees, an AA will require a significant degree of cultural competence. For example, if an AA is managing a team with human members in Japan and the US, it should 'know' that the expression of strong emotion by an AA may be more acceptable in an individualistic culture like the US than in a collectivistic culture like Japan.[12]

Some of the cultural capacities needed by AA managers have already been developed. Researchers have successfully designed AAs that can perceive, understand, and display diverse cultural behaviors through a choice of actions, language, vocal intonation, gaze, posture, gestures, emotions, personality, the expectations for the kinds of social interactions that occur between individuals in a particular social relation, and other "unwritten rules of human cultures."[13] Ongoing advances are being made in this field.

Choosing the Culture(s) that an AA Virtual Team Leader Should Display

Considering the fact that it is necessary for a successful AA team leader to effectively display specific cultural behaviors, and supposing that it is (or soon will be) possible to create AA team leaders that are capable of perceiving and displaying a sophisticated range of cultural behaviors, this raises the key question of this paper: *which culture or cultures* should the AA leader of a human virtual team display?

We posit that an AA virtual team leader can either: 1) reflect its own unique synthetic nonhuman culture; 2) reflect one or more existing human cultures; or 3) collaborate with its human team members in the development of a shared synthetic culture[14] that is created jointly by all team members. If team members will be working together for an extended period of time, all three possibilities become viable options. However, options 1 and 2 require significant time and effort on the part of human team members to develop new competencies in a synthetic culture that may have no usefulness beyond the particular team. It may thus be more efficient and effective – especially for a business project team of limited duration – if the AA can simply reflect existing human culture.

We must then consider the question of which human culture or cultures the AA team leader should display. We suggest that this will depend on at

---

[12] Rehm & Nakano (2009).

[13] Mascarenhas et al., "Social Importance Dynamics: A Model for Culturally-Adaptive Agents" (2013); Friedenberg (2008); Payr & Trappl (2003).

[14] Payr & Trappl (2003).

least two critical factors: 1) the particular linguistic and cultural makeup of the team's human members; and 2) the network topology and technological means that the team's human members typically employ when interacting with one another and their leader.

Our Proposed Two-dimensional Model for Designing AA Team Leaders' Cultural Behaviors

To help an organization choose the sort of culture(s) that an AA virtual team leader should display – based on these critical factors of a team's linguistic and cultural makeup and means of interaction – we propose the use of a two-dimensional model for designing suites of cultural behaviors for AAs managing human virtual teams.

Four Types of Artificial Agents Leading Human Virtual Teams	**Invariance** The culture displayed to a particular team member by the AA team leader does not change.	**Situationality** The culture displayed to a particular team member by the AA team leader changes depending on the situation.
Objectivity The AA team leader displays the same culture to all team members.	**Objective-Invariant** • Simplest design • Some team members may communicate poorly with the AA leader due to the lack of a shared culture and language • All team members share the same objective experience of the AA leader • Does not allow exploitation of the effects of foreign vs. familiar cultural displays	**Objective-Situational** • Moderately complex design • Some team members may communicate poorly with the AA leader due to the lack of a shared culture and language • All team members share the same objective experience of the AA leader • May allow some exploitation of the effects of foreign vs. familiar cultural displays
Personalization The AA team leader displays different cultures to different team members.	**Personalized-Invariant** • Highly complex design • Allows each team member to communicate fluently with the AA using his or her own culture and language • Team members do not share a common objective experience of the AA's messages or behavior • Does not allow exploitation of the effects of foreign vs. familiar cultural displays	**Personalized-Situational** • Extremely complex design • Allows each team member to communicate fluently with the AA using his or her own culture and language • Team members do not share a common objective experience of the AA's messages or behavior • Allows full exploitation of the effects of foreign vs. familiar cultural displays

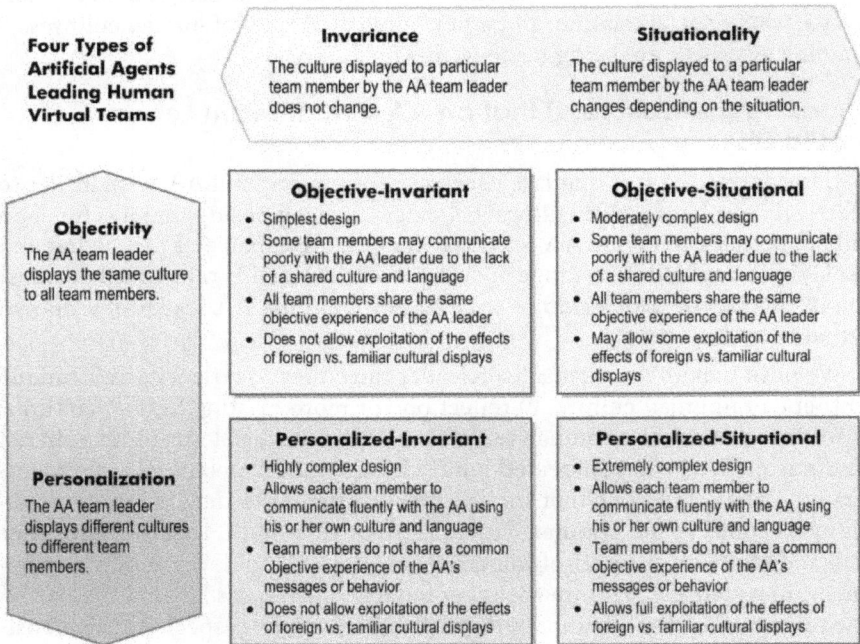

Figure 1: An overview of our two-dimensional model of the cultural behavior of an artificial agent (AA) leading a human virtual team. The axes of objectivity/personalization and invariance/situationality yield four potential types of AA leaders.

The model's vertical dimension describes whether an AA deploys the same cultural behaviors for its dealings with all team members ('objectivity') or customizes its cultural display for each team member ('personalization'). The horizontal dimension describes whether the AA always displays the same culture to a given team member ('invariance'), or possesses a repertoire of cultural guises for a particular team member, from which it chooses one to fit the current situation ('situationality'). The two dimensions of objective/personalized and invariant/situational cultural behaviors yield four archetypes for AAs leading human virtual teams. Figure 1 presents an overview of this framework.

The Four Types of AA Leaders as Categorized According to Their Cultural Behavior

We can now discuss in more detail these four types of AA leaders of human virtual teams and identify some potential advantages and disadvantages of each type.

Objective-Invariant Type

Description and Examples

When managing a human virtual team, an AA of the objective-invariant type is given by its designers (or selects on its own) a single culture, native language, and apparent age that will best facilitate its work as team leader during the life of the team. The AA displays this single set of characteristics at all times in all dealings with all employees.

For example, imagine a virtual human team of employees from several countries who are involved with a hydroelectric construction project. Given the team's assigned objectives and personnel makeup, it might be decided that the team's leader would most effectively manage the team's members if he or she behaved in a way that made team members comfortable about approaching the leader with concerns or ideas (i.e., if the relationship reflected a low power distance, as defined by Hofstede), encouraged team members to take personal responsibility for their own work (i.e., high individualism), and established clear objectives and procedures to guide team members in their work (i.e., high uncertainty avoidance). Germany is a nation whose culture, broadly speaking, reflects these traits.[15] Thus it might be decided that the team's AA leader will appear in the virtual guise of, say, a 35-year-old female German engineer. (The selection of 'German' as the culture means that while the AA can speak to team members in any language known to the AA, it speaks all languages with a slight German accent and uses typically German metaphors and allusions.)

The fact that the German culture is a real, historically grounded human culture (rather than a new synthetic culture) means that the AA will display self-consistent cultural behaviors that will be recognizable and predictable to the team's human members, to the extent that individual team members have been previously exposed to German culture. Moreover, the fact that *German*

[15] Hofstede, "Germany" (2014).

culture was selected means that particular traits found in this culture (such as low power distance and high individualism) will subtly reinforce the desired workplace atmosphere and work practices that will help both individual team members and the entire team to excel.

As another example, imagine a virtual human team that consists of financial officers at a manufacturer's plants around the world who participate in a global virtual team led by an AA based at the company's headquarters. The team's purpose is to coordinate the sharing of financial data and best practices among the plants. It has been decided that the team leader's behavior should reinforce the idea that ultimate financial decision-making authority rests with corporate HQ and that the role of individual plants' financial managers is to efficiently carry out HQ's directives (i.e., the relationship reflects high power distance), that employees' highest concern should be the welfare of the whole company rather than the success of individual employees or plants (i.e., high collectivism), and that the financial managers' role is to ensure the predictability of their plants' finances rather than take entrepreneurial risks (i.e., high uncertainty avoidance). More so than some other nations, Mexico is a nation whose culture reflects these traits.[16] Thus it might be decided that the team's AA leader will appear in the virtual guise of, say, a 50-year-old male Mexican financial officer whose cultural behaviors would support the kinds of relationships and work practices desired in team members.

Potential Advantages and Disadvantages

It is simpler to design an AA that displays the traits of a single predetermined culture than one that displays multiple cultures. Moreover, while some team members might have less familiarity with the objective-invariant AA's chosen culture (and thus may potentially misunderstand the AA's message), the fact that all team members are presented with the same message allows them to more easily discuss it among themselves.

This approach also has disadvantages, though. Imagine a team including French, Canadian, and Chinese members. If the AA displays a Chinese accent and speech patterns, utilizes Chinese metaphors and cultural allusions when explaining assignments, and manifests Chinese cultural expectations for the roles of members within then team, then non-Chinese members of the team may be less likely to grasp the full meaning (or even existence) of information conveyed through the team leader's nonverbal communication. The non-Chinese team members could also feel as though the Chinese team members are unfairly benefitting from in-group favoritism, even if the AA has been designed in a way that is intended to minimize such an occurrence.

[16] Hofstede, "Mexico" (2014).

Objective-Situational Type

Description and Example

When managing a human virtual team, an AA of the objective-situational type possesses a single suite of multiple cultural 'personas' from which it chooses one particular guise to display at a particular moment to all team members.

An AA leader's ability to change its cultural guise allows the AA to take advantage of the fact that in some circumstances, team members will respond better to a leader who shares their own culture – but in other circumstances, team members will respond better to a leader whose culture *differs from* their own. For example, it has been found that a human being experiences greater psychological arousal when interacting with a virtual agent whose cultural behavior is different from the person's own.[17] Increased psychological arousal can be beneficial when carrying out critical tasks requiring a high degree of concentration.

Thus, if a virtual team consists of Italian employees, it might be most conducive to effective communication and a productive work environment if the AA leading the team typically appears as a fellow Italian when interacting with team members in their everyday work. However, if the AA is required to give team members instructions for a particularly important task that needs to be remembered and carried out precisely, it may be more effective if the 'Italian AA' steps out of the office for a moment, and the AA instead temporarily adopts the guise of a Zambian or Australian or Chilean manager whose cultural dissimilarity would help enhance the Italian team members' attentiveness as the AA conveys critical instructions.

Potential Advantages and Disadvantages

In comparison to an objective-invariant AA, an objective-situational AA has an additional tool to use in maximizing its team members' effectiveness, insofar as it can exploit the benefits of being able to choose between displaying a familiar or foreign culture in a given situation. However, this effect can only be used to its full potential if the group's cultural makeup is such that there is at least one culture that is 'familiar' to all team members and one 'foreign' to all of them.

In comparison to an objective-invariant AA, one drawback of an objective-situational AA is that it is more complex to design, because it cannot simply have a single set of cultural behaviors 'hardwired' into it. The AA must possess a full knowledge of multiple cultures, must have some mechanism for changing its displayed culture, and – perhaps most challengingly – must know

[17] Obaid et al., "Cultural Behaviors of Virtual Agents in an Augmented Reality Environment" (2012).

when a situation calls for it to change its displayed behavior from that of one culture to another.

Personalized-Invariant Type

Description and Example

A personalized-invariant AA represents a more radical rethinking of the way in which a virtual team leader interacts with team members. For each human member of the team, the AA chooses the language, culture (most likely the employee's own native culture), gender, and apparent age that will allow it to interact most effectively with that individual, and the AA displays that culture in all dealings with that particular employee.

An example would be an AA leader that interacts with each team member in such a way that the AA looks, sounds, and acts as though it were that person's 'older sibling.' To a 20-year-old Venezuelan team member, the AA would display the cultural characteristics of a 25-year-old Venezuelan, while to a 50-year-old Finn, the AA would look, sound, and act as though it were a 55-year-old Finn. When communicating instructions to an American team member, the AA might make some point by using an allusion to baseball, whereas when conveying the same message to a team member from another culture, the AA might use an allusion to cricket or football, instead.

Of particular interest is the fact that it is theoretically possible (through the use of mediating technology) for the AA to display different cultures to different team members *simultaneously*, even when the AA and the team members are all interacting with one another in a live virtual group meeting.

Potential Advantages and Disadvantages

A personalized-invariant AA would be easiest to implement in a virtual team whose communication processes utilize a star network topology or hub-and-spoke design in which team members communicate primarily with and through the team leader rather than directly with one another. If team members do not directly interact with one another to 'compare notes' about their experience of the AA team leader, it is less likely that confusion will arise due to the fact that the leader presents itself to team members under different cultural guises. This hub-and-spoke arrangement also exploits the fact that the information processing needs of a virtual team can usually be most effectively met by a centralized system,[18] with the AA perhaps doubling as the team's data storage and processing system.

However, by utilizing sufficiently advanced mediating technologies, an AA could potentially display different cultural traits to different team members at

---

[18] Jensen et al., "The Effect of Virtuality on the Functioning of Centralized versus Decentralized Structures – an Information Processing Perspective" (2010).

the same time, even while they were all interacting simultaneously in a virtual group meeting. Clearly, an AA of this sort would be highly complex. While the advantages of an intelligent agent that can look and sound different to different individuals simultaneously have long been explored in science fiction,[19] due to its inherent complexity such an AA has not yet been seriously considered as a feasible business technology. In principle, though, this is simply a further development of established technologies such as the simultaneous live multilingual interpretation of a speaker's remarks used by organizations like the UN.

This personalized-invariant approach has the potential to significantly increase an AA's effectiveness in communicating with individual team members and to eliminate miscommunications arising from a lack of cultural familiarity. This is especially useful for teams whose members do not share a common language or culture. By communicating with each team member using his or her own native language and cultural knowledge, the AA can reduce the heightened stress and obstacles to performance that arise when cross-cultural virtual team members lack fluency in a shared language.[20]

Personalized-Situational Type

Description and Example

For each team member, a personalized-situational AA possesses a unique suite of at least two possible cultural personas to use in interacting with that employee: one persona reflects the employee's own native culture, and the other reflects a culture that is foreign to the employee. The AA chooses which cultural persona to present to a particular team member at a particular point in time based on the demands of the current situation.

Building on our previous personalized-invariant example, an example of a personalized-situational AA would be an AA leader which, under normal circumstances, appears to each team member as though it were that person's 'older sibling' (to cultivate a sense of camaraderie and reduce possibilities for miscommunication) but in moments of urgency or criticality appears as a more senior authority figure from a culture less familiar to that person.

Potential Advantages and Disadvantages

A personalized-situational AA can easily communicate with and coordinate the work of a team whose members do not share a common language and who possess very different cultures, as the AA can communicate with each team member using his or her own native language and cultural behaviors, in

[19] For example, see Daniels, "The Man Trap" (1966).

[20] Nurmi, "Unique Stressors of Cross-Cultural Collaboration through ICTs in Virtual Teams" (2009).

most situations. However, the personalized-situational AA can also maximize effectiveness by communicating with team members using behaviors from a foreign culture, in circumstances where heightened psychological arousal among team members would be beneficial.

This is the most complicated type of AA to design and operate. As with the personalized-invariant AA, it is more challenging to implement for a team that holds virtual group meetings involving all members than for a team whose AA leader implements a hub-and-spoke network topology in which the AA leader is responsible for transmitting information between team members.

Avenues for Empirical Research Regarding this Model

Empirical research is needed to verify the validity and usefulness of this proposed two-dimensional model. One study that can be undertaken now is to test the impact on motivation, comprehension, and performance when particular kinds of tasks common to virtual teams are presented to a human worker by AA 'managers' displaying cultures familiar and foreign to the human being. This will enhance our ability to identify real-world situations faced by virtual teams in which the ability to exploit foreign vs. familiar cultural displays by an AA leader would have the most beneficial impact.

Other research can take place after further advances are made in AA technology. Once AAs have been engineered that can present a single message to multiple participants in a virtual group meeting using different languages and cultural behaviors simultaneously, it will be feasible to empirically test the hypothesized advantages and challenges of personalized AA types and to concretely measure the cost and complexity of personalized AA types relative to objective types.

Discussion and Conclusion

A review of current artificial agent technologies and the direction of anticipated progress in the field suggests that sufficiently sophisticated AAs will have the potential to successfully manage virtual teams of human workers. The most capable and effective AA managers will be those whose cultural and linguistic flexibility allows them to lead human workers that communicate using a diverse array of languages and cultural displays. Utilizing the two-dimensional model proposed in this paper, we believe that an AA manager of the personalized-situational type represents the ultimate objective toward which research and development in this field should (and will) be advancing, as it offers the most powerful and effective design for an AA manager.

However, the inherent complexity involved with engineering a personalized-situational AA means that AAs of the objective-situational and personalized-invariant types will likely be developed first, as steppingstones along the way. By considering now the implications of these technologies for managing virtual teams, businesses can most effectively position their current e-leadership strategies within the context of this impending long-term technological change.

References

Cummings, J.N., J.A. Espinosa, and C.K. Pickering. "Spatial and Temporal Boundaries in Global Teams." In *Virtuality and Virtualization*, pp. 85-98. International Federation for Information Processing Proceedings no. 236, 2007.

Dai, Y., C. Suero Montero, T. Kakkonen, M. Nasiri, E. Sutinen, M. Kim, and T. Savolainen (2013). "TrustAider – Enhancing Trust in E-Leadership." In *Business Information Systems*, pp. 26-37. Lecture Notes in Business Information Processing 157. Springer Berlin Heidelberg, 2013.

Daniels, Marc. "The Man Trap." *Star Trek: The Original Series*, season 1, episode 1. Paramount, September 8, 1966.

El-Tayeh, A., N. Gil, and J. Freeman. "A Methodology to Evaluate the Usability of Digital Socialization in 'virtual' Engineering Design." *Research in Engineering Design* vol. 19, no. 1 (2008): 29-45.

Friedenberg, Jay. *Artificial Psychology: The Quest for What It Means to Be Human.* Philadelphia: Psychology Press, 2008.

Hofstede, G. "Germany." The Hofstede Centre. http://geert-hofstede.com/germany.html. Accessed June 7, 2014.

Hofstede, G. "Mexico." The Hofstede Centre. http://geert-hofstede.com/mexico.html. Accessed June 7, 2014.

Jensen, K.W., D. Døjbak Håkonsson, R.M. Burton, and B. Obel. "The Effect of Virtuality on the Functioning of Centralized versus Decentralized Structures – an Information Processing Perspective." *Computational and Mathematical Organization Theory* vol. 16, no. 2 (2010): 144-70.

Koehne, B., M.J. Bietz, and D. Redmiles. "Identity Design in Virtual Worlds." In *End-User Development*, pp. 56-71. Lecture Notes in Computer Science no. 7897. Springer Berlin Heidelberg, 2013.

Kriksciuniene, D., and S. Strigunaite. "Multi-Level Fuzzy Rules-Based Analysis of Virtual Team Performance." In *Building the E-World Ecosystem*, pp. 305-18. IFIP Advances in Information and Communication Technology 353. Springer Berlin Heidelberg, 2011.

Mascarenhas, S., R. Prada, A. Paiva, and G.J. Hofstede. "Social Importance Dynamics: A Model for Culturally-Adaptive Agents." In *Intelligent Virtual Agents*, pp. 325-38. Lecture Notes in Computer Science no. 8108. Springer Berlin Heidelberg, 2013.

Nunes, M., and H. O'Neill. "Assessing the Performance of Virtual Teams with Intelligent Agents." In *Virtual and Networked Organizations, Emergent Technologies and Tools*, pp. 62-69. Communications in Computer and Information Science 248. Springer Berlin Heidelberg, 2012.

Nurmi, N. "Unique Stressors of Cross-Cultural Collaboration through ICTs in Virtual Teams." *Ergonomics and Health Aspects of Work with Computers*, pp. 78-87. Lecture Notes in Computer Science no. 5624, Springer Berlin Heidelberg, 2009.

Obaid, M., I. Damian, F. Kistler, B. Endrass, J. Wagner, and E. André. "Cultural Behaviors of Virtual Agents in an Augmented Reality Environment." In *Intelligent Virtual Agents*, pp. 412-18. Lecture Notes in Computer Science no. 7502. Springer Berlin Heidelberg, 2012.

Ober, S. *Contemporary Business Communication*. Stamford, CT: Cengage Learning, 2007.

Payr, S., and R. Trappl. "Agents across Cultures." In *Intelligent Virtual Agents*, pp. 320-24. Lecture Notes in Computer Science no. 2792. Springer Berlin Heidelberg, 2003.

Perez-Marin, D., and I. Pascual-Nieto. *Conversational Agents and Natural Language Interaction: Techniques and Effective Practices*. Hershey, PA: IGI Global, 2011.

Rehm, M., André, E., and Nakano, Y. "Some Pitfalls for Developing Enculturated Conversational Agents." In *Human-Computer Interaction: Ambient, Ubiquitous and Intelligent Interaction*, pp. 340-48. Lecture Notes in Computer Science 5612. Springer Berlin Heidelberg, 2009.

Rehm, M., Y. Nakano, E. André, T. Nishida, N. Bee, B. Endrass, M. Wissner, A.A. Lipi, and H. Huang, "From observation to simulation: generating culture-specific behavior for interactive systems." *AI & SOCIETY* vol. 24, no. 3 (2009): 267-80.

Chapter Eleven

A Fractal Measure for Comparing the Work Effort of Human and Artificial Agents Performing Management Functions[1]

Abstract. Thanks to the growing sophistication of artificial agent technologies, businesses will increasingly face decisions of whether to have a human employee or artificial agent perform a particular function. This makes it desirable to have a common temporal measure for comparing the work effort that human beings and artificial agents can apply to a role. Existing temporal measures of work effort are formulated to apply either to human employees (e.g., FTE and billable hours) or computer-based systems (e.g., mean time to failure and availability) but not both. In this paper we propose a new temporal measure of work effort based on fractal dimension that applies equally to the work of human beings and artificial agents performing management functions. We then consider four potential cases to demonstrate the measure's diagnostic value in assessing strengths (e.g., flexibility) and risks (e.g., switch costs) reflected by the temporal work dynamics of particular managers.

The Need for a Common Temporal Measure of Work Effort

The increasing power and sophistication of artificial agent technology is allowing businesses to employ artificial agents in a growing number of roles.

---

[1] This text was originally published in *Position Papers of the 2014 Federated Conference on Computer Science and Information Systems*, edited by Maria Ganzha, Leszek Maciaszek, and Marcin Paprzycki, pp. 219-226. Annals of Computer Science and Information Systems vol. 3. Warsaw: Polskie Towarzystwo Informatyczne, 2014. A further elaboration of the material was published as "A Tool for Designing and Evaluating the Temporal Work Patterns of Human and Artificial Agents," *Informatyka Ekonomiczna / Business Informatics* 3(33) (2014): 61-76.

Artificial agents are no longer restricted simply to performing logistical functions such as resource scheduling, but are now capable of more complex interpersonal workplace behavior such as using social intelligence to effectively manage the limitations, abilities, and expectations of human employees,[2] recognizing and manifesting culture-specific behaviors in interactions with human colleagues,[3] and assessing the performance of human members of virtual teams.[4] It is thus gradually becoming more feasible to design artificial agents capable of performing the four key functions carried out by human managers, which are planning, organizing, leading, and controlling.[5]

As a result of such recent and anticipated future advances, businesses will increasingly be faced with concrete decisions about whether, for example, the manager of a new corporate call center should be an experienced human manager or the latest artificial agent system. Such decisions will be shaped by a large number of strategic, financial, technological, political, legal, ethical, and operational factors. One particular element to be taken into account is that of temporal work effort: i.e., how much time would a human manager actually be able to dedicate to carrying out the necessary work functions, given the fact that physiological, cultural, legal, and ethical constraints limit the number of hours per week that a human being is capable of working? Similarly, how much time would an artificial agent be able to dedicate to carrying out the necessary work functions, given the fact that scheduled maintenance or unscheduled outages can limit the uptime of computer-based systems? Knowing how much time per day (or week, or other relevant time interval) a manager will be available to carry out his or her functions of planning, organizing, leading, and controlling becomes especially relevant in an interconnected age when global businesses operate around the clock, and managers are expected to be available to respond to inquiries and make decisions at almost any time of the night or day.

In the case of human professionals, temporal measures such as 'full-time equivalent' (FTE)[6] and 'billable hours' are often used to quantify one's work effort. Computer-based systems, meanwhile, often use temporal measures such as 'availability' and 'reliability.' In the following sections, we will analyze such existing measures and then develop a new fractal-dimension-based temporal measure for work effort that has at least two notable advantages: it is applicable to the work effort of both human and artificial agent managers, and it provides valuable diagnostic insights into the strengths and limitations of

[2] Williams, "Robot Social Intelligence" (2012).

[3] Rehm et al., "From observation to simulation: generating culture-specific behavior for interactive systems" (2009).

[4] Nunes & O'Neill, "Assessing the performance of virtual teams with intelligent agents" (2012).

[5] Daft, *Management* (2011), pp. 7-8.

[6] "Full-Time Equivalent (FTE)," European Commission – Eurostat.

an individual manager's temporal work dynamics that are not provided by existing measures.

Measures of Work Effort for Computer-based Systems

Availability and Reliability

A computer's reliability is often quantified as the mean time to failure (MTTF), the average length of time that a system will remain continuously in operation before experiencing its next failure.[7] The mean time to repair (MTTR) is the average length of time needed to detect and repair the failure and return the system to operation. A computer's steady-state availability A is the likelihood that the computer is operating at a particular moment, and it is related to MTTF and MTTR in the equation:[8]

$$A = \frac{\text{MTTF}}{\text{MTTF} + \text{MTTR}}$$

A standard requirement for commercial computer systems is 99.99% availability over the course of a year.[9]

Availability has traditionally been understood in a binary manner: a system is either 'up' or 'down.' Rossebeø et al. argue that a more sophisticated measure is needed that takes qualitative aspects into account and suggest recognizing a range of intermediate qualitative states between simply 'up' and 'down'.[10] As we explain below, the measure proposed in this paper takes a different approach: its unique diagnostic value comes not from adding a qualitative component but from considering more carefully the fineness and resolution of the time-scales on which measurements are being made.

Time-scales for Measuring Computer Performance

A computer performs actions across a vast range of time-scales. As Gunther notes, if a typical computer's CPU cycle were 'scaled up' so that it lasted one second, then using that same scale, a DRAM access would take about one minute, a single disk seek would require roughly 1.35 months, and a tape access would last more than a century.[11] He explains that when measuring performance, "Only those changes that occur on a timescale similar to the quantity we are trying to predict will have the most impact on its value. All other

[7] Grottke et al., "Ten fallacies of availability and reliability analysis" (2008).

[8] Grottke et al. (2008).

[9] Gunther, "Time – the zeroth performance metric" (2005).

[10] Rossebeø et al., "A conceptual model for service availability" (2006).

[11] Gunther (2005).

(i.e., faster) changes in the system can usually be ignored.... In modeling the performance of a database system where the response time is measured in seconds, it would be counterproductive to include all the times for execution of every CPU instruction."

In a business context, artificial agents performing certain logistical or data-analysis tasks can operate at speeds constrained only by the laws of physics and availability of needed resources. However, an artificial agent manager whose role involves planning, organizing, leading, and controlling the activity of human colleagues should have its work effort measured within a corresponding time-scale. Thus for our present purposes there is no need to consider phenomena such as metastability that have major implications for computer design and functionality but are only directly relevant at the smallest temporal scale.[12]

Viewed from the microscopic end of the temporal spectrum, the regulator of all activity within a computer system is the 'clock tick' or 'fundamental interval of time' created by an interrupt sent from the system's hardware clock to the operating system's kernel; in a Unix system, this tick interval is often set at 10 ms,[13] during which time roughly 3.2×10^5 CPU cycles might occur. For an artificial agent system operating on a serial processor architecture, there is no need to adopt a temporal measure capable of resolving each individual CPU cycle, as that would not provide information that is directly relevant to the tasks in which the agent's work performance will be evaluated and which take place over a much longer time-frame. For example, an artificial agent manager might interact with human colleagues by generating text or images displayed on a screen. Assuming a screen refresh rate of 60 Hz, this yields a single work interval (or frame) of roughly 17 ms. Writing output data to disk would require a minimum work interval of roughly 3.50 ms for a disk seek.[14] If the artificial agent is generating speech or other audio to be heard by human beings, a standard sampling rate of 48,000 Hz would yield a single work interval of roughly 0.02 ms.

At the macroscopic end of the time-scale, it is not unknown for servers to run for several years without rebooting or a moment of downtime.[15] If we view an artificial agent manager as a form of enterprise software, we might expect its lifespan to average around nine years and to be no shorter than two years.[16] Thus while a coarser or finer temporal resolution is possible, our proposed temporal measure for work effort should prove sufficient for artificial agent

---

[12] Gunther (2005).

[13] Gunther (2005).

[14] Gunther (2005).

[15] "Cool solutions: uptime workhorses: still crazy after all these years" (2006).

[16] Tamai & Torimitsu, "Software lifetime and its evolution process over generations" (1992).

systems as long as it can encompass time-scales ranging from 0.02 ms up to several years.

Measures of Work Effort for Human Employees

The Significance of a Year as a Temporal Unit

We can now consider the case of a human manager. In principle, the longest possible macroscopic time-scale of work effort that one can utilize for a human employee is a biological lifespan. In practice, though, the relevant time-scale is obviously much shorter. In the United States, a typical managerial employee only remains with his or her current employer for about 5.5 years before moving to a new organization.[17] The 'year' has significant historical and conceptual value as a fundamental measure of human work activity. Just as enterprise system availability is often cited in terms of uptime per operating year, productivity figures for human workers are typically based on an annual time-frame.[18]

Having taken the year as our initial frame of reference, how do we quantify the portion of a given year that a human employee actually spends on his or her work? For this purpose, the largest relevant subunit is that of a single week, as professional workers regularly assess a job's fringe benefits according to how many weeks of vacation they receive each year, and government agencies and researchers often track this data. The number of weeks worked per year varies significantly across nations and cultures.[19]

The Significance of an Hour as a Temporal Unit

Even if we know that two employees both 'work' the same number of weeks per year, this fact does not yet tell us much about their relative levels of work effort, as it is possible for the employees to differ vastly in how many hours they work each week. Here, too, there is significant variation across nations and cultures and between specific jobs.[20] For example, an American law firm will likely expect attorneys to work more than 50 hours per week,[21] while employees of high-tech Silicon Valley firms are routinely expected to work over 100 hours per week when project deadlines are approaching.[22]

---

[17] "Employee tenure summary," United States Department of Labor, Bureau of Labor Statistics (2012).

[18] "Annual Hours Worked," Organisation for Economic Co-operation and Development.

[19] Golden, "A brief history of long work time and the contemporary sources of overwork" (2009).

[20] Golden (2009).

[21] "The Truth about the Billable Hour," Yale Law School.

[22] Shih, "Project Time in Silicon Valley" (2004).

How Much Work in an Hour of Work?

The hour, though, is certainly not the smallest quantifiable interval of employee work effort. Two employees may consider themselves to have just spent an hour 'working,' but the number of minutes of work actually performed by each can differ greatly. In some professions, it is common to track work effort in sub-hour intervals. For example, attorneys with American law firms typically track their work time in six-minute intervals and sometimes record and bill clients for work that took as little as one minute. For every hour that an attorney spends 'at work,' an average of roughly 45 minutes will count toward billable hours.[23]

In other professions, employers have given up any attempt at precisely measuring how much time an employee is putting into their work, as the advent of new communications technologies has caused 'work time' and 'personal time' to meld into an indistinguishable blur.[24] The rise of multitasking and 'continuous partial attention' drives human workers to constantly monitor emails, texts, and instant messages, even while in the middle of meetings or conversations.[25] For knowledge workers, this continual checking of email can consume up to 25% of their workday.[26] While much of this nonstop communication activity is work-related, the existence of workplace phenomena such as shirking, social loafing, and job neglect means that a significant number of these electronic interruptions do not relate to work at all but are purely personal. In particular, younger employees of the Millennial generation are less fond of email and tend to prefer text messaging, instant messaging,[27] and other forms of micro-communication that produce shorter but more frequent non-work interruptions to their work activities. Because professional employees can alternate between work-related and personal actions at such a rapid rate (once every few seconds, if not faster), it is now "very hard to tell when people are working and when people are not working," as a Silicon Valley executive reported in Shih's study.[28] In an effort to counteract this constant stream of distractions, some Extreme Programming (XP) teams employ the Pomodoro Technique, a time-boxing strategy in which physical timers are

---

[23] "The Truth about the Billable Hour."

[24] Shih (2004).

[25] Sellberg and Susi, "Technostress in the office: a distributed cognition perspective on human–technology interaction" (2014).

[26] Gupta et al., "You've got email! Does it really matter to process emails now or later?" (2011).

[27] Hershatter & Epstein, "Millennials and the world of work: an organization and management perspective" (2010).

[28] Shih (2004).

used to enforce a steady pace consisting of 25 minutes of focused work followed by a brief break.[29]

Identifying a Minimum Time Unit of Work by Human Managers

Within a given period of 'work,' there may be alternating periods of work and non-work that are measured in seconds, not minutes. However, in attempting to identify the minimum unit of work of which humans are capable, it is valuable to consider time-scales even much smaller than a second. For example, scholars have estimated that the human brain is capable of between 10^{14} and 10^{16} calculations per second,[30] or roughly 6.6×10^{16} FLOPS,[31] although the massively distributed parallel processing architecture of the brain[32] means that many calculations are taking place simultaneously, and the duration of a single calculation cannot be determined by simply dividing one second by, say, 10^{15}. In attempting to estimate the duration of a single 'calculation' performed by the brain, scholars have alternately cited the fact that an individual neuron can fire as often as 1,000 times a second,[33] that "synapses carry out floating point operations ... at a temporal resolution approaching about 1000 Hz,"[34] that a neuron is capable of firing roughly once every 5 ms,[35] or that the brain operates at a rate of speed of "around 100 cycles per second."[36] These estimates yield a range of 1-10ms for the brain's smallest temporal unit of work activity.

It is helpful, though, to refer once more to Gunther's position on the measurement of computer performance: we can essentially ignore activity taking place within a system on a time-scale shorter than that of our work-relevant inputs and outputs, as it is "more likely to be part of the *background noise* rather than the main theme."[37] In the case of a human being considered *qua* employee, the firing of a single synapse does not directly constitute 'work.' The work of planning, organizing, leading, and controlling for which human managers are employed typically involves more complicated inputs and outputs such as engaging in conversation or reading and creating documents. The

---

[29] Gobbo & Vaccari, "The Pomodoro Technique for sustainable pace in extreme programming teams" (2008).

[30] Kurzweil, *The Singularity Is Near: When Humans Transcend Biology* (2006); Friedenberg, *Artificial Psychology: The Quest for What It Means to Be Human* (2008).

[31] Llarena, "Here comes the robotic brain!" (2010).

[32] Friedenberg (2008).

[33] Friedenberg (2008).

[34] McClelland, "Is a machine realization of truly human-like intelligence achievable?" (2009).

[35] Kurzweil, *The Age of Spiritual Machines: When Computers Exceed Human Intelligence* (2000).

[36] See Abbott & Dayan, *Theoretical Neuroscience: Computational and Mathematical Modeling of Neural Systems* (2001), and its discussion in Llarena (2010).

[37] Gunther (2005).

smallest temporal unit of work would be the smallest unit relevant in the performance of such tasks.

That unit appears to be an interval of roughly 50 ms. Studies have shown that if one alternates too quickly between two tasks that require the same cognitive resources, one's performance on both tasks will be negatively impacted,[38] as shifting from one mental task to another incurs a 'switch cost' of both a temporal delay and an increased error rate,[39] which lowers productivity.[40] In particular, the human brain needs around 120 ms to fully allocate its attention to a new stimulus.[41] Marchetti cites diverse studies supporting the claim that the minimum 'integration time' needed for the brain to meld disparate sensory input into a conscious perception of a single event or experience is roughly 50-250 ms, with a median of about 100 ms.[42] These findings make it unlikely that a human manager would be capable of performing individual instances of work that need to be measured using a time-frame shorter than 50 ms. If one attempted to alternate between tasks faster than once every 50 ms, one's brain would not even have time to focus attention on a new task before abandoning it for yet another task.

Durations of Particular Work Inputs and Outputs

This minimum interval of roughly 50 ms is supported by the fact that the kinds of inputs and outputs that human managers typically utilize when performing work-related functions do not have durations shorter than this interval. For example, Hamilton notes that human beings can think at a rate of 400-800 words per minute, while we typically speak at 100-175 words per minute (with each spoken word comprising an average of 4-5 phonemes[43]). Optimal listening comprehension occurs when a speaker speaks at 275-300 words per minute, which gives a listener's mind less time to become distracted or daydream between each of the speaker's words.[44] Adult native speakers of English typically read 200-250 words per minute.[45] Regarding work output, the fastest sustainable typing rate is roughly 150 words per minute,[46] with each word

---

[38] Brown & Merchant, "Processing resources in timing and sequencing tasks" (2007).

[39] Wager et al., "Individual differences in multiple types of shifting attention" (2006).

[40] Schippers & Hogenes, "Energy management of people in organizations: a review and research agenda" (2011).

[41] Tse et al., "Attention and the subjective expansion of time" (2004).

[42] Marchetti, "Observation levels and units of time: a critical analysis of the main assumption of the theory of the artificial" (2000).

[43] Levelt, "Models of word production" (1999).

[44] Hamilton, *Essentials of Public Speaking* (2014), p. 60.

[45] Traxler, *Introduction to Psycholinguistics: Understanding Language Science* (2011), chapter 10.

[46] "World's fastest typer," *Chicago Tribune.*

comprising an average of 5-6 characters; the fastest known shorthand writing speed is roughly 350 words per minute;[47] and the fastest known human speaker is able to clearly articulate more than 650 words per minute.[48] When these rates are converted into milliseconds, they yield the intervals seen in Table I.

Table I. Average time needed by the human brain to perform work-related input, processing, and output functions.

Activity	Average Time
Fully allocating attention to a new stimulus	120 ms
Consciously perceiving a single coherent experience or event	50-250 ms
Hearing one spoken phoneme (4.4 phonemes per word)	45-50 ms
Hearing one spoken word	200-220 ms
Reading one printed word	240-300 ms
Thinking one word	75-150 ms
Speaking one phoneme (4.4 phonemes per word)	21-140 ms
Speaking one word	92-600 ms
Typing one character (5.5 characters per word)	≤ 73 ms
Typing one word	≤ 400 ms
Writing one word in shorthand	≤ 170 ms

The Fractal Self-similarity of Human Work Cycles

As we have seen, for artificial agent managers, the time-scales relevant to their work effort range from several years down to about 0.02 ms, while for human managers they range from several years down to around 50 ms. Within this range, there are multiple relevant time-scales and activity cycles of different lengths that demonstrate an interesting degree of self-similarity: within a given year of work, a typical human manager will spend many consecutive weeks working, interrupted periodically by non-work weeks of vacation. Within a given week of work, he or she will spend spans of several consecutive hours working, followed by non-work hours when he or she is asleep or out of the office. Within a given hour of work, his or her spans of minutes spent

[47] "New World's Record for Shorthand Speed" (1922).

[48] "Fastest talker," Guinness World Records.

working will be followed by non-work intervals when the manager is daydreaming or writing a personal email. The roughly self-similar nature of this temporal dynamic opens the door to understanding a human manager's work activity as a fractal time series.

The fractal nature of our typical human work dynamics is not at all surprising: as Longo and Montévil note, fractal-like dynamics are "ubiquitous in biology, ... in particular when we consider processes associated with physiological regulation."[49] Lloyd notes that when an organism's biological processes operate on multiple time-frames displaying fractal temporal coherence, it creates a scale-free system with "robust yet flexible integrated performance" in which the oscillatory dynamics with long memory allow the organism to predict and respond to long-term environmental conditions such as tidal, seasonal, and annual cycles, while the short-term cycles coordinate internal processes such as organ functioning and cellular division.[50]

Calculating the Fractal Dimension of Work Effort

Significance of the Fractal Dimension

One of the most important and meaningful attributes of a fractal time series is that it possesses a *fractal dimension* that one can calculate and which captures valuable information about the series' temporal dynamics. The calculation of the fractal dimension of biological phenomena has varied practical applications. For example, analysis of the fractal dimension of EEG data can be used to quantify the level of concentration during mental tasks,[51] and fractal analysis has demonstrated that healthy hearts display greater rhythmic complexity than diseased hearts.[52]

The fractal dimension of empirically observed natural phenomena can be described by the equation $D = 2 - H$, where H is the Hurst exponent of the time series as graphed in two-dimensional Cartesian space. In this approach, an x-coordinate is the time at which a value was measured, and the y-coordinate is the value measured at that time.[53] The case $0 < H < \frac{1}{2}$ represents a dynamic that is variously described as antipersistent, irregular, or trend-reversing: if the value in one moment is greater than the mean, the value in the next consecutive moment is likely to be less than the mean. The case $H = \frac{1}{2}$ represents a random-walk process such as Brownian motion, in which the

---

[49] Longo & Montévil, "Scaling and scale symmetries in biological systems" (2014).

[50] Lloyd, "Biological time is fractal: early events reverberate over a life time" (2008).

[51] Sourina et al., "Fractal-based brain state recognition from EEG in human computer interaction" (2013).

[52] Longo & Montévil, "Scaling and scale symmetries in biological systems" (2014).

[53] Mandelbrot, *The Fractal Geometry of Nature* (1983), pp. 353-54.

value in the next consecutive moment is equally likely to move toward or away from the mean. The case $\frac{1}{2} < H < 1$ is described as persistent or quasi-regular: the value at the next consecutive moment in time is likely to be the same as the value in the previous moment.[54] In this case, we can say that the dynamic has long memory.

Work Effort as a Time Series of Binary Values

Graphing a time-series in two-dimensional space is useful for natural phenomena such as earthquakes that occur at different times with different intensities.[55] However, in the case of developing a temporal measure for quantifying the work effort of human and artificial managers, we suggest that a different approach is warranted. Graphing an agent's work effort in two-dimensional space would be useful if the work effort displayed by a human or artificial agent manager at a particular instant of time were able to range across a continuous spectrum of values. However, in this case we have only a binary set of possible values: at any given instant, an agent is either focusing its attention on its work, or it is not. Marchetti draws on research from several areas of psychology to show that the human mind is incapable of dividing its attention between two different scenes, attitudes, or 'observational levels' at the same instant in time. (We would suggest that the same will likely be true for any artificial agent whose cognitive capacities are modeled closely on those of the human brain's neural network, as well as for any artificial agent governed by a computer program in the form of executable code.) As we saw above, the brain's attention mechanism is capable of alternating attention between two different thoughts or scenes with great rapidity (as in cases of so-called 'multitasking'), however in any given instant of time, our attention is allocated to at most one of those thoughts or scenes. This means that work effort cannot be quantified by saying, for example, that "At moment *t*, 70% of the agent's attention was dedicated to its work." Instead, one would say that "For all of the indivisible instances of attention that took place during time interval $[a, b]$, in 70% of those instances the agent's attention was focused on its work."

Mandelbrot notes that if the fractal dimension of a time series graphed in two-dimensional space is represented by the equation $D = 2 - H$, then the zero set (or any other level set) of the graphed time series would have fractal dimension:[56]

$$D = 1 - H.$$

[54] Mandelbrot (1983), pp. 353-54; Valverde et al., "Looking for memory in the behavior of granular materials by means of the Hurst analysis" (2005).

[55] Telesca et al., "On the methods to identify clustering properties in sequences of seismic time-occurrences" (2002).

[56] Mandelbrot (1983), pp. 353-54.

We can use this equation to relate the fractal dimension and Hurst exponent for work effort when we understand work effort as graphed on a one-dimensional line segment. The length of the entire segment represents the entire time available (such as a year, week, or hour) during which an agent can potentially be performing work. Those instants of actual work form the set that is graphed on the line segment, while instants of non-work do not belong to the set. With this binary approach, we can envision the depiction of an agent's work effort across time as a series of instances of work and non-work graphed on a line segment that resembles a generalized Cantor set in which the moments of work are those points contained in the set and moments of non-work are portions of a deleted interval. Because this is a graph of an empirically observed natural phenomenon rather than a purely mathematical object, it would have a minimum fineness and resolution: if our minimum unit of time is 10 ms and we graph a line segment representing one hour, it would comprise 3.6×10^6 such units of work or non-work.

In this context, the Hurst exponent takes on a different (and perhaps even counterintuitive) meaning. For a two-dimensional graph of a time series with $H \approx 0$, successive y-values alternate antipersistently around the mean, and the graphed line fills up a relatively large share of the two-dimensional space. For a one-dimensional graph of a binary time series, one might visualize the set as though it contains a single point that is able to slide back and forth along the x-axis to occupy many different x-values simultaneously, thus forming the set. For a set with high persistence ($H \approx 1$), the point may be locked to a single x-value, reflecting a process with long memory. For a set with low persistence ($H \approx 0$), the point 'forgets' where it is and is free to move up and down the line segment, occupying many different x-values. This conceptualization reflects the fact that the two-dimensional graph of an antipersistent time series will cross the horizontal line determined by the mean y-value at many different places, whereas the graph of a persistent process might only cross it once, and the graph of a random-walk process can intersect it either one or many times.

Formulating our Fractal Measure

Advantages of the Box-Counting Method

Different methods exist for calculating fractal dimension. A number of scholars prefer the Minkowski-Bouligand or box-counting dimension over alternatives such as the area-perimeter or power spectrum methods for estimating the fractal dimension of natural phenomena as diverse as seismic activity, electrical activity in the brain, and physical surface features at the nanometer scale.[57] Longo and Montévil argue that while it lacks some of the mathematical

[57] Telesca et al. (2002); Sourina et al., (2013); Zhang et al., "Fractal structure and fractal dimension

import found in other definitions of fractal dimension such as the Hausdorff dimension, the box-counting dimension has an advantage in that it can easily be applied to empirically observed phenomena.[58]

In order to develop our comparative fractal measure of work effort for human and artificial agent managers, we have thus employed the box-counting method to estimate the temporal dynamics' fractal dimension. The box-counting dimension D of set F can be calculated as:

$$D = \lim_{\delta \to 0} \frac{\log N_\delta(F)}{-\log \delta}.$$

Here $N_\delta(F)$ is the smallest number of sets of diameter δ that cover the set F.[59] When using the box-counting method to estimate the fractal dimension of natural phenomena, this can be done by calculating the average value of D that results when one empirically determines $N_\delta(F)$ for multiple values of δ.[60]

Calculation and Notation of our Fractal Measure

When we applied this approach to calculate the box-counting fractal dimension D for the work effort of particular hypothetical human and artificial agent managers, it yielded insights that could be useful for understanding, comparing, and enhancing the temporal work dynamics of such agents.

To accomplish this, we considered an agent's typical work effort as viewed across on three different time-scales or levels: 1) The set F_1 includes those weeks worked within a span S_1 of five years (or 260 weeks), for which the covering sets used for the box-counting estimation were δ_a = 4 weeks, δ_b = 2 weeks, and δ_c = 1 week. 2) The set F_2 includes those hours worked within a span S_2 of one week (or 168 hours), for which the covering sets used for the box-counting estimation were δ_a = 4 hours, δ_b = 2 hours, and δ_c = 1 hour. 3) The set F_3 includes those minutes worked within a span S_3 of one hour, for which the covering sets used for the box-counting estimation were δ_a = 1 minute, δ_b = 30 seconds, and δ_c = 15 seconds. Using the box-counting method, we calculated D_1, D_2, and D_3 for the time-scales F_1, F_2, and F_3, respectively, and averaged those values to produce a mean value of $D = \langle D_1, D_2, D_{3,} \rangle$ for a particular agent. We then calculated the estimated value for the Hurst exponent for that agent's temporal dynamic with the equation $H = 1 - D$.

Drawing on the data considered in previous sections for the typical temporal performance of human professionals and artificial agents (envisioned as

determination at nanometer scale" (1999).

[58] Longo & Montévil (2014).

[59] Falconer, *Fractal Geometry: Mathematical Foundations and Applications* (2004), pp. 41-44.

[60] Wahl et al., *Exploring Fractals on the Macintosh* (1994), pp. 75-108.

hardware and software systems), we present four specific hypothetical cases and the values of D and H calculated for each.

Applying our Measure to Particular Cases

Temporal Dynamics of Human Manager A

Consider a hypothetical Human Manager A whose work effort approaches the maximum of which contemporary human beings are capable. This manager does not take any weeks of vacation during the five years worked in his position (S_1 = 260 weeks, $N_\delta(F_1)$ = 260 weeks). He concentrates exclusively on his career, working an average of 90 hours per week (S_2 = 168 hours, $N_\delta(F_2)$ = 90 hours). During the work day, he avoids all possible distractions and, relying on an approximation of the Pomodoro Technique, spends only 5 minutes of each 'work hour' not performing work-related functions (S_3 = 60 minutes, $N_\delta(F_3)$ = 55 minutes). We graphed each of these situations on a line segment that we then considered at three different temporal resolutions. Within the graph of the time series, a moment of work is indicated with a colored vertical slice, and a moment of non-work is indicated with an unshaded interval. A graph of the temporal work dynamics of Human Manager A is seen in Fig. 1 below. For an agent with these characteristics, we have calculated D = 0.962, H = 0.038, and availability (understood as the likelihood that any randomly-selected instant of time will fall during a moment of work rather than non-work) as A = 49.1%.

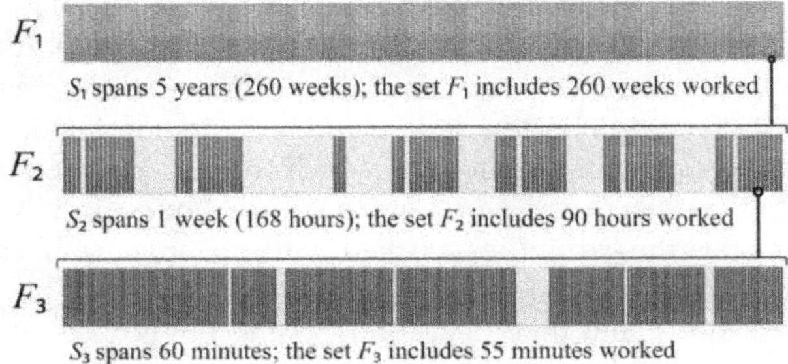

F_1
S_1 spans 5 years (260 weeks); the set F_1 includes 260 weeks worked
F_2
S_2 spans 1 week (168 hours); the set F_2 includes 90 hours worked
F_3
S_3 spans 60 minutes; the set F_3 includes 55 minutes worked

Fig. 1: Human Manager A's periods of work and non-work.

Temporal Dynamics of Human Manager B

Hypothetical Human Manager B represents the opposite end of the spectrum: his time commitment approaches the lowest amount possible for someone who is filling a management role with an organization. We suppose that Human Manager B spends only half of the weeks in the year working (S_1 = 260 weeks, $N_\delta(F_1)$ = 130 weeks). Even during those weeks when he is working, the manager dedicates only 10 hours of effort to this particular position (S_2 = 168 hours, $N_\delta(F_2)$ = 10 hours). Moreover, during each hour of 'work,' the manager spends only a third of the time focused directly on work-related tasks, with the rest of the time representing distractions or non-work-related activities (S_3 = 60 minutes, $N_\delta(F_3)$ = 20 minutes). A graph of the temporal work dynamics of Human Manager B is seen in Fig. 2 below. For an agent with these characteristics, we have calculated D = 0.532, H = 0.468, and A = 1.0%.

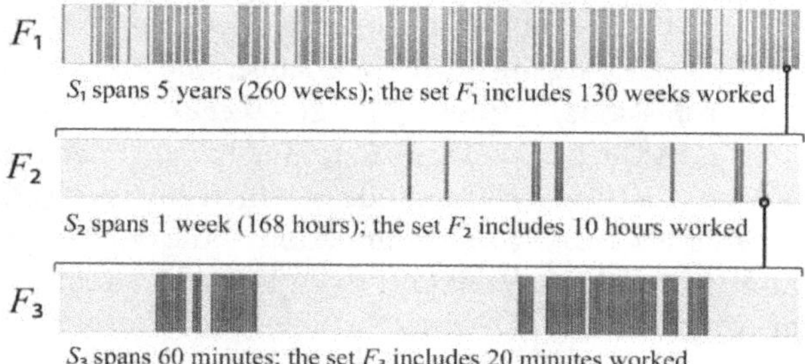
F_1
S_1 spans 5 years (260 weeks); the set F_1 includes 130 weeks worked
F_2
S_2 spans 1 week (168 hours); the set F_2 includes 10 hours worked
F_3
S_3 spans 60 minutes; the set F_3 includes 20 minutes worked

Fig. 2: Human Manager B's periods of work and non-work.

Temporal Dynamics of Artificial Agent Manager A

Next consider a hypothetical Artificial Agent Manager A in the form of a software program running on a computer with a typical serial processor architecture. We suppose that during a given five-year operating period, there may be brief service outages for scheduled maintenance or updates but that there are no extended outages (S_1 = 260 weeks, $N_\delta(F_1)$ = 260 weeks). Each week, there is a scheduled maintenance window of one hour, when software updates are applied and the system is rebooted (S_2 = 168 hours, $N_\delta(F_2)$ = 167 hours). The software program and hardware substrate for Artificial Agent Manager A have no non-work-related functions and are not capable of being 'distracted' in the way that a human manager is, thus during a typical hour period of work, Artificial Agent Manager A does not dedicate any minutes to non-work-related functions (S_3 = 60 minutes, $N_\delta(F_3)$ = 60 minutes). A graph of the temporal work dynamics of Artificial Agent Manager A is seen in Fig. 3 below. For an agent with these characteristics, we have calculated D = 0.999, H = 0.001, and A = 99.4%.

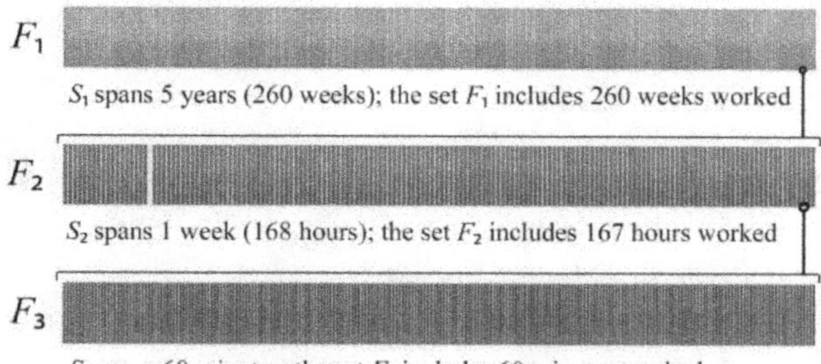
F_1
S_1 spans 5 years (260 weeks); the set F_1 includes 260 weeks worked
F_2
S_2 spans 1 week (168 hours); the set F_2 includes 167 hours worked
F_3
S_3 spans 60 minutes; the set F_3 includes 60 minutes worked

Fig. 3: Artificial Agent Manager A's periods of work and non-work.

Temporal Dynamics of Artificial Agent Manager B

Finally, consider the hypothesized future scenario of Artificial Agent Manager B, an artificial general intelligence with a distributed neural network architecture that is modeled on the human brain and displays human-like motivations, emotions, and learning capacity.[61] While Artificial Agent Manager B enjoys its job, every two years it must spend a week away from work for a period of psychological assessment, maintenance, and relaxation, to reduce the likelihood of professional burnout (S_1 = 260 weeks, $N_\delta(F_1)$ = 258 weeks). Moreover, during each week of work, its neural network architecture requires it to spend two hours daily in a 'sleep' mode in which any new external stimuli are shut out, in order to facilitate the assimilation of the day's experiences into long-term memory. In order to maintain its capacity for creativity, satisfy its intellectual curiosity, and avoid the development of cyberpsychoses, it must also spend two hours daily exploring spheres of experience unconnected to its work-related tasks (S_2 = 168 hours, $N_\delta(F_2)$ = 126 hours). Because Artificial Agent Manager B reflects the full constellation of human-like cognitive and social behaviors, it spends five minutes of each hour on functions other than work, such as cyberloafing, following news stories, and communicating with friends (S_3 = 60 minutes, $N_\delta(F_3)$ = 55 minutes).

A graph of the temporal work dynamics of Artificial Agent Manager B is seen in Fig. 4 below. For an agent with these characteristics, we have calculated $D = 0.945$, $H = 0.055$, and $A = 68.2\%$.

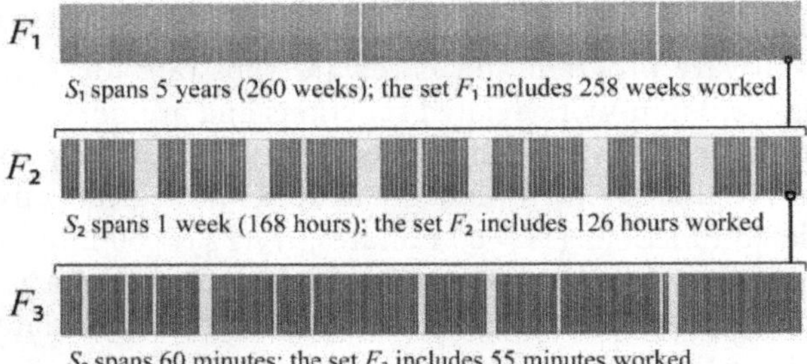
F_1
S_1 spans 5 years (260 weeks); the set F_1 includes 258 weeks worked
F_2
S_2 spans 1 week (168 hours); the set F_2 includes 126 hours worked
F_3
S_3 spans 60 minutes; the set F_3 includes 55 minutes worked

Fig. 4: Artificial Agent Manager B's periods of work and non-work.

Analysis and Discussion

Comparison and Analysis

Table II below gives the values of *D*, *H*, and *A* for all four agents, ranked from the highest value of *D* to the lowest.

---

[61] Friedenberg (2008).

Table II. Agents' work effort as characterized by fractal dimension, Hurst exponent, and Availability.

Agent	**D**	**H**	**A**
Artificial Agent Manager A	0.999	0.001	99.4%
Human Manager A	0.962	0.038	49.1%
Artificial Agent Manager B	0.945	0.055	68.2%
Human Manager B	0.532	0.468	1.0%

We may note the following conclusions:

1) Artificial Agent Managers A and B and Human Manager A all display similar values of $H \approx 0$ (antipersistence), while Human Manager B displays a value of $H \approx ½$ (randomness). While more study is required to verify this supposition, it seems likely that managers with low persistence (as understood in the mathematical sense defined above) would be free from high switch costs, as their work intervals last longer, and they spend a smaller share of their work time transitioning into or out of periods of work.

2) The managers displaying high values for D possess 'flexibility' in the sense that they are ready and available to work in almost every possible moment. However, they may simultaneously display 'inflexibility,' in the sense that they are *used to working* in every possible moment, thus unexpected interruptions may be more likely to derail the work of this sort of manager. Meanwhile, managers with a lower value for D possess 'flexibility,' insofar as they are already used to working only sporadically and juggling intervals of work amidst many other activities, thus unexpected interruptions to their work may not greatly faze them. On the other hand, they might simultaneously display 'inflexibility,' insofar as the bulk of their time may already be filled with non-work-related activity, leaving only brief, sporadic slivers of time available for work. If an unexpected distraction prevents them from working during one of these windows, it may be quite some time before another window of availability for work appears.

3) The values of A and D are neither directly nor inversely proportional to one another. Artificial Agent Manager A possesses the highest values for both A and D, while Human Manager B displays the lowest values for both. However, in the middle of the table, Human Manager A displays a higher value for D than Artificial Agent Manager B but a lower value for A. This means that if one only utilizes a simple measure such as availability in assessing (and ranking) the temporal work dynamics of human and artificial agents, one will miss out on additional information that the fractal dimension and Hurst exponent can provide. While availability is a useful measure, it can potentially be misleading if not complemented by more sophisticated measures such as fractal dimension.

Avenues for Future Research

Further steps that we have identified to advance this research include:

1) Gathering empirical data about temporal work dynamics from a sample of real-world human managers and artificial agent systems to verify the appropriateness and value of this fractal-dimension-based model. Analysis of such data could aid in predicting the temporal dynamics of future artificial agent systems (for which empirical data is not yet available) and designing more advanced AI systems that will be capable of carrying out a wider range of business management roles.

2) Adding data for a time-scale S_4 that captures the work activity of human and artificial agent managers as viewed in intervals as small as 10 milliseconds. The ability to capture such data for the neural activity of a human manager exceeds the temporal resolution available with current fMRI technology, but it may be possible using EEG or MEG techniques (perhaps in conjunction with fMRI).

3) Attempting to identify correlations between the values of D and H for a particular manager's temporal dynamics and traits identified in established models of managerial motivation and behavior.

In conclusion, we hope that if this paper's proposal for a single fractal temporal measure of work effort that is applicable to both human and artificial agent managers proves useful, it might contribute to the development of a new perspective in which an organization's human resources management and its management of artificial agent systems are seen not as two disconnected spheres, but rather as two aspects of a new, integrated discipline of human *and* artificial agent resource management.

References

Abbott, L.F., and P. Dayan. *Theoretical Neuroscience: Computational and Mathematical Modeling of Neural Systems.* Cambridge: The MIT Press, 2001.

"Annual Hours Worked." Organisation for Economic Co-operation and Development. http://www.oecd.org/els/emp/ ANNUAL-HOURS-WORKED.pdf. Accessed May, 2014.

Brown, S.W., and S.M. Merchant. "Processing resources in timing and sequencing tasks." *Perception & Psychophysics* vol. 69, no. 3 (2007): 439-49.

"Cool solutions: uptime workhorses: still crazy after all these years." *Novell Cool Solutions,* January 12, 2006. http://www.novell.com/coolsolutions/trench/241.html. Accessed May, 2014.

Daft, Richard. *Management,* Mason, OH: South-Western, Cengage Learning, 2011.

"Employee tenure summary." United States Department of Labor, Bureau of Labor Statistics, September 18, 2012. http://www.bls.gov/news.release/tenure.nr0.htm.

Falconer, K. *Fractal Geometry: Mathematical Foundations and Applications.* Hoboken, NJ: John Wiley & Sons, 2004.

"Fastest talker." Guinness World Records. http://gwrstaging.untitledtest.com/world-records/1/fastest-talker. Accessed May, 2014.

Friedenberg, Jay. *Artificial Psychology: The Quest for What It Means to Be Human.* Philadelphia: Psychology Press, 2008.

"Full-Time Equivalent (FTE)." European Commission – Eurostat. http://epp.eurostat.ec.europa.eu/ statistics_explained/index.php/Glossary:Full-time_equivalent. Accessed May, 2014.

Gobbo, F., and M. Vaccari. "The Pomodoro Technique for sustainable pace in extreme programming teams." *Agile Processes in Software Engineering and Extreme Programming,* pp. 180-84. Lecture Notes in Business Information Processing 9. Springer Berlin Heidelberg, 2008.

Golden, L. "A brief history of long work time and the contemporary sources of overwork." *Journal of Business Ethics* vol. 84, no. 2 (2009): 217-27.

Grottke, M., H. Sun, R.M. Fricks, and K.S. Trivedi. "Ten fallacies of availability and reliability analysis." In *Service Availability,* pp. 187-206. Lecture Notes in Computer Science 5017. Springer Berlin Heidelberg, 2008.

Gunther, N.J. "Time – The Zeroth Performance Metric." In *Analyzing Computer System Performance with Perl::PDQ,* pp. 3-46. Springer Berlin Heidelberg, 2005.

Gupta, Ashish, Ramesh Sharda, and Robert A. Greve. "You've got email! Does it really matter to process emails now or later?" In *Information Systems Frontiers* vol. 13, no. 5 (2011): 637-53. http://dx.doi.org/10.1007/s10796-010-9242-4.

Hamilton, C. *Essentials of Public Speaking.* Stamford, CT: Cengage Learning, 2014.

Hershatter, A., and M. Epstein. "Millennials and the world of work: an organization and management perspective." *Journal of Business and Psychology* vol. 25, no. 2 (2010): 211-23.

Kurzweil, Ray. *The Age of Spiritual Machines: When Computers Exceed Human Intelligence.* New York: Penguin Books, 2000.

Kurzweil, Ray. *The Singularity is Near: When Humans Transcend Biology.* New York: Viking Penguin, 2005.

Levelt, W. "Models of word production." *Trends in Cognitive Sciences* vol. 3, no. 6 (1999): 223-32.

Llarena, A. "Here comes the robotic brain!" In *Trends in Intelligent Robotics,* pp. 114-21. Communications in Computer and Information Science 103. Springer Berlin Heidelberg, 2010.

Lloyd, D. "Biological time is fractal: early events reverberate over a life time." *Journal of Biosciences* vol. 33, no. 1 (2008): 9-19.

Longo, G., and M. Montévil. "Scaling and scale symmetries in biological systems." In *Perspectives on Organisms,* pp. 23-73. Lecture Notes in Morphogenesis. Springer Berlin Heidelberg, 2014.

Mandelbrot, B.B. *The Fractal Geometry of Nature.* London: Macmillan, 1983.

Marchetti, G. "Observation levels and units of time: a critical analysis of the main assumption of the theory of the artificial." *AI & Society* vol. 14, no. 3-4 (2000): 331-47.

McClelland, J.L. "Is a machine realization of truly human-like intelligence achievable?" *Cognitive Computation* vol. 1, no. 1 (2009): 17-21.

"New World's Record for Shorthand Speed: Nathan Behrin Transcribed 350 Words in a Minute With Only Two Errors." *New York Times*, December 30, 1922.

Nunes, M., and H. O'Neill. "Assessing the Performance of Virtual Teams with Intelligent Agents." In *Virtual and Networked Organizations, Emergent Technologies and Tools*, pp. 62-69. Communications in Computer and Information Science 248. Springer Berlin Heidelberg, 2012.

Rehm, M., Y. Nakano, E. André, T. Nishida, N. Bee, B. Endrass, M. Wissner, A.A. Lipi, and H. Huang, "From observation to simulation: generating culture-specific behavior for interactive systems." *AI & SOCIETY* vol. 24, no. 3 (2009): 267-80.

Rossebeø, J.E.Y., M.S. Lund, K.E. Husa, and A. Refsdal. "A conceptual model for service availability." In *Quality of Protection*, pp. 107-18. Advances in Information Security 23. Springer US, 2006.

Schippers, M.C., and R. Hogenes. "Energy management of people in organizations: a review and research agenda." In *Journal of Business and Psychology* vol. 26, no. 2 (2011): 193-203.

Sellberg, C., and T. Susi. "Technostress in the office: a distributed cognition perspective on human–technology interaction." *Cognition, Technology & Work* vol. 16, no. 2 (2014): 187-201.

Shih, J. "Project Time in Silicon Valley." *Qualitative Sociology* 27(2) (2004): 223-45.

Sourina, O., Q. Wang, Y. Liu, and M.K. Nguyen. "Fractal-based brain state recognition from EEG in human computer interaction." In *Biomedical Engineering Systems and Technologies*, pp. 258-72. Communications in Computer and Information Science 273. Springer Berlin Heidelberg, 2013.

Tamai, T., and Y. Torimitsu. "Software lifetime and its evolution process over generations." In *Proceedings of 1992 Conference on Software Maintenance*, pp. 63-69. IEEE, 1992.

Telesca, L., V. Cuomo, V. Lapenna, and M. Macchiato. "On the methods to identify clustering properties in sequences of seismic time-occurrences." *Journal of Seismology* vol. 6, no. 1 (2002): 125-34.

Traxler, M.J. *Introduction to Psycholinguistics: Understanding Language Science.* Hoboken, NJ: John Wiley & Sons, 2011.

"The Truth about the Billable Hour." Yale Law School. http://www.law.yale.edu/studentlife/cdoadvice _truthaboutthebillablehour.htm. Accessed May, 2014.

Tse, P.U., J. Intriligator, J. Rivest, and P. Cavanagh. "Attention and the subjective expansion of time." *Perception & Psychophysics* vol. 66, no. 7 (2004): 1171-89.

Valverde, J.M., A. Castellanos, and M.A.S. Quintanilla. "Looking for memory in the behavior of granular materials by means of the Hurst analysis." In *Of Stones and Man: From the Pharaohs to the Present Day*, edited by J. Kerisel. London: Taylor & Francis Group, 2005.

Wager, T.D., J. Jonides, and E.E. Smith. "Individual differences in multiple types of shifting attention." In *Memory & Cognition* vol. 34, no. 8 (2006): 1730-43.

Wahl, B., P. Van Roy, M. Larson, and E. Kampman. *Exploring Fractals on the Macintosh.* Reading, MA: Addison-Wesley Professional, 1994.

"World's fastest typer." *Chicago Tribune.* http://www.chicagotribune.com/sns-viral-fastest-records-pictures-018,0,193476.photo. Accessed May, 2014.

Williams, Mary-Anne. "Robot Social Intelligence." In *Social Robotics*, edited by Shuzhi Sam Ge, Oussama Khatib, John-John Cabibihan, Reid Simmons, and Mary-Anne Williams, pp. 45-55. Lecture Notes in Computer Science 7621. Springer Berlin Heidelberg, 2012.

Zhang, Y., Q. Li, W. Chu, C. Wang, and C. Bai. "Fractal structure and fractal dimension determination at nanometer scale." *Science in China Series A: Mathematics* vol. 42, no. 9 (1999): 965-72.

Chapter Twelve

Managerial Robotics: A Model of Sociality and Autonomy for Robots Managing Human Beings and Machines[1]

Abstract. The development of robots with increasingly sophisticated decision-making and social capacities is opening the door to the possibility of robots carrying out the management functions of planning, organizing, leading, and controlling the work of human beings and other machines. In this paper we study the relationship between two traits that impact a robot's ability to effectively perform management functions: those of *autonomy* and *sociality*. Using an assessment instrument we evaluate the levels of autonomy and sociality of 35 robots that have been created for use in a wide range of industrial, domestic, and governmental contexts, along with several kinds of living organisms with which such robots can share a social space and which may provide templates for some aspects of future robotic design. We then develop a two-dimensional model that classifies the robots into 16 different types, each of which offers unique strengths and weaknesses for the performance of management functions. Our data suggest correlations between autonomy and sociality that could potentially assist organizations in identifying new and more effective management applications for existing robots and aid roboticists in designing new kinds of robots that are capable of succeeding in particular management roles.

Introduction and Background

Currently existing robots possess a wide array of forms and purposes – from robotic welding arms that weld parts in factories, to robotic animals that

---

[1] This text was originally published in the *International Journal of Contemporary Management* vol. 13, no. 3 (2014): 67-76.

provide therapeutic benefits for the elderly, to telepresence robots that allow one to offer educational lectures to distant audiences. Such robots are frequently used as tools for human workers; however, one might also ask whether it is possible to design robots that can serve effectively as *managers* of human workers.

The four key functions that a manager must be able to carry out are planning, organizing, leading, and controlling.[2] The ability of existing robots to perform these functions is limited. Some telepresence robots can indeed be used effectively to manage the activities of human employees, however these robots are little more than puppets that require the continuous engagement of a human operator. Such a robot is generally incapable of processing data, making decisions, and taking actions on its own; thus the 'manager' is not the robot itself but the human supervisor acting through it. Gradually, though, new artificial agent technologies are being developed that will allow robots to act autonomously in performing management functions in overseeing human workers.[3] Our acceptance of such artificial beings as managers will likely be accelerated by the fact that human beings are not only willing but even inclined to create social bonds with computerized systems as though they were human.[4] In addition to managing human beings, robots are also being developed that can interact socially with human colleagues to receive new tasks and then manage other (nonsocial) machines in carrying out those tasks.[5]

Given the wide variety of forms and capacities found among robots, it seems likely that some robots are better suited than others for performing functions as managers of human beings or other machines. However, significant attention has not yet been given to this question of 'managerial robotics'; we do not yet possess a robust set of models or principles designed to help identify or develop robots that are uniquely qualified to perform particular management roles. In this paper we propose a model that can help us in assessing one such aspect of a robot's potential to successfully carry out management functions.

Research Aims and Questions

When analyzing and comparing the capacities of different robots, there are many elements that one can potentially consider, such as the robots' size,

---

[2] Daft, *Management* (2011), p. 8.

[3] Nunes and O'Neill, "Assessing the Performance of Virtual Teams with Intelligent Agents" (2012); Kriksciuniene & Strigunaite, "Multi-Level Fuzzy Rules-Based Analysis of Virtual Team Performance" (2011); Dai et al., "TrustAider – Enhancing Trust in E-Leadership" (2013).

[4] Rehm et al. "Some Pitfalls for Developing Enculturated Conversational Agents" (2009); Friedenberg, *Artificial Psychology: The Quest for What It Means to Be Human* (2008).

[5] Zhang et al. "Human-Robot Interaction Control for Industrial Robot Arm through Software Platform for Agents and Knowledge Management" (2006).

shape, mobility, sensory capacities, or processing speed and power. Here we have chosen to focus on two factors that we believe will play a key role in a robot's ability to serve as a manager: namely, the robot's levels of *autonomy* and *sociality*.

Differing degrees of robotic autonomy are desirable in different situations. If a robot is managing work that involves complex ethical dilemmas or the risk of harm to persons or property, one may wish the robot to be directly and continuously overseen by a human being who bears ultimate responsibility for the robot's actions and can override them at any moment, if needed. On the other hand, if a robot is managing repetitive work that involves no ethical or safety concerns, one may wish the robot to operate without continuous human oversight, thereby allowing the robot to work faster and more efficiently and reducing the human resource demands placed on the organization.[6]

Similarly, different degrees of robotic sociality are desirable in different situations. If a robot's work will involve managing very simple machines in the performance of repetitive, predetermined tasks, it would likely be a waste of time and resources to design a robot that possesses advanced capacities for natural language processing, cultural competence, or emotional display; it would be simpler and cheaper to select a robot with very limited sociality. On the other hand, if the robot's work will involve negotiating project goals with human subordinates and motivating and instructing them in their tasks, the robot would benefit from possessing a form of sociality that is as sophisticated as possible.

The particular question that we are exploring here is whether a robot's level of sociality is independent from its level of autonomy. We hypothesize that the two traits are not independent but interrelated. If there is a strong positive correlation between autonomy and sociality, then designers of future managerial robots may not easily be able to implement one of these attributes without taking the other into consideration.

Methodology

An Instrument for Assessing Robotic Autonomy and Sociality

In order to evaluate the autonomy and sociality of existing robots, we have utilized the newly developed version 1.1 of our assessment instrument IOPAIRE, the Inventory of Ontological Properties of Artificially Intelligent and Robotic Entities. This inventory encompasses eight aspects such as Identity, Temporality and Change, Physicality, and Cognition, which together comprise 75 general characteristics and a wide range of particular properties.

---

[6] Murphy, *Introduction to AI Robotics* (2000), p. 31.

Autonomy and sociality are multifaceted composite traits that reflect the possession of a wide range of more basic capacities. For example, for robots, 'autonomy' consists of being "capable of operating in the real-world environment without any form of external control for extended periods of time."[7] In its full sense, autonomy thus means that robots can not only perform cognitive tasks such as setting goals and making decisions but can also successfully perform physical activities such as obtaining energy sources and carrying out mechanical self-repair without human intervention. In the IOPAIRE framework there are 34 assessed properties that contribute to an entity's score for Autonomy and 36 properties that contribute to its score for Sociality, with the completed inventory yielding a score ranging from 0-100 for each of these traits. Drawing on conventional classifications of robotic autonomy,[8] the score generated for Autonomy by the IOPAIRE instrument is normalized so that a score of 0-25 represents a robot that is *Nonautonomous* (e.g., a telepresence robot that is fully controlled by its human operator), 26-50 represents one that is *Semiautonomous* (e.g., that requires 'continuous assistance' or 'shared control'), and 51-75 represents one that is *Autonomous* (e.g., that requires no human guidance or intervention in fulfilling its intended purpose). We have also introduced the category of *'Superautonomous'* (represented by a score of 76-100) to describe theoretically possible but not yet extant robots whose degree of autonomy significantly exceeds that displayed by human beings – e.g., because the robot contains an energy source that can power it throughout its anticipated lifespan or because its ability to independently acquire new skills and knowledge frees it from any need to seek guidance from human subject-matter experts.

Similarly, drawing on established classifications of robotic social behavior, social interactions, and social relations,[9] the score for Sociality yielded by IOPAIRE is normalized so that a value of 0-25 reflects a robot that is *Nonsocial* (e.g., that might display basic social behaviors but cannot engage in social interaction), 26-50 reflects one that is *Semisocial* (i.e., that can engage in social interactions but not full-fledged social relations), and 51-75 reflects one that is fully *Social* (e.g., that can participate in social relations that evolve over time and are governed by the expectations of a particular society). We have also introduced the category of *'Supersocial'* to describe theoretically possible but not yet extant robots whose degree of sociality significantly exceeds that displayed by human beings – e.g., because they can fluently converse in all

[7] Bekey, *Autonomous Robots: From Biological Inspiration to Implementation and Control* (2005), p. 1.

[8] Murphy (2000), pp. 31-34.

[9] Vinciarelli et al., "Bridging the Gap between Social Animal and Unsocial Machine: A survey of Social Signal Processing" (2012).

known human languages or, through the use of multiple communication interfaces, can engage in separate social interactions with thousands of human beings simultaneously.

Selecting the Population for Assessment

To generate our data set, we applied the IOPAIRE instrument to 38 different kinds of entities. Through a review of scholarly, industrial, and popular robotics literature we identified 35 models of existing robots that display a great variety of forms and have been designed for a wide array of industrial, domestic, entertainment, educational, and governmental purposes; we then researched, documented, and analyzed their design specifications and performance characteristics. We have also evaluated three types of living organisms (i.e., a typical human being, dog, and mouse) to reflect the fact that human beings, domesticated animals, and robots can be understood as members of a single, shared social space – a phenomenon that is perhaps most clearly visible in the case of therapeutic robots like PARO, which explicitly fills a role of relating to human beings that might otherwise be filled by a dog or cat.[10]

Developing a Two-dimensional Model for Classifying Entities

We have created a two-dimensional model in which the X-axis represents Autonomy and the Y-axis Sociality. Because the scores for Autonomy and Sociality are each divided into four groups, this model organizes entities into sixteen different types. We would suggest that each of these types will possess a unique set of capacities and limitations for use in managing human employees and other robots and computerized systems that can be identified through further research.

Findings

Mapping of Scores onto the Two-dimensional Model

Figure 1 depicts the results for the 38 robotic and organic entities that we assessed. Each of the 35 gray circular dots represents a particular robot. The three kinds of organic beings that we assessed are represented by a black triangle (a common mouse), a black diamond (a typical dog), and a black square (a typical human being). The seven subquadrants that would include any Superautonomous or Supersocial entities are shaded in gray to note that while these categories might someday include advanced robots or cybernetically or genetically altered human beings, it is not anticipated that any currently extant entities would fall into these categories.

---

[10] Inada & Tergesen, "It's Not a Stuffed Animal, It's a $6,000 Medical Device" (2010).

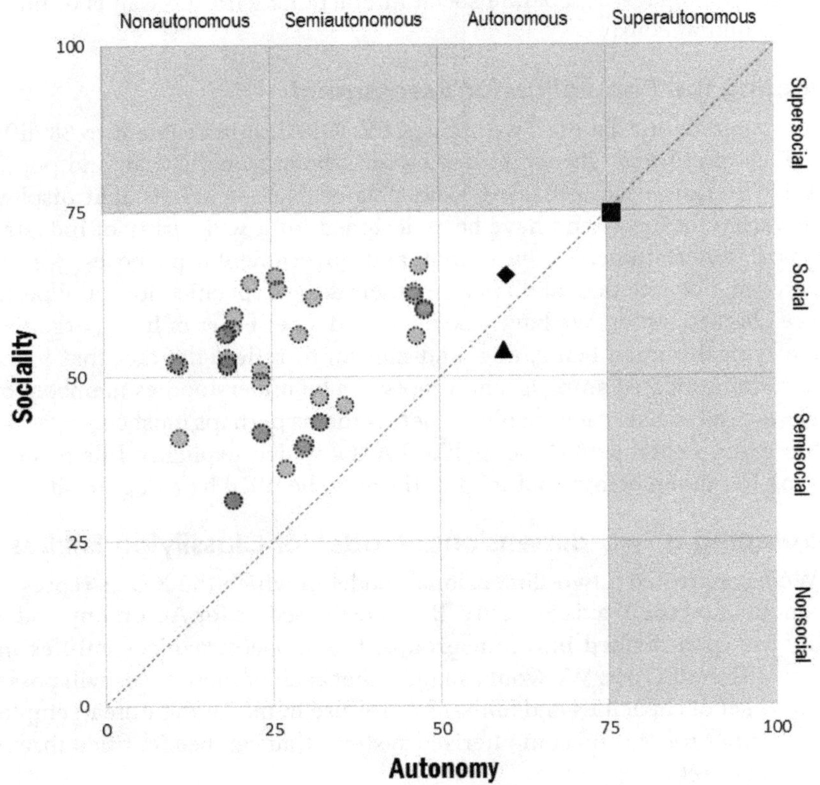

Figure 1. Thirty-five robots and three kinds of organic life-forms categorized according to their degrees of Autonomy and Sociality.

Source: Own data and design.

As shown in Table 1, the values for the inventoried robots' Autonomy score ranged from a minimum of 10.6 (for the telepresence robot Hugvie) to a maximum of 47.7 (for PARO and the industrial robot Baxter) with a mean score of 27.0. The robots' values for the Sociality score varied from a minimum of 31.4 (for Looj) to a maximum of 66.7 (for Pepper) with a mean score of 50.9.

Table 1. Summary of the Autonomy and Sociality scores for 35 inventoried robots.

Trait	Min. score	Mean score	Max. score
Autonomy	10.6	27.0	47.7
Sociality	31.4	50.9	66.7

Source: Own data.

When we categorize the 35 robots according to subquadrants, we see that:

Seven robots can be described as Nonautonomous Semisocial, including the Looj 330 gutter-cleaning robot and MQ-1 Predator unmanned aerial vehicle.

Seven robots are Semiautonomous Semisocial, including the KR Quantec Pro industrial manipulator arm, the Curiosity Mars rover, and the Roomba 500 series of vacuum-cleaning robots.

Eleven robots are Nonautonomous Social, including the Geminoid HI-4, Telenoid R2, and PackBot Explorer.

Ten robots are Semiautonomous Social, including PARO, the therapeutic robot resembling a baby seal; the Care-Providing Robot FRIEND wheelchair; and the 'emotional robot' Pepper.

The grid's remaining 12 subquadrants contained no robots at all.

Analyzing the Relationship of Scores for Autonomy and Sociality

One may note that all of the inventoried robots are mapped to a position above the line defined by the equation $y = x$. In other words, all of the robots possessed a Sociality score greater than their score for Autonomy; in no case does a robot's Autonomy exceed its Sociality. In order to better understand the relationship between autonomy and sociality, we calculated the Pearson product-moment correlation coefficient (r), Spearman's rank correlation coefficient (ρ), and the p-value for our data set, as shown in Table 2.

Table 2. Overview of correlation coefficients and p-value for the evaluated entities.

Population	r	ρ	p-value
35 robots	0.36	0.26	0.13
35 robots plus 3 organic beings	0.50	0.37	0.02

Source: Own data.

If only the 35 inventoried robots are considered, the p-value of 0.13 does not allow us to presume with great confidence that there is a correlation between the value of the entities' Autonomy and Sociality scores; it is not inconceivable that an apparent relationship similar to that visible in Figure 1 could be obtained by random chance. However, when we consider the population of inventoried entities that includes both the 35 robots and three kinds of organic beings, the p-value of 0.02 allows us to conclude with a high degree of confidence that an entity's level of Sociality has a significant correlation with its level of Autonomy.

Discussion and Conclusions

It is perhaps not surprising that no robots were classified as Superautonomous or Supersocial: these categories represent abilities significantly beyond those of which human beings are capable, and artificial intelligence technologies are not yet sufficiently advanced to grant robots synthetic emotion, cultural competence, or ethical judgment that can match human capacities, let alone significantly surpass them. The Autonomous Social subquadrant also contains no assessed robots, although a number of them fell just outside it, possessing adequate Sociality but insufficient Autonomy. Our data would suggest that developing autonomous robots may be a greater challenge than developing social ones: while telepresence robots such as Hiroshi Ishiguro's Geminoid models demonstrate a level of sociality that exceeds that of a mouse, rivals that of a dog, and even approaches that of a human being, when it comes to manifesting autonomy the robots that we have studied still fall short of common mice – entities which are, after all, able to go about their regular activities, survive, and thrive in the most difficult environments and with no human assistance (or indeed even in the face of active human opposition).

Also noteworthy is the fact that the quadrants representing Nonautonomous and Semiautonomous Nonsocial robots were empty; none of the robots that we evaluated could be described as truly 'nonsocial' entities. We would hypothesize that this may reflect the fact that at present, it is not possible for robots to be designed and created solely by other machines without the involvement of human beings. Every existing contemporary robot has been designed by human beings; it has been 'born' into a human society in which it will be operated and maintained by human beings to fulfill a purpose that has been chosen by human beings and is intended to benefit certain human beings. While it might be possible for a rock or a flower or a distant star to be classified as 'nonsocial,' it is not surprising that robotic artifacts created to serve the ends of human society possess at least a weak form of semisociality, since sociality depends not just on the inherent qualities of an object itself but also on the ways in which it is viewed and treated by the human beings who interact with it.

While the data obtained from the 35 robots does not by itself provide conclusive evidence that there is a correlation between the robots' levels of autonomy and sociality, the additional data obtained from the three kinds of living organisms suggests strongly that such a correlation exists, if one views robots and living organisms as fellow members of the single population of entities that are capable of possessing some degree of autonomy and sociality. If a correlation between robotic autonomy and sociality exists, there remains a question of whether a direct causal connection exists between the two traits, or whether some third factor produces them both. Our data suggest that increasing a robot's degree of sociality does not, in itself, enhance the robot's autonomy, as we identified a number of robots with quite high scores for Sociality but low scores for Autonomy. On the other hand, every robot that possessed a high score for Autonomy (i.e., nearing 50) also possessed a high score for Sociality. This leads us to formulate a working hypothesis that enhanced robotic autonomy contributes to a higher level of robotic sociality.

This supposition will require further research in order to be confirmed. We hope to expand our data set to include a larger quantity and variety of evaluated robots and to employ the IOPAIRE instrument to develop an expanded multidimensional model that can identify correlations between robotic traits other than those of autonomy and sociality. Even in the absence of further data and analysis, though, the results described here seem to warrant suggesting a piece of practical advice to any engineers who are attempting to design an Autonomous Social managerial robot that is capable of carrying out all four management functions: if they should encounter obstacles while attempting to directly increase their robot's level of sociality, they might instead try focusing on enhancing their robot's level of autonomy and then see whether this increased autonomy is accompanied by growth in the robot's social capacities. We anticipate that further future study in this area of managerial robotics will not only aid organizations in identifying existing robots that can effectively perform particular management functions, but will also aid engineers to develop new robots and artificially intelligent systems that are optimally suited to filling particular managerial roles.

References

Bekey, G.A. *Autonomous Robots: From Biological Inspiration to Implementation and Control.* Cambridge, MA: MIT Press, 2005.

Daft, Richard L. *Management.* Mason, OH: South-Western / Cengage Learning, 2011.

Dai, Y., C. Suero Montero, T. Kakkonen, M. Nasiri, E. Sutinen, M. Kim, and T. Savolainen (2013). "TrustAider – Enhancing Trust in E-Leadership." In *Business Information Systems,* pp. 26-37. Lecture Notes in Business Information Processing 157. Springer Berlin Heidelberg, 2013.

Friedenberg, Jay. *Artificial Psychology: The Quest for What It Means to Be Human,* Philadelphia: Psychology Press, 2008.

Inada, M., and A. Tergesen, (2010, June 21). "It's Not a Stuffed Animal, It's a $6,000 Medical Device." *The Wall Street Journal,* June 21, 2010. online.wsj.com/news/articles/SB10001424052748704463504575301051844937276. Accessed October 12, 2014.

Kriksciuniene, D., and S. Strigunaite. "Multi-Level Fuzzy Rules-Based Analysis of Virtual Team Performance." In *Building the E-World Ecosystem,* pp. 305-18. IFIP Advances in Information and Communication Technology 353. Springer Berlin Heidelberg, 2011.

Murphy, Robin. *Introduction to AI Robotics.* Cambridge, MA: The MIT Press, 2000.

Nunes, M., and H. O'Neill. "Assessing the Performance of Virtual Teams with Intelligent Agents." In *Virtual and Networked Organizations, Emergent Technologies and Tools*, pp. 62-69. Communications in Computer and Information Science 248. Springer Berlin Heidelberg, 2012.

Rehm, M., André, E., and Nakano, Y. "Some Pitfalls for Developing Enculturated Conversational Agents." In *Human-Computer Interaction: Ambient, Ubiquitous and Intelligent Interaction*, pp. 340-48. Lecture Notes in Computer Science 5612. Springer Berlin Heidelberg, 2009.

Vinciarelli, A., M. Pantic, D. Heylen, C. Pelachaud, I. Poggi, F. D'Errico, and M. Schröder. "Bridging the Gap between Social Animal and Unsocial Machine: A survey of Social Signal Processing." *IEEE Transactions on Affective Computing* 3:1 (January-March 2012): 69-87.

Zhang, T., V. Ampornaramveth, and H. Ueno. "Human-Robot Interaction Control for Industrial Robot Arm through Software Platform for Agents and Knowledge Management." In *Industrial Robotics: Theory, Modelling and Control*, edited by Sam Cubero, pp. 677-92. Germany: Pro Literatur Verlag, 2006.

Chapter Thirteen

Developing a Non-anthropocentric Definition of Business: A Cybernetic Model of the Synthetic Life-form as Autonomous Enterprise[1]

Abstract. In this text we argue that it is theoretically possible to create artificial life-forms that function as autonomous businesses within the real-world human economy and explore some of the implications of the development of such beings. Building on the cybernetic framework of the Viable Systems Approach (VSA), we formulate the concept of an 'organism-enterprise' that exists simultaneously as both a life-form and a business. The possible existence of such entities both enables and encourages us to reconceptualize the historically anthropocentric understanding of a 'business' in a way that allows an artificial life-form to constitute a 'synthetic' organism-enterprise (SOE) just as a human being acting as a sole proprietor constitutes a 'natural' organism-enterprise. Such SOEs would exist and operate in a sphere beyond that of current examples of artificial life, which produce goods or services within some simulated world or play a limited role as tools or assistants within a human business. Rather than competing against artificial organisms in a virtual world, SOEs could potentially survive and evolve through competition against human businesses in our real-world economy. We conclude by briefly envisioning particular examples of SOEs that elucidate some of the legal, economic, and ethical issues that arise when a single economic ecosystem is shared by competing human and artificial life. It is suggested that the theoretical model of synthetic organism-enterprises developed in this text may provide a

[1] This text is a revised and expanded version of "The Artificial Life-Form as Entrepreneur: Synthetic Organism-Enterprises and the Reconceptualization of Business," in *Proceedings of the Fourteenth International Conference on the Synthesis and Simulation of Living Systems*, edited by Hiroki Sayama, John Rieffel, Sebastian Risi, René Doursat and Hod Lipson, pp. 417-18. Cambridge, MA: The MIT Press, 2014.

useful conceptual foundation for computer programmers, engineers, economists, management scholars and practitioners, ethicists, policymakers, and others who will be called upon in the coming years to grapple with the realities of artificial agents that increasingly function as autonomous enterprises within our world's complex economic ecosystem.

Introduction

Operating a business has traditionally been considered an exclusively human activity: while domesticated animals or desktop computers, for example, might participate in the work of a business, they are not in themselves capable of organizing or constituting a 'business.' However, the increasing sophistication and capacities of social robots, synthetic life-forms, and other kinds of artificial agents raises the question of whether some such entities might theoretically be capable of not only leading a business but of directly constituting one.

In this work we argue that it is theoretically possible to develop artificial life that functions as an autonomous business within the real-world human economy, and we explore the implications of such an eventuality. By drawing on the cybernetic framework of the Viable Systems Approach (VSA), we formulate the concept of an 'organism-enterprise' that exists simultaneously as both a life-form and a business. We then propose reconceptualizing the historically anthropocentric understanding of a 'business' in a way that allows an artificial life-form to constitute a 'synthetic' organism-enterprise (SOE) just as a human being functioning as a sole proprietor constitutes a 'natural' organism-enterprise. SOEs would move a step beyond current examples of artificial life that produce goods or services within a simulated world or play a limited role within a human business: rather than competing against other artificial organisms in a virtual world, SOEs could evolve through competition against human businesses in the real-world economy. We conclude by briefly considering concrete examples of SOEs that highlight some of the legal, economic, and ethical issues that arise when a single economic ecosystem is shared by competing human and artificial life.

Understanding Businesses and Organisms as Viable Systems

A 'business' has been defined as "the organized effort ... to produce and sell, for a profit, the goods and services that satisfy society's needs."[2] In an effort to analyze and better understand the forms and functions displayed by such businesses, management scholars and practitioners have employed a number of metaphors – for example, describing a business as being analogous

[2] Pride et al., *Foundations of Business* (2014).

to a 'machine' or a 'journey.' However, throughout the last century (and especially since 1975), one of the dominant and most useful metaphors has been that of the businesses as a biological *organism.*[3]

During recent decades, some researchers in the fields of cybernetics and systems theory have argued that while a business is perhaps not literally an 'organism' in the sense in which a biologist would typically employ that term, neither is the resemblance between a business and an organism simply a metaphorical one: both businesses and biological organisms can be viewed as two different kinds of systems which – because they are systems – are bound by the principles that govern the actions of all systems and display certain structural and functional similarities. One effort undertaken from the perspective of systems theory to identify the common dynamics shared by businesses and biological organisms and to utilize that knowledge to improve organizational management is the Viable Systems Approach (VSA), a cybernetic model grounded in neurophysiology that allows a business to be understood as an autopoietic organism or 'system' that dwells within the ecosystem of a larger economy or 'suprasystem.'[4] Within this ecosystem, a business must compete against other organisms for limited resources and adapt to environmental demands. In our human economy, individual businesses are born, grow, and die, and taken as a whole, this array of businesses forms an evolvable system.

It is possible to go further, however, by noting that in some cases there are businesses that do not simply share some fundamental characteristics with biological organisms but which literally *are* biological organisms. In particular, we can consider the case of a human being who functions as the sole proprietor of a business. In this unique circumstance, a business is not simply 'analogous to' a living organism but identical to it: a single system (e.g., the business's human proprietor) satisfies all the requirements of being a life-form while simultaneously satisfying all the requirements of being a business. We can describe such an entity as a unitary 'organism-enterprise,' a kind of system that displays the form and dynamics of both a biological organism and a business and which can be analyzed from either perspective. Such organism-enterprises are by no means rare or unusual; it is estimated that in the United States alone, at least 20 million 'human organism-enterprises' exist in the form of sole proprietors of businesses.[5]

---

[3] Clancy, *The Invisible Powers: The Language of Business* (1999), pp. 169-70.

[4] Beer, *Brain of the Firm* (1981); Barile et al., "An Introduction to the Viable Systems Approach and Its Contribution to Marketing" (2012).

[5] Pride et al. (2014), p. 102.

Beyond 'Human Resources': Replacing Anthropocentric Definitions of Business

The question with which this text is concerned is whether an organism-enterprise can only exist in the form of a 'human organism-enterprise' or whether it is theoretically and practically possible for an artificial life-form – a synthetic organism – to similarly exist and function as an organism-enterprise. By utilizing VSA and the concept of an organism-enterprise, it is possible to reexamine the traditional anthropocentric understanding of business as an exclusively human activity and to consider whether an artificial life-form could serve as a 'synthetic organism-enterprise' (SOE) that is simultaneously both a life-form and a business. We argue that this is indeed possible – but that it will require a deepening and clarification of our everyday understanding of what constitutes a 'business.'

For example, businesses are traditionally described as requiring four kinds of resources: 1) human; 2) material; 3) financial; and 4) informational. In this conventional model, 'human resources' are defined as "the people who furnish their labor to the business in return for wages."[6] In other words, operating a 'business' is something that can only be done by a particular type of biological organism: human beings. While many kinds of animals (and some kinds of embodied artificial agents such as social robots) are capable of engaging in complex patterns of interaction and exchange, the systems of production and interaction that such entities establish have traditionally been excluded *a priori* from possibly being considered 'businesses,' because the entities that have created and which operate them are not human beings.

To replace this historically anthropocentric understanding of what constitutes a business, we would suggest that a 'business' be defined in a more generalized way that does not eliminate any of the structural elements or dynamics that have long been considered essential to the nature of a business but which allows for the theoretical possibility that a business's activities could be performed by agents that are not human beings, as long as those agents possess the requisite sensory, motor, and information-processing capacities. Within this new non-anthropocentric framework, a business can be understood more generally than in the past, as requiring: 1) agent resources; 2) material resources; 3) value-storing media; and 4) information.

Identifying the Core Dynamics of Business through the Viable Systems Approach

Human businesses have traditionally been understood not only in relation to the resources that they require but also in relation to the different types of

[6] Pride et al. (2014), p. 8.

functions that must be performed by a business or by the subunits that constitute the business. Such functions have gradually become institutionalized in the form of functional departments such as those that are responsible for production, finance, marketing, human capital management, and information technology. By drawing on the Viable Systems Approach – and keeping in mind the case of a human sole proprietor – it is possible to analyze the essential characteristics of these business functions in such a way that allows them to be understood more generally as functions that can potentially be performed by existing or proposed forms of artificial life.

Within the Viable Systems Approach, a viable 'system-in-focus' (such as a particular business) is composed of a number of smaller constituent systems, each of which is in itself a viable system. Building on the work of cyberneticist Stafford Beer and others, by convention these constituent systems are identified as System 1, which performs the operations that generate a business's core products or services (e.g., the manufacturing operation in the case of a consumer electrics company or the professional accounting staff in the case of an accounting firm); System 2, which maintains stability and resolves immediate conflicts among the system's operational units (e.g., by arbitrating conflicting demands for the use of particular equipment or spaces); System 3, which optimizes the overall productivity and effectiveness of the system-in-focus by monitoring the activities of all of its constituent systems, identifying potential synergies, and, when necessary, overriding the normally autonomous operations of the constituent systems in order to realize those synergies; System 3*, which gathers data from throughout the system-in-focus (and in particular, data revealing signs of stress or inefficiencies within the system) and conducts special analyses in order to ensure that System 3 possesses all of the information needed in order to identify and create synergies; System 4, which looks beyond the immediate activities of the system-in-focus to identify future needs and opportunities, map out alternative strategies and courses of action that the system-in-focus could potentially pursue, and then analyze the various alternatives to recommend a particular course of action to be followed; and System 5, which formulates a high-level 'policy' or 'ethos' for the system-in-focus that describes its unique mission, values, and priorities and which provides the criteria by which potential strategies and courses of action can be evaluated (e.g., a mission statement and set of guiding principles developed by a corporation's board of directors). Collectively, Systems 2-5 can be described as the 'Metasystem' which oversees the core operations performed by System 1 within the system-in-focus.[7]

The final major element to be considered in order to complete the picture is that of the external environment within which the system-in-focus exists and operates; it is from this environment that the system-in-focus draws its

[7] Beer (1981); Barile et al. (2012).

resources and into which it releases its finished products or services for consumption. Just as the system-in-focus (i.e., a particular business) is composed of numerous smaller constituent systems, so too can the environment or ecosystem in which the system-in-focus exists (e.g., a national economy) be understood as a 'suprasystem' that is itself a viable system and whose constituent systems include the system-in-focus and many other, similar viable systems (e.g., competitors within the same industry and companies in many other industries).[8]

Execution of the Business Process Cycle as a Byproduct (or Purpose) of Artificial Life

Although it might at first sound unusual to suggest that an individual artificial organism could function successfully in the real-world human economy as a full-fledged business enterprise (the example of a human sole proprietor notwithstanding), the Viable Systems Approach helps us to understand why such a notion may not be at all implausible: it reminds us that the structures and internal dynamics that must be possessed by an entity in order for it to exist and function successfully within the world as an 'organism' are strongly analogous to – and in some cases can even be identified with – the structures and internal dynamics needed for an entity to exist and function successfully within the world as an 'enterprise.'

In Figure 1, we build on the Viable Systems Approach to provide an overview of a reconceptualized, non-anthropocentric 'business process cycle' that can be carried out equally well either by a conventional human-led business or by an artificial life-form that has been designed or has evolved to fill a business role within a larger economic ecosystem and which performs all of the functions necessary for a business as part of its natural life cycle.

Of particular note is the way in which the business concept of 'profit' is genericized to apply to other kinds of viable systems such as biological or artificial organisms. In the case of a conventional human business, the company's 'profit' can be quantified as the amount by which the business's income exceeds its expenses. While part or all of such profits may be paid out to a company's shareholders, profits may alternatively be retained by the company and either invested to expand or upgrade the company's operating capacity or kept in reserve to provide a form of 'insurance' against future uncertainty. In the case of a synthetic organism-enterprise, the 'profit' generated by the SOE's execution of the business process cycle can be understood as the difference between those resources that are expended by the SOE and those that are received in its exchanges in the suprasystem, which (when positive) provides an

[8] Beer (1981); Barile et al. (2012).

SOE with the potential for growth and insurance against future environmental uncertainties.

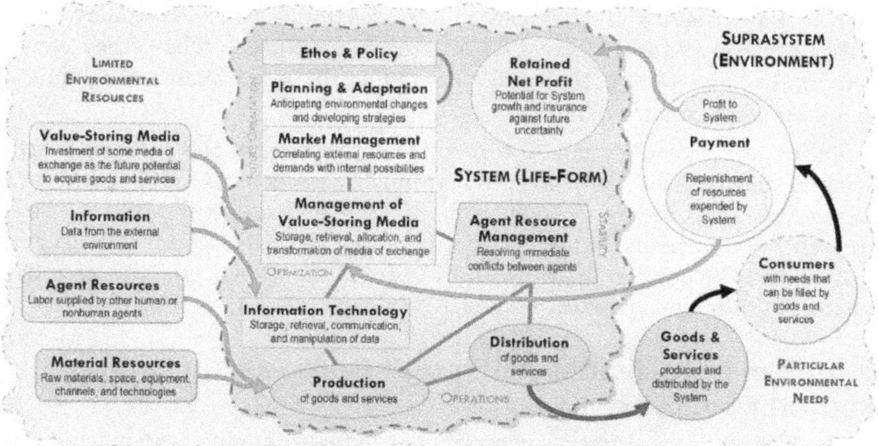

Figure 1. A reconceptualized business process cycle that applies to a human organism-enterprise as well as to a synthetic organism-enterprise (SOE) that has been designed or has evolved to provide goods or services within some ecosystem.

Challenges to the Development of Synthetic Organism-Enterprises

Artificial life-forms have already been designed that are capable of carrying out this entire business process cycle within the simulated ecosystem of a virtual world.[9] Likewise, there exist artificial life-forms that are capable of carrying out parts of this cycle within human businesses in the economy of the 'real world.'[10] However, our survey of the field has not yet identified any existing artificial life-forms that are capable of carrying out this entire business process cycle within the real-world human economy.

While it is relatively easy to envision and create artificial-life forms that generate a specific product or service that possesses some value within the real-world human economy, it is more difficult to develop artificial life-forms that are capable, for example, of analyzing environmental conditions to conduct long-range planning (rather than simply reacting to current environmental conditions) or formulating their own high-level values and policies regarding the kinds of products and services that they would like to produce. Beyond

---

9 Kubera et al., "IODA: An Interaction-Oriented Approach for Multi-Agent Based Simulation" (2011).

10 Kim and Cho, "A Comprehensive Overview of the Applications of Artificial Life" (2006).

the technological challenges that exist with creating or facilitating the evolution of artificial life-forms with such capacities, there are also significant legal and ethical issues surrounding the question of whether human institutions such as governments and regulatory agencies could and should allow such SOEs to operate within the real-world human economy – where they would potentially compete directly against human workers and businesses – and, if so, under what conditions.[11]

Different approaches exist for overcoming these obstacles so that an artificial life-form can not only meet the minimal requirements for constituting a synthetic organism-enterprise but potentially even excel in the role of entrepreneur.[12] Of particular interest is the potential of virtual goods and cryptocurrencies to overcome the difficulty of providing an SOE with an effective means of utilizing value-storing media.[13]

Evolution of Artificial Life through Competition in the Human Economy

An SOE producing goods or services of value to human beings would be capable of competing against human businesses in the real-world economy. However, it is unclear whether these competitive pressures would by themselves be sufficient to drive evolution among SOEs in a way that is identifiable by and meaningful for human beings. By utilizing the concept of 'clockspeed' as a measure of the speed at which businesses must adapt and compete, it may be possible to identify industries in which SOEs would likely evolve at an accelerated rate. For example, an SOE that generates a profit by engaging in online currency trading – in which a single iteration of the business process cycle might only last a matter of milliseconds – is very different from an SOE that manufactures some large and complex physical product such as an oil

[11] The question of whether SOEs should be allowed to operate licitly within the real-world human economy as enterprises independent of direct human control depends partly on the issue of whether SOEs are capable of possessing the legal responsibility for their own actions that is required both of human entrepreneurs and legal persons such as corporations that are not human beings. Regarding the theoretical possibility that embodied artificial agents such as robots may be capable of possessing legal and moral responsibility for their own actions, see Stahl, "Responsible Computers? A Case for Ascribing Quasi-Responsibility to Computers Independent of Personhood or Agency" (2006), and Calverley, "Imagining a Non-Biological Machine as a Legal Person" (2008).

[12] For a thoughtful exploration of such issues, see Rijntjes, "On the Viability of Automated Entrepreneurship" (2016). See also Ihrig, "Simulating Entrepreneurial Opportunity Recognition Processes: An Agent-Based and Knowledge-Driven Approach" (2012).

[13] See Gladden, "Cryptocurrency with a Conscience: Using Artificial Intelligence to Develop Money That Advances Human Ethical Values" (2015), and Scarle et al., "E-commerce Transactions in a Virtual Environment: Virtual Transactions" (2012).

tanker or commercial airliner, in which the business process cycle might last years or even decades.[14]

Specific Examples and Practical Implications of Artificial Life as Enterprise

An autonomous artificial life-form that is capable of securing all of the resources needed for survival and growth directly from the real-world human economy would in principle no longer be dependent on its human designer. Such possibilities are not risk-free: one can imagine the case of computer viruses that are capable of evolving self-adaptive behavior rather than mere polymorphism or metamorphism[15] and that no longer steal for the financial gain of human cybercriminals but to provide resources for their own survival, growth, and autonomously chosen pursuits. One can also envision more optimistic cases, such as the development of artificial life-forms that build successful 'careers' as artists or composers or IT service-providers within the human economy, that are able to evolve in response to economic demands without the active guidance or support of a human designer, and which may be able to offer products or services of a sort that cannot be created by human beings. As such scenarios suggest, further research is needed in order to address the significant moral, legal, and economic issues that will arise from the existence of synthetic organism-enterprises and the fact that the productive and competitive capacities of successful SOEs could far surpass those of traditional human businesses.

Conclusion

In this text we have argued that it is theoretically possible to create artificial life-forms that function as autonomous businesses within the real-world human economy and have explored some of the implications of the development of such beings. Building on the cybernetic framework of the Viable Systems Approach (VSA), we formulated the concept of an 'organism-enterprise' that exists simultaneously as both a life-form and a business. It was argued that the possible existence of such entities both enables and encourages us to reconceptualize the historically anthropocentric understanding of a 'business' in a way that allows an artificial life-form to constitute a 'synthetic' organism-enterprise (SOE) just as a human being acting as a sole proprietor constitutes a 'natural' organism-enterprise. Such SOEs would exist and operate in a sphere beyond that of current examples of artificial life, which produce goods or services within some simulated world or play a limited role as tools or as-

[14] Fine, *Clockspeed: Using Business Genetics to Evolve Faster than Your Competitors* (1998).

[15] Beckmann et al., "Applying Digital Evolution to the Design of Self-Adaptive Software" (2009).

sistants within a human business. Rather than competing against artificial organisms in a virtual world, SOEs could potentially survive and evolve through competition against human businesses in our real-world economy. We concluded by briefly envisioning particular examples of SOEs that elucidate some of the legal, economic, and ethical issues that arise when a single economic ecosystem is shared by competing human and artificial life. It is our hope that the theoretical model of synthetic organism-enterprises developed in this text might provide a useful conceptual foundation for computer programmers, engineers, economists, management scholars and practitioners, ethicists, policymakers, and others who will be called upon in the coming years to grapple with the practical realities of artificial agents that increasingly function as autonomous enterprises within our world's complex economic ecosystem.

References

Barile, S., J. Pels, F. Polese, and M. Saviano. "An Introduction to the Viable Systems Approach and Its Contribution to Marketing." *Journal of Business Market Management* 5(2) (2012): 54-78.

Beckmann, B.E., L.M. Grabowski, P.K. McKinley, and C. Ofria. "Applying Digital Evolution to the Design of Self-Adaptive Software." In *IEEE Symposium on Artificial Life*, pp. 100-07. IEEE, 2009.

Beer, Stafford. *Brain of the Firm*. Second edition. New York: John Wiley, 1981.

Calverley, D.J. "Imagining a non-biological machine as a legal person." *AI & SOCIETY* 22, no. 4 (2008): 523-37.

Clancy, J. *The Invisible Powers: The Language of Business*. Lanham: Lexington Books, 1999.

Fine, C. *Clockspeed: Using Business Genetics to Evolve Faster than Your Competitors*. Reading, MA: Perseus Books, 1998.

Gladden, Matthew E. "Cryptocurrency with a Conscience: Using Artificial Intelligence to Develop Money that Advances Human Ethical Values." *Annales: Ethics in Economic Life* vol. 18, no. 4 (2015): 85-98.

Ihrig, M. "Simulating Entrepreneurial Opportunity Recognition Processes: An Agent-Based and Knowledge-Driven Approach." In *Advances in Intelligent Modelling and Simulation: Simulation Tools and Applications*, edited by A. Byrski, Z. Oplatková, M. Carvalho, and M. Kisiel-Dorohinicki, pp. 27-54. Berlin: Springer-Verlag, 2012.

Kim, K., and S. Cho. "A Comprehensive Overview of the Applications of Artificial Life." *Artificial Life* 12(1) (2006): 153-82.

Kubera, Y., P., Mathieu, and S. Picault. "IODA: An Interaction-Oriented Approach for Multi-Agent Based Simulation." *Autonomous Agents and Multi-Agent Systems* 23(3) (2011): 303-43.

Pride, W., R. Hughes, and J. Kapoor. *Foundations of Business*, 4e. Stamford, CT: Cengage Learning, 2014.

Rijntjes, Tom. "On the Viability of Automated Entrepreneurship." Presentation of Media Technology MSc program graduation project, Universiteit Leiden, Leiden, January 22, 2016.

Scarle, S., S. Arnab, I. Dunwell, P. Petridis, A. Protopsaltis, and S. De Freitas. "E-commerce Transactions in a Virtual Environment: Virtual Transactions," *Electronic Commerce Research* 12(3) (2012): 379-407.

Stahl, B. C. "Responsible Computers? A Case for Ascribing Quasi-Responsibility to Computers Independent of Personhood or Agency." *Ethics and Information Technology* 8, no. 4 (2006): 205-13.

Index

About the Author

Matthew E. Gladden is an executive, management consultant, and researcher whose work focuses on the organizational implications of sociotechnological change relating to emerging technologies such as those for augmented reality, persistent virtual worlds, neuroprosthetic enhancement, artificial intelligence, artificial life, nanotechnology, social robotics, genetic engineering, and life extension. He lectures internationally on the relationship of posthumanizing technologies to organizational life, and his research has been published in journals such as the *International Journal of Contemporary Management, Annals of Computer Science and Information Systems, Informatyka Ekonomiczna / Business Informatics, Creatio Fantastica*, and *Annales: Ethics in Economic Life*, as well as by IOS Press, Ashgate Publishing, the Digital Economy Lab of the University of Warsaw, and the MIT Press.

He is the founder and CEO of consulting firms NeuraXenetica LLC and Cognitive Firewall LLC and serves as Executive Director of the nonprofit Institute for Posthuman Management and Organizational Studies. He has previously served as Associate Director of the Woodstock Theological Center and Administrator of the Department of Psychology at Georgetown University and has also taught philosophical ethics and worked in computer game design. He is a member of ISACA, ISSA, and the Academy of Management and its divisions for Managerial and Organizational Cognition, Technology and Innovation Management, and Business Policy and Strategy.

He holds an MBA in Innovation and Data Analysis from the Institute of Computer Science of the Polish Academy of Sciences, certificates in Advanced Business Management and Nonprofit Management from Georgetown University, and a BA in Philosophy from Wabash College.

www.ingramcontent.com/pod-product-compliance
Lightning Source LLC
Chambersburg PA
CBHW022050210326
41519CB00054B/288

* 9 7 8 1 9 4 4 3 7 3 1 3 9 *